84

Fortschritte der Chemie organischer Naturstoffe

Progress in the Chemistry of Organic Natural Products

Founded by
L. Zechmeister

Edited by
W. Herz, H. Falk,
and G. W. Kirby

Authors:
M. Glasenapp-Breiling, P. G. Jagtap,
D. G. I. Kingston, F.-P. Montforts,
L. Samala, H. Yuan

SpringerWienNewYork

Prof. W. Herz, Department of Chemistry,
The Florida State University, Tallahassee, Florida, U.S.A.

Prof. H. Falk, Institut für Chemie,
Johannes-Kepler-Universität, Linz, Austria

Prof. G. W. Kirby, Chemistry Department,
The University of Glasgow, Glasgow, Scotland

ISBN 978-3-7091-3228-9 ISBN 978-3-7091-6160-9 (eBook)
DOI 10.1007/978-3-7091-6160-9
Softcover reprint of the hardcover 1st edition 2002

Library of Congress Catalog Card Number AC 39-1015

Typesetting: Thomson Press (India) Ltd., Noida

Printed on acid-free and chlorine-free bleached paper

SPIN: 10845363

With 12 Figures

ISSN 0071-7886

Contents

List of Contributors

Glasenapp-Breiling, Dr. M., Institut für Organische Chemie, FB 2 Biologie/Chemie, Universität Bremen, Postfach 330440, D-28334 Bremen, Germany

Jagtap, Dr. P. G., Department of Chemistry, M/C 0212, Virginia Polytechnic Institute and State University, Blacksburg, VA 24061, U.S.A. (current address: Laboratory for Drug Discovery in Neurodegeneration, Harvard Center for Neurodegeneration and Repair, 65 Landsdowne Street, Cambridge, MA 02139, U.S.A.)

Kingston, Univ. Dist. Prof. D. G. I., Department of Chemistry, M/C 0212, Virginia Polytechnic Institute and State University, Blacksburg, VA 24061, U.S.A.

Montforts, Prof. Dr. F.-P., Institut für Organische Chemie, FB 2 Biologie/Chemie, Universität Bremen, Postfach 330440, D-28334 Bremen, Germany

Samala, Dr. L., Department of Chemistry, M/C 0212, Virginia Polytechnic Institute and State University, Blacksburg, VA 24061, U.S.A. (current address: Combinatorial and Medicinal Chemistry, Pharmacia Corporation, 7000 Portage Road, Kalamazoo, MI 49001, U.S.A.)

Yuan, Dr. H., Department of Chemistry, M/C 0212, Virginia Polytechnic Institute and State University, Blacksburg, VA 24061, U.S.A. (current address: Allergan Inc., 2525 Dupont Dr., Mail Code RD-3D, Irvine, CA 92623, U.S.A.)

Naturally Occurring Cyclic Tetrapyrroles*

F.-P. Montforts and M. Glasenapp-Breiling
Institut für Organische Chemie, Universität Bremen,
Bremen, Germany

Contents

---

* Dedicated to Professor L. F. Tietze on the occasion of his 60th birthday.

1. Introduction

Of the porphyrinoid structures occurring in nature the most important and most widespread are the red blood pigment heme (**1**), the green pigment of plant photosynthesis chlorophyll *a* (**2**), the bacterial photosynthetic pigment bacteriochlorophyll *a* (**3**), and the "antipernicious" red pigment vitamin B_{12} (**4**). The basic function of these cofactors are determined by the incorporation of the different metal ions into the macrotetracycles. The different oxidation levels of the macrocyclic ligand system regulate the fine tuning of these functions. The final adaptation of the cofactors to their special molecular environments in the cell compartments is effected by variation of the substitution patterns of the chromophores.

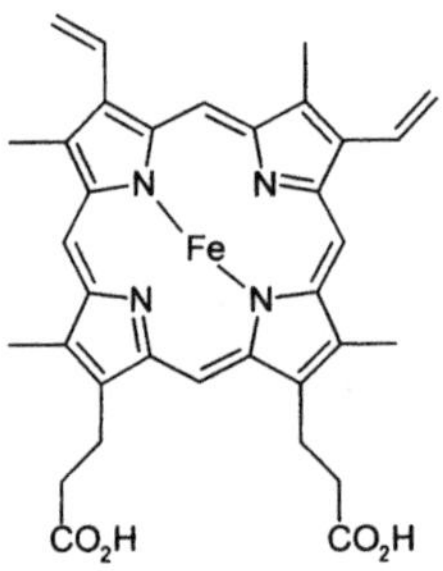

1

Heme

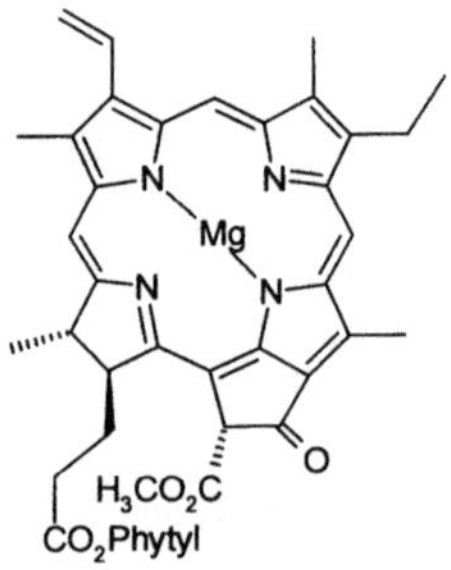

2

Chlorophyll *a*

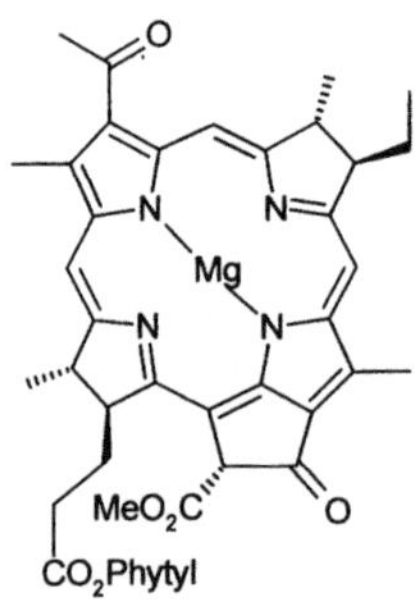

3

Bacteriochlorophyll *a*

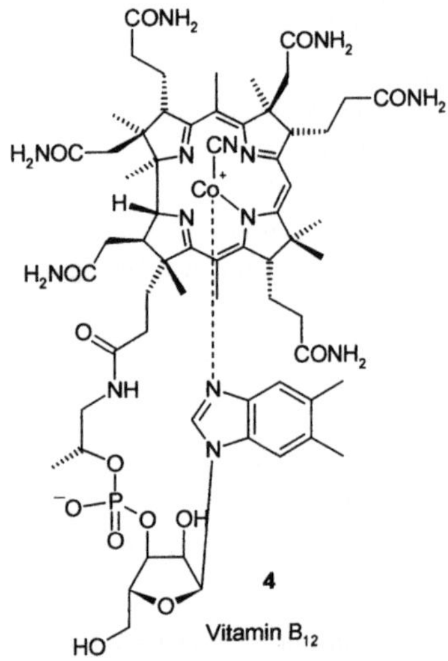

4

Vitamin B_{12}

References, pp. 42–51

Until the mid-1970s the four classic cyclic tetrapyrrolic structures with their porphyrin, chlorin, bacteriochlorin, and corrin skeletons were almost the only representatives in the class of porphyrinoid natural products (*1–10*). Although other partially reduced porphyrins were conceivable, none of these partially saturated porphyrinoid structures had hitherto been found in nature.

Over the past 25 years, two developments have extended the range of naturally occurring porphyrinoids with novel, interesting structures. The first route of development originates from the investigation of vitamin B_{12} biosynthesis (*10–12*). Novel hydroporphyrinoid structures have been discovered in the search for biosynthetic intermediates leading to vitamin B_{12}. At the same time some of these structures of vitamin B_{12} biosynthesis were identified as cofactors of redox enzymes of micro-organisms and plants (*8, 13, 14*). The second route was characterized by the deliberate search for new porphyrinoid structures in marine organisms and micro-organisms, which resulted in the discovery of novel porphyrinoid and hydroporphyrinoid compounds (*13–18*). In both cases isolation and structural elucidation of novel structures, which often occur as traces in natural sources, had become possible due to the simultaneous development of new separation techniques and modern NMR spectroscopic methods. The identification of intermediates in the biosynthesis especially of vitamin B_{12} was also decisively facilitated by application of methods from molecular biology. Last but not least the novel structures with unusual biological activities have also attracted the attention of organic synthetic chemists, who have developed synthetic pathways and have thus been able to support structure elucidation, to produce sufficient amounts of material, and to modify the natural lead structures.

2. Classification

Due to their extended conjugated π-system, porphyrins and their derivatives as well as corrins are deeply coloured and show characteristic electronic absorption spectra which vary with the saturation of the chromophoric system, with the central metal ions of the macrocycles, and the various substituents attached to the chromophores (*1, 19*). Thus, the different porphyrin derivatives can be easily distinguished from each other. The most intense absorption of prophyrins occurs at $\lambda_{max} \approx 400\,nm$ with a molar extinction coefficient of $\varepsilon \approx 400000$. This band is named after its discoverer (1883) the Soret band. Other bands with lower intensity than the Soret band occur in the long-wave length range of the absorption spectra ($\lambda = 500-800\,\text{nm}$). The most striking changes in colour and in the absorption spectra are observed with successive saturation of the completely unsaturated porphyrin chromophore (**5**). Reduction of one peripheral double bond of the porphyrin gives the green chlorin (**6**) which is named after chlorophyll. According to the IUPAC nomenclature (*20*) this structure is systematically named as 2,3-dihydroporphyrin. The reduced symmetry of the chromophore, which still contains like the porphyrin 18π-electrons in cyclic conjugation and is therefore an aromatic system, influences the electronic spectrum which shows a bathochromic shift and a higher molar extinction coefficient of the long-wavelength absorption bands compared to the porphyrin.

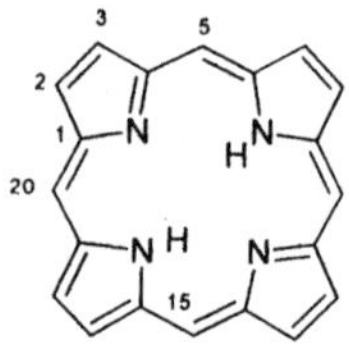

5
Porphyrin

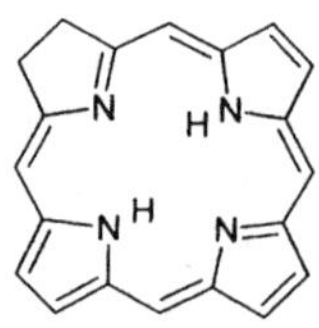

6
Chlorin

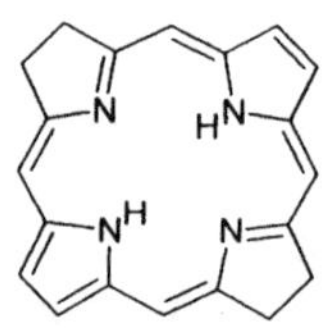

7
Bacteriochlorin

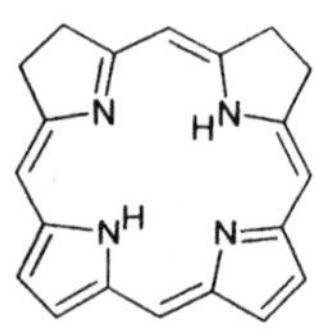

8
Isobacteriochlorin

The bacteriochlorin (**7**) (2,3,12,13-tetrahydroporphyrin) structural type named after the bacteriochlorophylls is formally derived from a porphyrin by saturation of two peripheral double bonds in opposite pyrrole rings. Again the 18π-electrons of the chromophore give rise to the aromatic character of the structure.

Isobacteriochlorins (**8**) (2,3,7,8-tetrahydroporphyrin) with two adjacent saturated pyrrole rings are the constitutional isomers of bacteriochlorins. Whereas in porphyrins, chlorins, and bacteriochlorins the 18π-perimeter includes two nitrogens excluding their lone pairs, the 18π-system of the isobacteriochlorin includes three nitrogens, one with its lone pair as part of the aromatic π-system. It is therefore less stabilized.

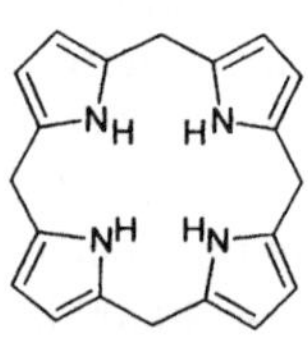

9
Porphyrinogen

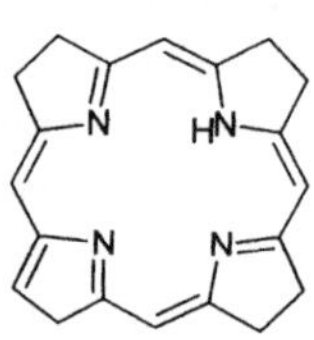

10
Corphin

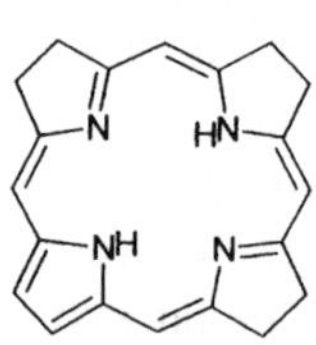

11
Pyrrocorphin

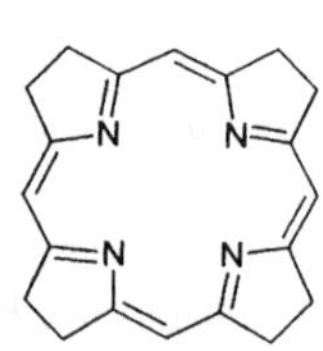

12
Tetraaza[16]annulene

Among the several hexahydroporphyrinoid structures which are tautomers on the same oxidation level, porphyrinogen (**9**), corphin (**10**), pyrrocorphin (**11**), and tetraaza[16]annulene (**12**) are especially distinguished by their structures and by their biological importance which will be discussed later in sections 3 and 8. Pyrrocorphin (**11**) has 18π-electrons in cyclic conjugation and is therefore aromatic whereas the tetraaza[16]annulene (**12**), as indicated by its name, has 16π-electrons in cyclic conjugation and is not a stabilized aromatic system.

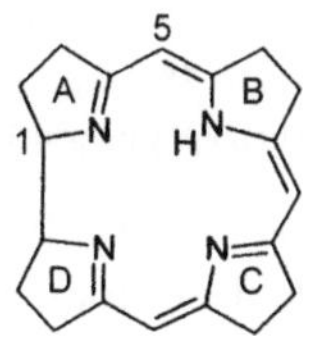

13

Corrin

Corrin (**13**) is the chromophoric skeleton of the naturally occurring vitamin B_{12} (**4**). The striking difference between porphyrinoids and corrins is the direct linkage of the pyrrole rings A and D in the latter structure originating from the loss of the 20-methine bridge. Another special feature of the corrin structure is the complete saturation of the β-periphery of the macrocycle and the interrupted cyclic conjugation. But at a first glance the corrin shows similar electronic and complexation properties as the porphyrinoid systems. The classification according to the oxidation level of the porphyrinoids will be used in this contribution as a principle to organize the different naturally occurring cyclic tetrapyrroles.

3. General Aspects of Cyclic Tetrapyrrole Biosynthesis

Key building block in the biosynthesis of all tetrapyrrolic pigments with the exception of few hexahydroporphyrins (see section 8, Figure 10) known at the present is uroporphyrinogen III (**14**) (*21a*). The framework of this key intermediate consists of four isolated pyrrole rings which are linked by methylene bridges. The prophyrinogen structure has the oxidation level of an hexahydroporphyrin (see section 2) and is therefore extremely sensitive to oxygen which easily oxidizes porphyrinogens to the porphyrins which are stabilized by aromatization.

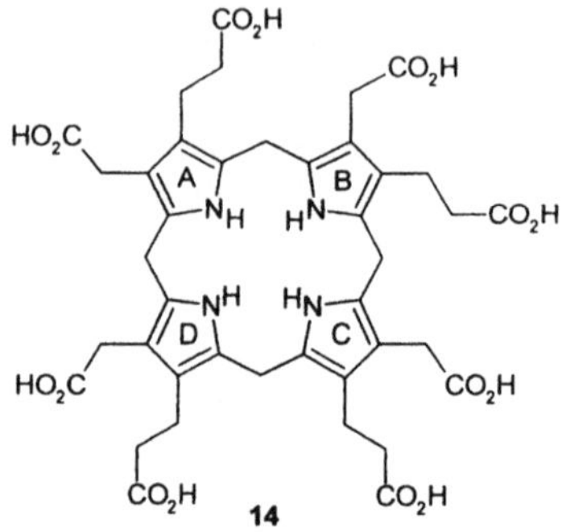

14

Uroporphyrinogen III

In the case that oxygen is not strictly excluded during isolation of porphyrinogens from biological sources the corresponding porphyrins are obtained as stable artifacts of the isolation procedure. At the β-positions each of the four pyrrole subunits bear acetic acid (A) and propionic acid (P) side chains. Three of the pyrrole rings are assembled in the order AP, AP, AP whereas the substitution pattern of pyrrole D is inverted. The character and the sequence of substituents in all naturally occurring tetrapyrroles reflect their origin from the acetic acid and propionic acid residues of uroporphyrinogen III (**14**). Especially for the biosynthesis of hemes and chlorophylls the mechanistic course of the chemical modification of carboxylic acid side chains is today well established. Uroporphyrinogen III (**14**) is built from eight molecules of δ-aminolevulinic acid (**18**), which itself is formed along two different pathways. Plants and many microorganisms produce **18** from glutamic acid (**15**) in three enzymatically controlled reaction steps, whereas animals use succinyl CoA (**16**) and glycine (**17**) as starting materials for the formation of **18**. But in many organisms both pathways, the glutamate route and the so called Shemin route, named after its discoverer, occur together.

Two molecules of δ-aminolevulinic acid (**18**) pair to yield porphobilinogen (**19**) (PBG). It is worth mentioning that the involved enzyme porphobilinogen synthase uses the same substrate (**18**) in two different modes to form the pyrrole PBG (**19**) (*21b*). Four porphobilinogen molecules are assembled through several reaction steps, controlled by two single enzymes, to generate uroporphyrinogen III (**14**) (*21*).

The porphobilinogen desaminase (**21**) which contains a sulfur linked dipyrrylmethane as prosthetic group takes up four PBGs (**19**) yielding a linear hexapyrrole. This enzyme linked hexapyrrole (**23**) is cleaved to give the hydroxybilane (**24**) which cyclizes cosynthase-controlled *via* the "spirointermediate" (**25**) leading to uroporphyrinogen III (**14**) or

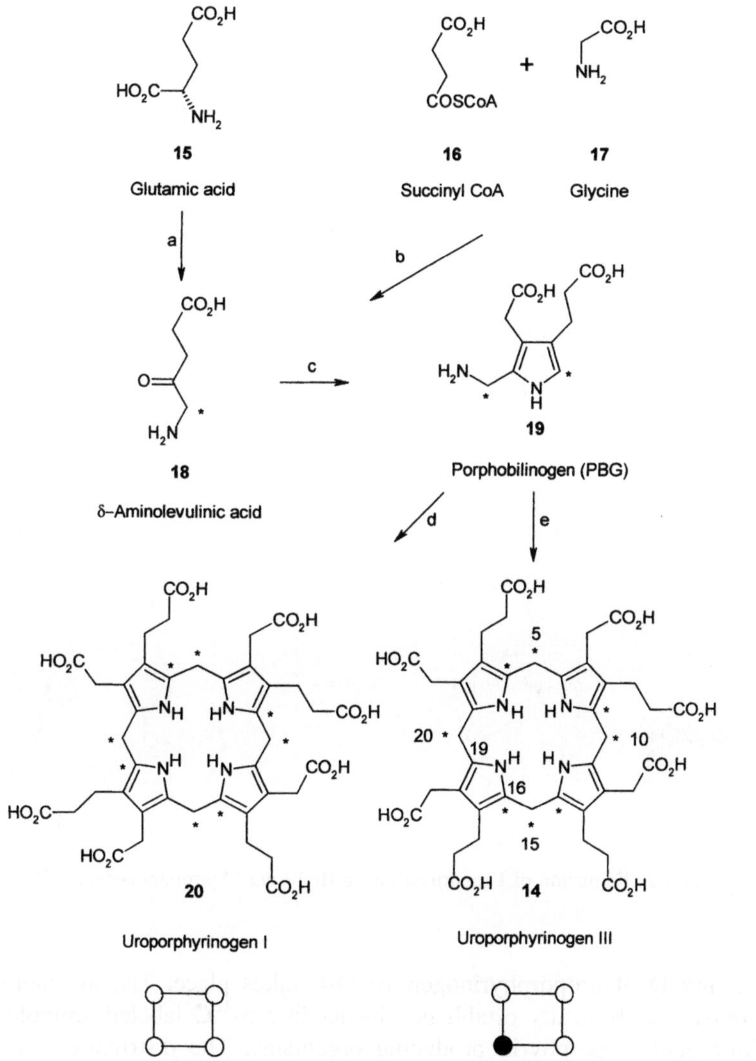

Fig. 1. Biosynthesis of Uroporphyrinogen III. **a** glutamate-*t*RNA ligase (Mg^{2+}, ATP); glutamyl-*t*RNA reductase (NADPH); glutamate-1-semialdehyde 2.1-aminomutase; **b** δ-aminolevulinate synthase (pyridoxal phosphate); **c** porphobilinogen synthase (Mg^{2+}, Zn^{2+}, K^+); **d** porphobilinogen deaminase; **e** porphobilinogen deaminase + cosynthase

alternatively in the absence of the second enzyme in a spontaneous chemical reaction to give uroporphyrinogen I (**20**). In the latter non-natural structure the acetic acid and propionic acid side chains are arranged in series. Through the "spirointermediate" (**25**) the inversion

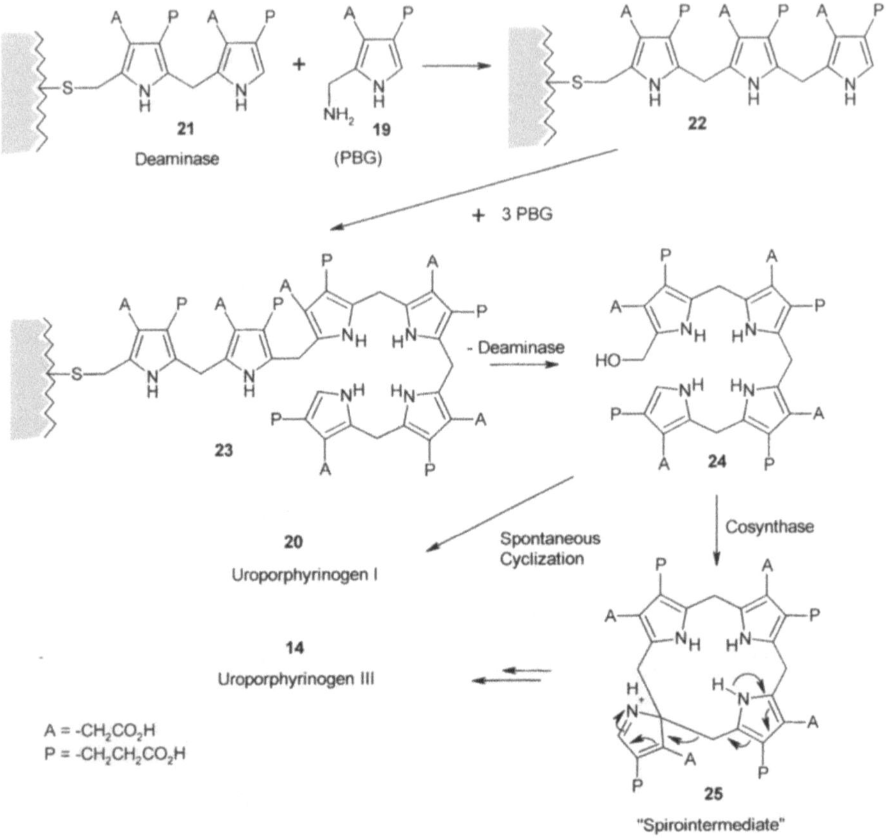

Fig. 2. Formation of Uroporphyrinogen III **14** *via* "Spirointermediate" **25**

of ring D of uroporphyrinogen III (**14**) takes place. The mechanistic course was basically established by feeding 5-^{13}C labeled aminolevulinic acid to porphyrin producing organisms. The distribution of the isotopic label shows always the same characteristic pattern for all the biosynthesized porphyrinogens/porphyrins namely the shift of one labelled C-atom from the 19-position into position 16, so that three labelled carbon atoms are linked together. It is still a mystery why nature makes use of the uroporphyrinogen III type from the four possible constitutional isomers of uroporphyrinogen (*21c*).

An answer could come from experiments which demonstrate that under "chemical" or "prebiotic" reaction conditions the cyclotetramerization of four PBG units gives a mixture of the four possible

References, pp. 42–51

uroporphyrinogen constitutional isomers I–IV in which the III-type is preferred due to statistical reasons (*7*).

4. Porphyrins

4.1. Occurrence, Structure, and Biological Function

Hemes are the prosthetic groups of different proteins like hemoglobin, myoglobin, cytochromes, catalases, and peroxidases. The central

Table 1. Naturally occurring Hemes and Porphyrins

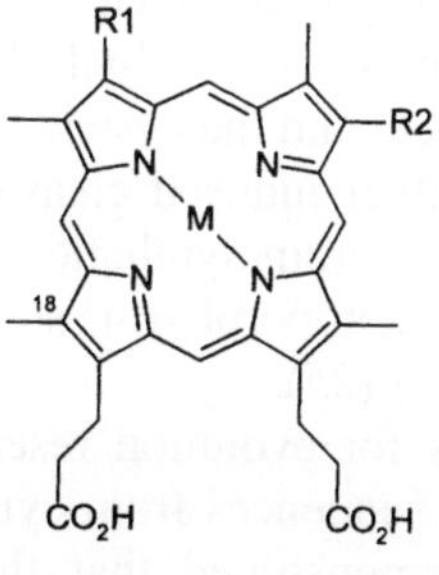

No.	R1	R2	M	Name
1	-CH=CH_2	-CH=CH_2	Fe	Heme
26	-CH=CH_2	-CH=CH_2	2H	Protoporphyrin IX
27	-CH=CH_2	-CHO	Fe	Chlorocruoroheme (Spirographis Heme)
28	OH -CH-CH_2-Farnesyl	-CH=CH_2	Fe	Heme o
29	OH -CH-CH_2-Farnesyl	-CH=CH_2	Fe 18^1-oxo	Heme a
30	S-$CH_2CH(NH_2)$-CO_2H -CH-CH_3	S-$CH_2CH(NH_2)$-CO_2H -CH-CH_3	Fe	Heme c
31	OH -CH-CH_3	OH -CH-CH_3	2H	Hematoporphyrin
32	-CH_2-CH_2-CO_2H	-CH_2-CH_2-CO_2H	2H	Coproporphyrin III
33	-CH=CH_2	-CH_2-CH_2-CO_2H	2H	Harderoporphyrin
34	-CH=CH-CO_2H(*E*)	-CH_2-CH_2-CO_2H	2H	Porphyrin S-411
35	-H	-CH=CH_2	2H	Pemptoporphyrin

iron atoms of the hemes are responsible for their functions, namely oxygen transport, oxygen activation, and electron transfer (*22*). Heme (**1**) the most common porphyrinoid pigment of myoglobin, hemoglobin and other hemoproteins is non-covalently linked to the proteins and basically involved in oxygen and electron transport along the respiratory chain. Chemically modified hemes occur in some organisms.

Chlorocruoro heme (**27**) is the prosthetic group of chlorocruorin, the oxygen-carrying pigment of certain polychete worms (*23*). Heme *a* (**29**) and heme *o* (**28**) contain a long lipophilic farnesyl residue at ring A of the porphyrin framework. Heme *a* (**29**) occurs in different tissues (*24*) and organisms for instance in the heart muscle of mammalians. Heme *o* (**28**) is found in the respiratory chain of *Escherichia coli* and other bacteria. C-type cytochromes, as well as proteins of the respiratory chain, contain heme *c* (**30**) as a prosthetic group which is covalently linked *via* thioether bridges to cystein residues. On acid hydrolysis heme *c* (**30**) consisting of the porphyrin macrocycle and the cystein residues is removed from the protein, whereas the silver induced cleavage yields optically active hematoporphyrin (**31**) (*23*). From synthetic studies the relative configuration of the formed hematoporphyrin stereoisomer could be assigned as (3′*S*, 8′*R*) or (3′*R*, 8′*S*) (*25*).

Cytochromes *c* are important proteins for evolution research. The comparison of more than 100 amino acid sequences from cytochromes *c* from very different organisms has demonstrated that the protein was only modified to a minor extent during evolution so that a genealogical tree of organisms could be established by that comparison (*25c*).

Naturally occurring metal free porphyrins are rare. Protoporphyrin IX (**26**), coproporphyrin III (**32**), harderoporphyrin (**33**), uroporphyrin I, and uroporphyrin III are all formed from the corresponding porphyrinogens on oxidation by air during the isolation procedure from natural sources (*23*). As will be demonstrated in section 4.2. these porphyrinogens are the "true" intermediates in the biosynthesis of hemes. Humans especially suffering from a disordered porphyrin metabolism caused by endogenous or exogenous factors produce large amounts of such porphyrins. The porphyrins are excreted but also deposited in bones and the skin, where they show severe phototoxic activities on exposure to light. Interesting is the occurrence of harderoporphyrin (**33**) (*23*) and porphyrin S-411 (**34**) (*23*) in certain tissues of mammalians. The harderian gland which is located in the orbital cavity and surrounds the optical nerve contains harderoporphyrin (**33**) in rather high concentration. Porphyrin S-411 (**34**) was detected in calf and foal meconium and in human bile. Pemptoporphyrin (**35**) (*23*) was isolated from feces of patients with intestinal absorption as one of five porphyrins [pempto (greek) = fifth]. It is assumed that it is formed by microorganisms in the intestinal tract following oral administration of antibiotics.

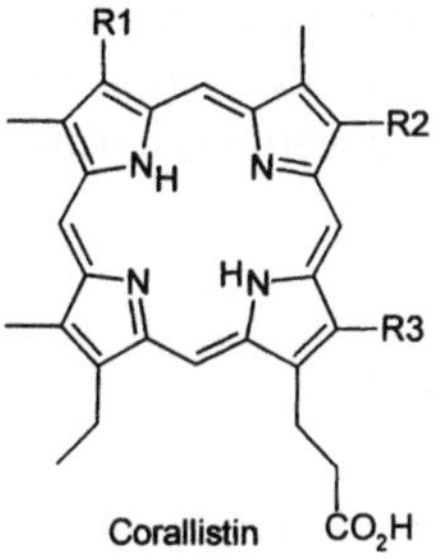

Corallistin

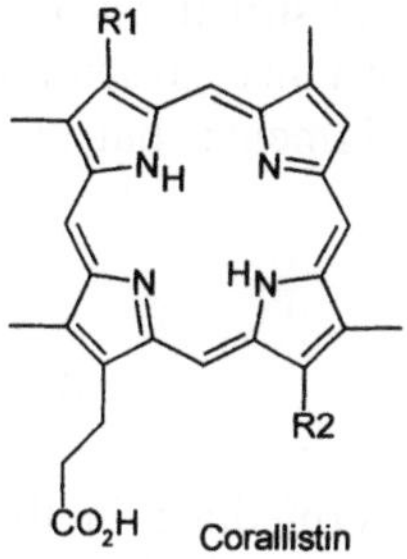

Corallistin

a R1= -H R2= -Et R3= $-CH_2-CO_2H$ **36**

b R1= -Et R2= -Et R3= $-CH_2-CO_2H$ **37**

c R1= -H R2= -H R3= $-CH_3$ **38**

d R1= -Et R2= $-CH_2-CH_2-CO_2H$ **39**

e R1= -H R2= -Et **40**

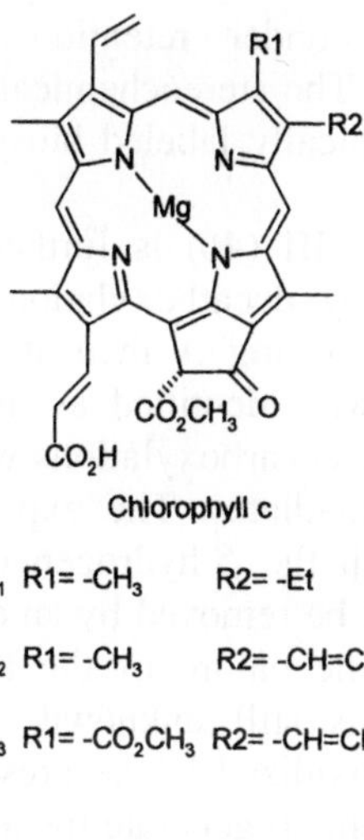

Chlorophyll c

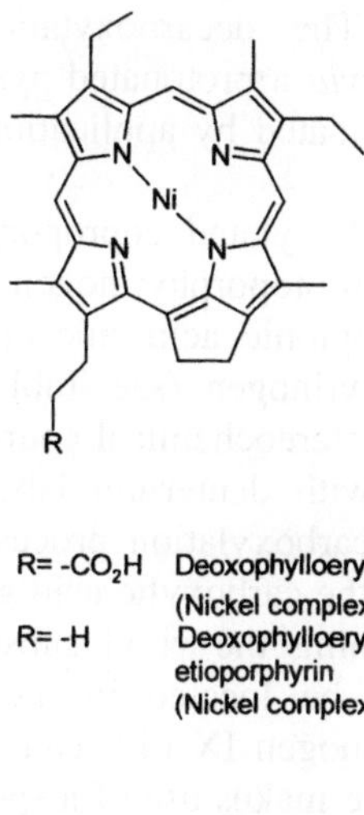

c_1 R1= $-CH_3$ R2= -Et **41**

c_2 R1= $-CH_3$ R2= $-CH{=}CH_2$ **42**

c_3 R1= $-CO_2CH_3$ R2= $-CH{=}CH_2$ **43**

R= $-CO_2H$ Deoxophylloerythrin (Nickel complex) **44**

R= -H Deoxophylloerythro-etioporphyrin (Nickel complex) **45**

Unique is the occurrence of different porphyrin derivatives (**36**–**40**) named corallistin a–e (*26*) after the marine sponge *Corallistes sp.* which was found near to New Caledonia. The chlorophylls c_1–c_3 (**41**–**43**) (*27*) of some algae also belong to the porphyrin class though chlorophyll pigments usually have the chlorin type framework. An interesting source of mainly nickel or other transition metal containing porphyrins are sediments. Out of numerous porphyrins isolated from sediments (*28, 31*) only nickel complexes of deoxophylloerythrin (**44**) and deoxophyllo-erythro-etioporphyrin (**45**) should be mentioned here, because they represent the most common structures. In most cases conclusions can be drawn from the structures of fossil porphyrins regarding the biological origin. For instance structures **44** and **45** are derived from different chlorophyll pigments, but also heme derived geoporphyrins are quite common.

4.2. Biosynthesis

A key intermediate in the biosynthesis of hemes as well as for chlorophylls is protoporphyrin IX (**26**) (*29*) (Fig. 3). The formation of protoporphyrin IX (**26**) from uroporphyrinogen III (**14**) starts with the successive enzymatic decarboxylation of the acetic acid side chains beginning at ring D and proceeding clockwise over ring A and B to ring C. All intermediates which are produced by a single enzyme have been identified by comparison with synthetically derived compounds. The decarboxylation occurs under retention of the configuration *via* a protonated pyrrole (**48**). The stereochemical course could be elucidated by application of isotopically labeled biosynthetic precursors.

Tetradecarboxylated coproporphyrinogen III (**46**) is further transformed into protoporphyrinogen IX (**47**) by decarboxylation of the (**50** → **51**) propionic acid side chains first at ring A then at ring B. Harderoporphyrinogen (see Table 1, **33**) was identified as an intermediate. The stereochemical course of these decarboxylations was also investigated with deuterium labelled intermediates. The experiments showed a decarboxylation process in which the β-hydrogen oriented antiplanar to the carboxylic acid group must be removed by an external acceptor forming the vinyl substituents. The nature of the acceptor which might be located in the enzyme is still unknown. Though protoporphyrinogen IX (**47**) can be easily oxidized in the presence of oxygen, nature makes use of a special enzyme to generate the aromatic protoporphyrin IX (**26**), whereby the enzymatic dehydrogenation proceeds stereoselectively. But the absolute stereochemical course could not yet been determined. The formation of chlorophylls requires further reaction steps at the protoporphyrin framework whereas the biosynthesis of hemes is completed by enzymatic insertion of the central iron ion.

References, pp. 42–51

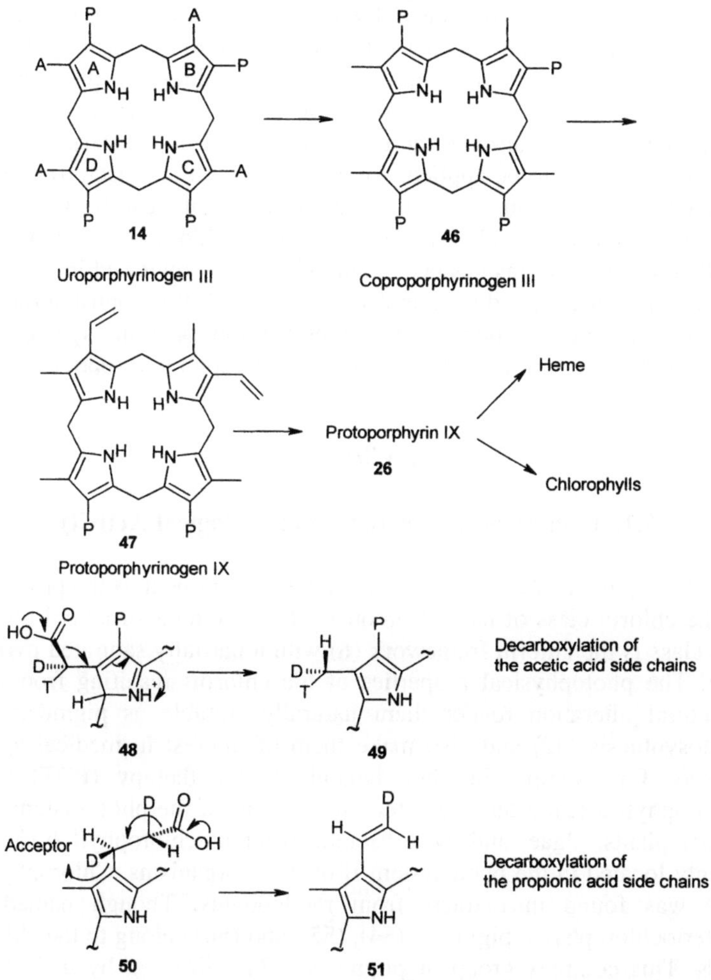

Fig. 3. Biosynthesis of Protoporphyrin IX (**26**) from Uroporphyrinogen III

4.3. Aspects of Synthesis

Synthesis of naturally occurring porphyrins is a well developed field in organic chemistry, so that today numerous synthetic strategies are available to prepare desired porphyrinoid targets. At the present new developments in porphyrin synthesis are mainly directed to the preparation of non-natural porphyrins with unusual substitution patterns, with distorted or extended chromophores and with new functions. Results from this research could have an impact on the synthesis of naturally occurring porphyrins in the future. Pyrroles, dipyrrylmethanes, dipyrrylmethenes and bilanes (linear tetrapyrroles) with different oxidation states are the most important building blocks for the synthesis of porphyrins. The kind of substitution pattern and symmetry of the target porphyrins determine the choice of the starting material and the synthetic strategy to be applied. Another important approach to prepare porphyrinoid compounds is the modification of readily accessible naturally occurring porphyrinoids like heme or chlorophyll *a*. Since the synthesis of porphyrins has been comprehensively reviewed in journals, monographs, series, and handbooks (*1, 19, 30, 31*) this contribution will not cover the field of porphyrin synthesis. Progress in the synthesis of hydroporphyrins will be addressed in the corresponding sections.

5. Chlorins

5.1. Occurrence, Structure, and Biological Activity

Chlorophyll *a* (**2**), the green photosynthesis pigment, is the prototype of the chlorin class of natural products. The common structural unit in this class is the chlorin framework (**6**) with a partially saturated pyrrole ring. The photophysical properties of the chlorins resulting from this structural alteration render them naturally suitable as pigments for photosynthesis (*32*) and also make them of interest in medical applications, for example in photodynamic tumor therapy (PDT) (*33*). Chlorophyll *a* (**2**) is the characteristic pigment of the photosystem I of higher plants, algae, and cyanobacteria whereas chlorophyll *b* (**52**) is mainly located in the photosystem II of these organisms. Chlorophyll *d* (**53**) was found in extracts from rhodophytes. Though named as bacteriochlorophylls, pigments (**54**), (**55**), and (**56**) belong to the chlorin class. This complex group of pigments (*27a*), differing by their alkyl substituents and by their alcohol components at the 17-propionic acid function, is present in green and brown or red sulfur bacteria Chlorobiaceae and in Chloroflexaceae. Compared to chlorophyll *a* and *b* these bacteriochlorophylls lack the 13^2-carbomethoxy group, and they bear an α-hydroxyethyl substituent at position C-3.

The absolute configuration at C-3′ has been determined by different chemical and physical methods (*27a*) showing that both configurations occur. Another interesting structural feature of the bacteriochlorophylls of the *c* and *e* series is the methyl group at the 20-methine position. The mentioned photosynthetic bacteria contain only one photosystem which makes use of hydrogensulfide or organic compounds as hydrogen donors

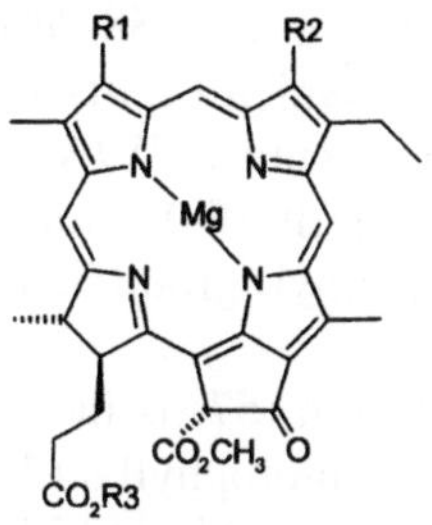

Chlorophyll

a R1= $-CH=CH_2$ R2= $-CH_3$ **2**

b R1= $-CH=CH_2$ R2= -CHO **52**

d R1= -CHO R2= $-CH_3$ **53**

R3= 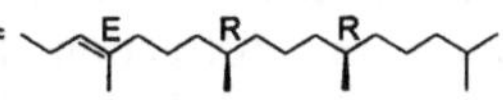

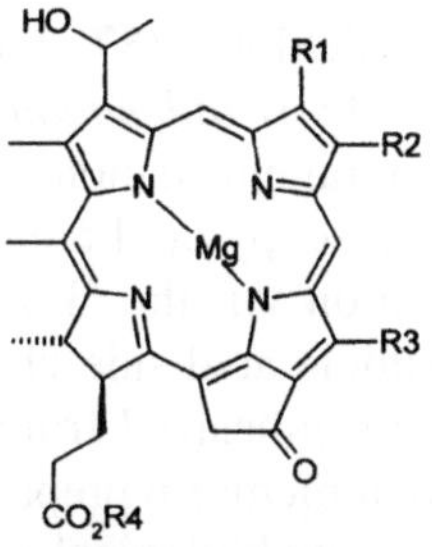

Bacteriochlorophyll

c R1= $-CH_3$ **54**

e R1= -CHO **55**

R2= Et, *n*-Pr, *i*-Bu or *neo*-Pent

R3= Me or Et

R4= farnesyl or geranyl

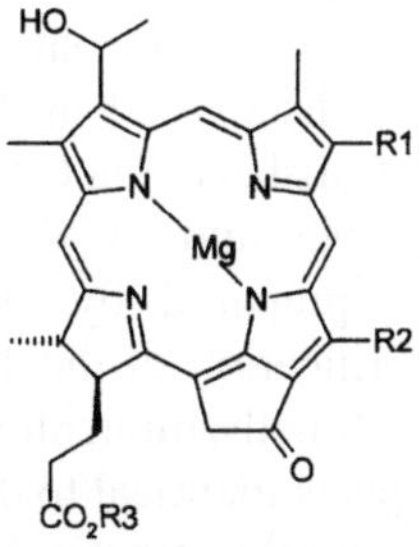

Bacteriochlorophyll *d* **56**

R1= Et, *n*-Pr, *i*-Bu or *neo*-Pent

R2= Me or Et

R3= farnesyl

for photosynthesis. A not yet fully structurally identified derivative of bacteriochlorophyll *c* or *d* was recently discovered as a visual pigment of the deep-sea dragon fish *Malacosteus niger* (*34*).

Novel chlorin structures have been isolated from various marine organisms and their substitution patterns suggest their origin from chlorophyll *a*.

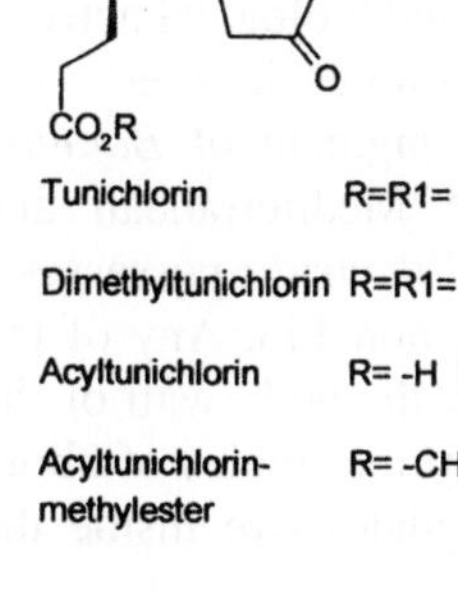

Tunichlorin	R=R1= -H	**57**
Dimethyltunichlorin	R=R1= $-CH_3$	**58**
Acyltunichlorin	R= -H R1= fatty acids	**59**
Acyltunichlorin-methylester	R= $-CH_3$ R1=fatty acids	**60**

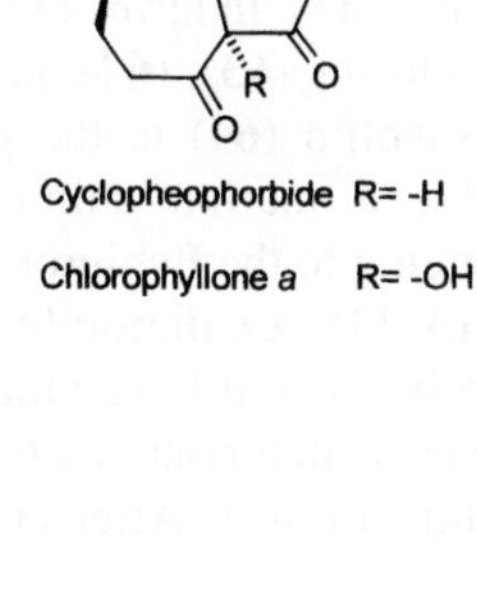

Cyclopheophorbide R= -H **61**

Chlorophyllone *a* R= -OH **62**

The nickel containing tunichlorin (**57**) and its derivatives (**58–60**) were discovered in the tunicate *Trididemnum solidum* which occurs in the Carribean Sea (*35*). Structural elucidation was carried out by UV/Vis, CD, MS, and ^{1}H NMR spectroscopy on dimethyltunichlorin (**58**), which was prepared by etherification of the 3-hydroxymethyl group and esterification of the 17-propionic acid side chain.

Dimethyltunichlorin (**58**) synthesized from tunichlorin (**57**) is in all respects identical to dimethyltunichlorin prepared from chlorophyll *a* (**2**) by partial synthesis. However, the biological function of tunichlorin (**57**) is still unknown. Cyclopheophorbide (**61**), which occurs in its enol structure with the C-C-double bond located in the seven membered ring, and chlorophyllone *a* (**62**), which are closely related to each other, were found in the sponge *Darwinella oxeata* and in the short-necked clam *Ruditapes philippinarum.* Cambie *et al.* (*36*) have isolated cyclopheophorbide (**61**) and determined its structure by X-ray diffraction. It is worth noting, that the structure of naturally occurring cyclopheophorbide, which was first isolated in 1986, is identical to that of the cyclopheophorbide prepared from chlorophyll *a* (**2**) by partial synthesis in Eschenmoser's laboratory (*37*) some years before. Structural elucidation of chlorophyllone *a* (**62**), an oxidation product of (**61**), was achieved by classical spectroscopic methods. Both compounds exhibit antioxidative properties and protect the animals containing the chlorin derivatives (**61**) and (**62**) against unwanted oxidation processes. Products of a further oxidation (*38*) of chlorophyllone *a* (**62**) named as chlorophyllone lactone *a*, and purpurin 18 were not only found in *Ruditapes philippinarum* but also in several other marine organisms. Chlorophyllone lactone *a* is the oxidation product of (**62**) in which the five-membered ring has undergone a Bayer-Villiger oxidation. From this lactone chlorophyllonic acid *a* and purpurin 18 are derived by hydrolysis and further oxidation. As can be deduced from their occurrence in animals, pigments (**61**) and (**62**) have no photosynthetic activity (*39*). Another group of chlorins, which are not related to the chlorophylls, is characterized by geminally disubstituted structural parts in the saturated pyrrole rings. In terms of their substitution pattern and biological activity the chlorins (**63**), (**64**), and (**65**) differ markedly from each other.

Bonellin (**64**) is the green sex-differentiating pigment of *Bonellia viridis*, a marine animal found throughout the Mediterranean and belonging to the Echiuroida class of animals. *Bonellia viridis* possesses a remarkable sex dimorphism, which is induced by bonellin. Any of the initially asexual larvae that come into contact with the body wall of the female, which contains the green bonellin, develop into males of about 1–3 mm in size. After contact with bonellin the males live inside the

References, pp. 42–51

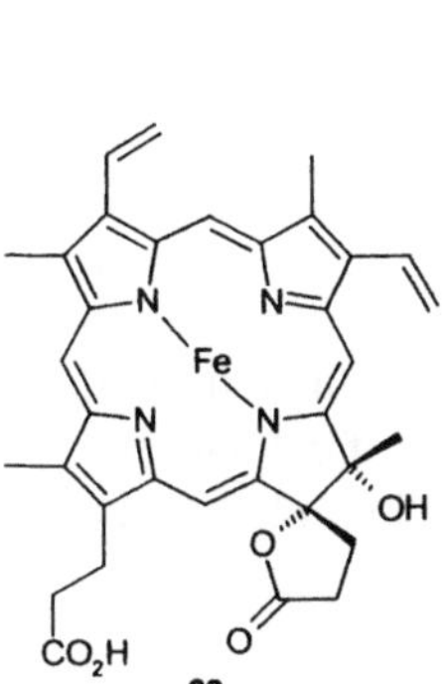

63

Heme d

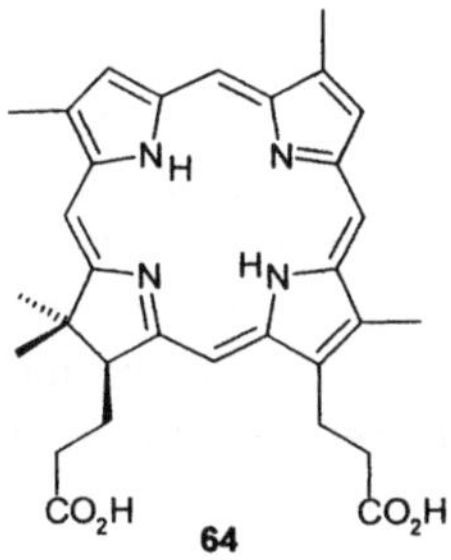

64

Bonellin

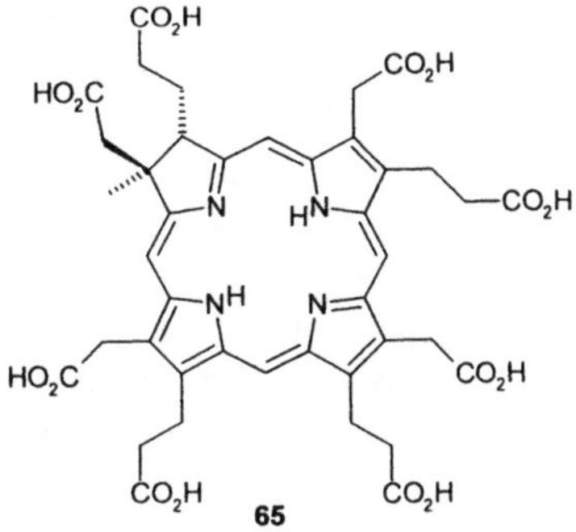

65

Factor I

body cavity of the larger females (15 cm). The female of the species develop from those larvae which have had no contact with **64** (*13*). Although pure crystalline bonellin (**64**) was first isolated by Lederer *et al.* in 1939 (*40*), it was not until 1976 that Pelter *et al.* determined the constitutional formula of bonellin by modern spectroscopic methods (*41*). Chemical degradation was used to elucidate the absolute configuration of bonellin (**64**) (*42*). The ring D degradation product of **64** was found to be constitutionally and configurationally identical to the already familiar ring C degradation product of vitamin B_{12} (**4**). Factor I (**65**) was first isolated as its octamethyl ester by Müller *et al.* (*43*) and by Arigoni *et al.* (*44*) from bacteria that produce vitamin B_{12}, such as *Propionibacterium shermanii* and *Clostridium tetanomorphum.* Its reduced form precorrin-1 is the first link in the biosynthetic chain between uroporphyrinogen III (**14**) and vitamin B_{12} (**4**) (*11, 12*). Apart from this, factor I (**65**) has no other biological function. The structural determination of factor I was greatly simplified by the fact that it originates during the biosynthetic pathway leading to vitamin B_{12}. Structural alternatives for factor I (**65**) were thus restricted.

In contrast to bonellin (**64**) and factor I (**65**), heme *d* (**63**) is bishydroxylated in the periphery of ring C, which lends it the characteristic geminally disubstituted structural part and the typical chlorin chromophore (*13, 45*). One of the hydroxyl groups is part of a spirolactone structure. But there is no certainty whether the spirocyclic lactone structure is the natural structural element or whether it is generated during isolation. X-ray structure investigations (*46*) on the catalase of *Penicillium vitale* and on a catalase hydroperoxidase of *Escherichia coli* demonstrated the *cis* configuration of the oxygen substituents with (17*S*,18*R*) absolute configuration.

As is typical in many bacteria, the respiratory chain of *Escherichia coli* contains two terminal oxidases, cytochrome *o* and cytochrome *d*. Heme *d* (**63**) is the prosthetic group of the terminal oxidase cytochrome *d* (*47*). Cytochrome *o* prevails in the early exponential phase during aerobic growth of the culture. Cytochrome *d* plays an important role in the late exponential phase, or when cells are grown with a limited oxygen supply. The K_m value (*48*) for the oxygen-binding of cytochrome *d* is about 8 times lower than that of cytochrome *o*. This allows the bacteria to grow under conditions of low oxygen supply.

5.2. Biosynthesis

Whereas elucidation of the biosynthesis of hemes and vitamin B_{12} can be considered as more or less completed, the investigation of chlorin biosynthesis is still awaiting a breakthrough. Only the formation of factor I (**65**), whose reduced form is itself an intermediate in the biosynthesis of vitamin B_{12} (see section 9) is well established (*43, 44*).

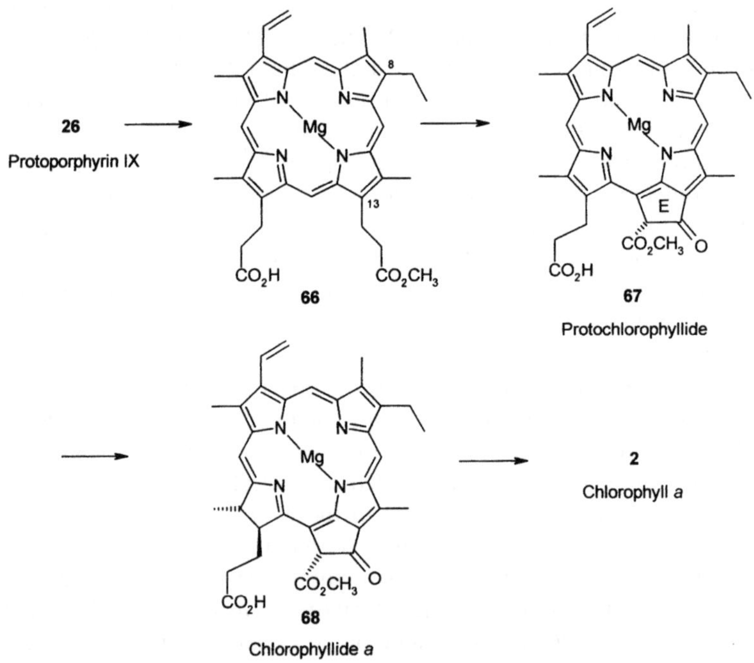

Fig. 4. Last steps in the Biosynthesis of Chlorophyll *a* (**2**)

Methylation of uroporphyrinogen III (**14**) by means of *S*-adenosylmethionine (SAM) forms directly the reduced form of (**65**). Factor I is obtained as an artefact by oxidation during the isolation procedure. Typically as for most of the methylation reactions with SAM the methyl group is transferred with inversion of configuration. Biosynthesis of heme (**1**) and chlorophyll *a* (**2**) as well as other chlorophyll derivatives branches out on the protoporphyrin IX (**26**) stage.

Protoporphyrin IX (**26**) is enzymatically generated in solution so that it is quite easy to isolate intermediates of the biosynthetic pathway (*49*). However the second part of chlorophyll biosynthesis takes place in the chloroplasts (*49, 50*). All enzymes participating in the formation of chlorophylls are located in the membrane and closely associated to each other so that the isolation of intermediates is very difficult. The identification of biosynthetic intermediates is also complicated by the fact that the structures become more and more lipophilic at the end of the biosynthetic chain making them insoluble in water. Another difficulty results from the lack of specificity of the enzymes involved in the late chlorophyll biosynthesis leading very often to an interchange of reaction steps in the biosynthetic sequence. Due to this very complicated situation it is intended here to give only a rough overview of the biosynthesis of chlorophyll *a* (**2**) (*49–51*).

In the first steps of the last section of chlorophyll biosynthesis magnesium is enzymatically introduced into protoporphyrin IX (**26**) followed by esterification of the 13-propionic acid residue by Mg-protoporphyrin-*O*-methyl transferase with SAM as a cofactor. At which point the 8-vinyl group is reduced is still unknown. The formed Mg-porphyrin (**66**) undergoes ring closure with the 17-propionic acid to give protochlorophyllide (**67**) with the typical isocyclic ring E of chlorophylls. The porphyrin macrocycle of (**67**) is reduced to yield the chlorin chromophore of chlorophyllide *a* (**68**).

Enzymatic esterification (*51*) by chlorophyllase and with phytylpyrophosphate completes the biosynthetic sequence and gives chlorophyll *a* (**2**). Phytylpyrophosphate itself is produced by NADPH reduction of geranyl geranylpyrophosphate. But it is also possible that esterification occurs before reduction. As can be expected from the esterification reagent and was demonstrated by application of isotopically labeled compounds the oxygen of the *O*-phytyl residue originates from the carboxylic group.

The ring E forming reaction step and the reduction yielding chlorophyllide *a* (**68**) are very complex from the mechanistic point of view. In addition, the reduction requires light. A detailed elucidation of the complex process is therefore very difficult. Though the substitution patterns of heme *d* (**63**) and bonellin (**64**) suggest a biosynthetic origin from uroporphyrinogen III (**14**), attempts to investigate the biosynthesis of these structures have not yet been made. Structures like tunichlorin (**57**) and cyclopheophorbide (**61**) are most likely formed from chlorophylls and finally from uroporphyrinogen III (**14**), but here also biosynthetic investigations are still lacking.

5.3. Aspects of Synthesis

Although methods for the preparation of chlorins were available when the discovery of novel chlorins like factor I (**65**) and bonellin (**64**) began, Woodward's total synthesis of chlorophyll *a* (**2**) (*52*) was the only selective pathway for the construction of chlorins. Woodward's ingenious synthesis of chlorophyll *a* (**2**), which is now published with full experimental details (*52c*), is based, in the first part of the synthetic route, on a classical porphyrin synthesis. The porphyrin intermediate is

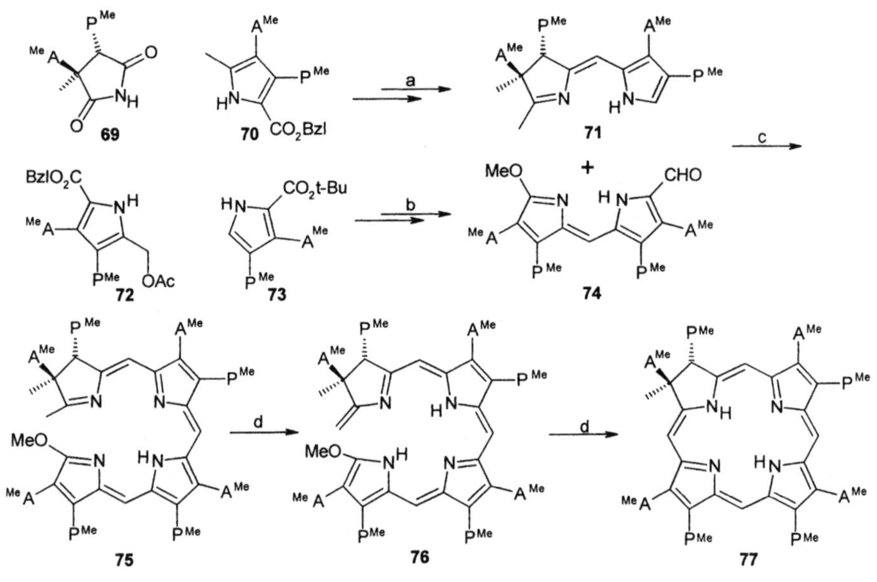

$P^{Me} = CH_2CH_2CO_2Me$

$A^{Me} = CH_2CO_2Me$

Fig. 5. **a**, **b** several steps (Ref. 55*a*); **c** TFA, rt, **d** hν (visible), TFA, *i*-Pr_2NEt, THF [34% rel (**71**) and (**74**)]

References, pp. 42–51

then transformed into a chlorin. But this transformation is restricted to the special substitution pattern found in chlorophyll *a* (**2**) and cannot be generalized for the synthesis of any other chlorin. Because Woodward's chlorophyll synthesis is well documented (*52*) it should not be reviewed in this contribution. The geminally dialkylated structural parts typical for factor I (**65**) and bonellin (**64**) cannot be created with Woodward's approach and require new synthetic concepts.

The laboratories of Battersby (*13, 53*) and Montforts (*13, 54*) developed selective syntheses for chlorins on model systems, which contain the typical dialkylated parts in the saturated rings of the chlorin structure. The experiences gained from these investigations were later used to synthesize naturally occurring chlorins.

Following the model studies, factor I octamethylester (**77**) (*53f, 55*) and bonellin dimethylester (**85**) (*53c, 56, 57*) were synthesized by the mentioned researchers. Battersby's synthesis of factor I (*53f, 55*) makes use of two dipyrrolic building blocks (**71**) and (**74**) which were prepared according to standard methods from pyrrole/porphyrin chemistry using the monopyrrolic starting compounds (**69**), (**70**), (**72**), and (**73**). The optically active AB component (**71**) is condensed at the α-free pyrrolic position under acidic reaction conditions with the aldehyde function of the DC part (**74**) to give the tetrapyrrolic biline intermediate (**75**). For the final cyclization step a new photochemical procedure was applied which also has been proved to be very useful for the synthesis of several hydroporphyrins (*53*). A plausible mechanism for the ring closure includes the formation of the enamine tautomer (**76**) which cyclizes in a photochemical 18π-σ isomerization followed by elimination of methanol yielding the target structure (**77**).

The total synthesis of bonellin dimethylester (**85**) could be achieved by a very similar approach as for (**77**) including the photochemical cyclization (*53c, 56*). The approach to bonellin dimethylester (**85**) (*57*) developed by Montforts *et al.* makes use of four monopyrrolic building blocks (**78**), (**79**), (**81**), and (**83**) which were assembled together by methods from porphyrin, bile pigment, and corrin chemistry to give the macrotetracycle.

Ring B pyrrole aldehyde (**78**) and ring C lactam (**79**) form under base induced condensation and subsequent introduction of sulfur the bicyclic BC thiolactam (**80**). This bicyclic lactam was coupled by a modification of Eschenmoser's sulfide contraction method (*58*) with the monocyclic chiral D building block (**81**) to yield after ester cleavage and complexation with nickel (II) the tricyclic intermediate (**82**). After hydrogenolytic cleavage of the benzylester in (**82**) the crude carboxylic acid formed with the ring A pyrrole aldehyde (**83**) the linear tetrapyrrole

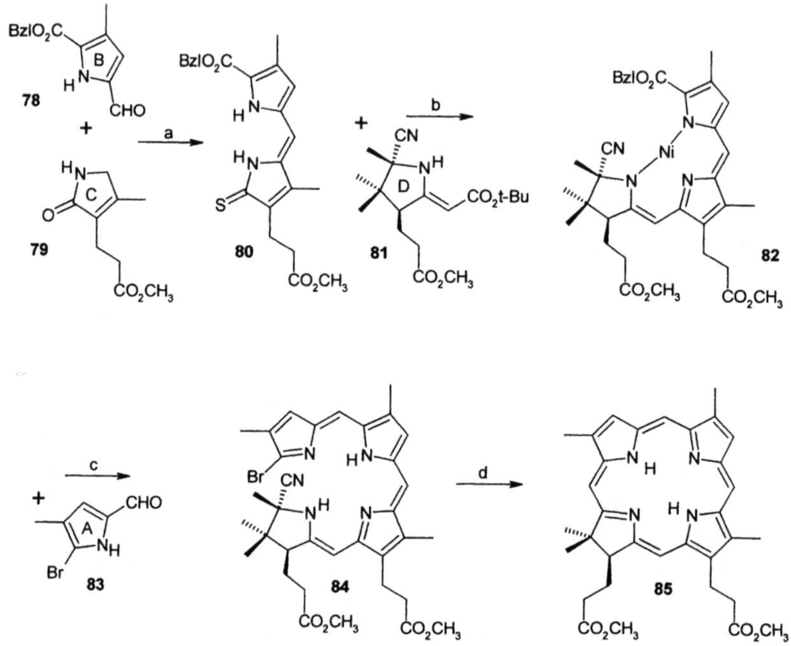

Fig. 6. **a** 2-(*tert*-Butylimino)-2-(diethylamino)-1,3-dimethylperhydro-1,2,3-diazophosphorine, benzene; Lawesson's reagent, THF; **b** NBS + **81**, CH_2Cl_2; DBU, CH_3CN; $P(CH_2CH_2CN)_3$, TFA, benzene; $Ni(OAc)_2 \times 4H_2O$, NaOAc, CH_2Cl_2; **c** Pd/C, H_2, CH_3OH; **83**, *p*-TsOH, $CHCl_3$; **d** $Zn(OAc)_2$, *t*-BuOK, *t*-BuOH; CH_2N_2, ether

(**84**). Reaction with *t*-butyl alkoxide in the presence of zinc (II), which acts as a template and can be readily removed with acid from the cyclized ligand system that is formed, finally results in the formation of bonellin dimethylester (**85**) from the linear tetrapyrrole (**84**).

Tunichlorin (**57**) (*13, 35a,b*) and cyclopheophorbide (**61**) (*13, 37*) as well as derivatives of both structures were obtained by partial synthesis starting from methylpheophorbide *a* as a common intermediate (*59*).

Methylpheophorbide *a* itself can be formed from chlorophyll *a* by removal of the central magnesium ion and transesterification with methanol. Partial synthesis was also used for the preparation of heme *d* (**63**) (*13, 60*) along a multistep reaction sequence starting from protoporphyrin IX dimethylester (**26**).

References, pp. 42–51

6. Bacteriochlorins

6.1. Occurrence, Structure, and Biological Activity

The bacteriochlorin structural type is formally derived from the porphyrin by saturation of two peripheral double bonds in opposite pyrrole rings. The term "bacteriochlorin" originates from the bacteriochlorophylls, which are widespread as pigments in bacterial photosynthesis (*1, 13, 27a*).

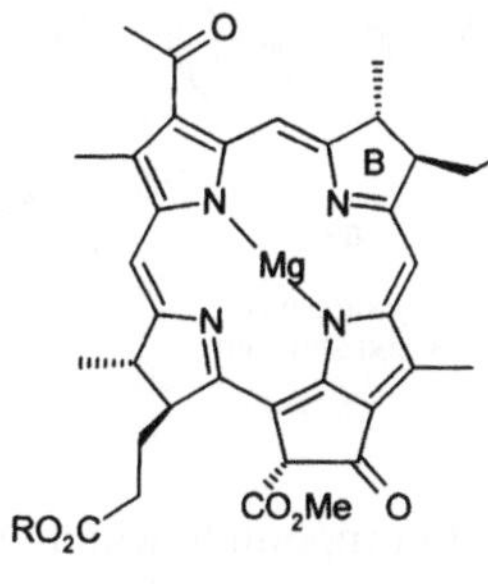

3

Bacteriochlorophyll *a*

R = Phytyl

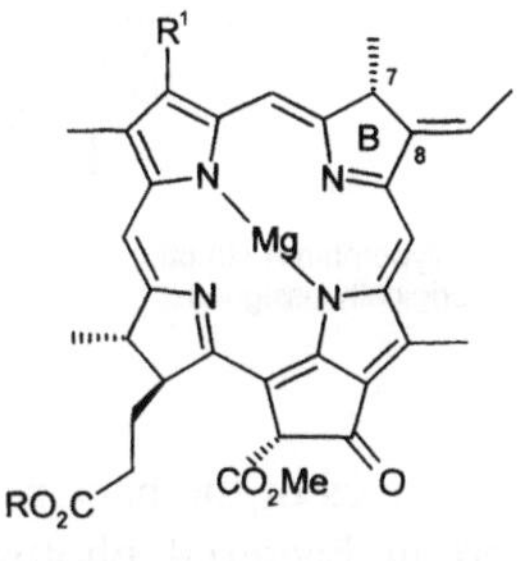

Bacteriochlorophyll *b* $R^1 = COCH_3$ R = Phytyl **86**

Bacteriochlorophyll *g* $R^1 = -CH{=}CH_2$ R = Farnesyl **87**

Bacteriochlorophyll *a* (**3**) was for a very long time considered as the only representative of the bacteriochlorophylls. But then bacteriochlorophylls *b* (**86**) and *g* (**87**) (*61*) were discovered, which are on the oxidation level of a chlorin because the peripheral double bond of ring B is only shifted by a tautomeric process into an exocyclic position and not hydrogenated. Nevertheless the structure should be considered as a bacteriochlorin because conjugation is interrupted and electronic properties match more those of typical bacteriochlorins. Also electronic absorption spectra of (**86**) and (**87**) resemble completely that of bacteriochlorophyll *a* (**3**). Configurations of the C-7 position as well as of the 8-double bond are still unknown. But it seems most likely that C-7 of (**86**) and (**87**) have the same configuration as C-7 of bacteriochlorophyll *a* (**3**) and that bacteriochlorophyll *b* (**86**) is transformed by hydrogenation of the exocyclic double bond at C-8 into **3**. Biosynthesis of the bacteriochlorophylls follows that of plant chlorophylls until formation of chlorophyllide *a* (**68**) (see Figure 4).

The transformation of chlorophyllide *a* (**68**) (see Figure 4) into the bacteriochlorophylls is only insufficiently understood (*50, 51*) and requires further investigations.

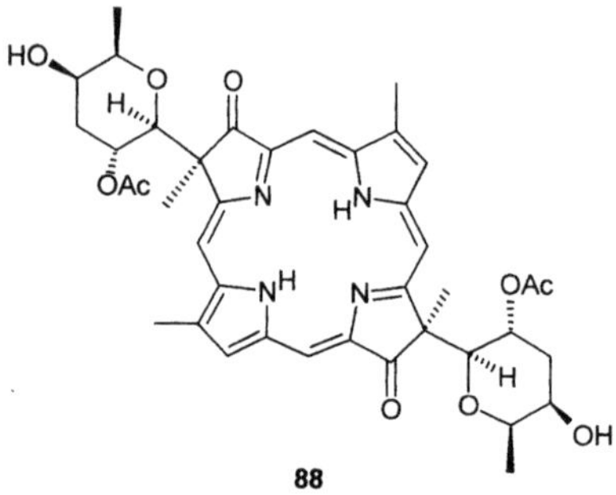

88

Tolyporphin A structure originally assigned

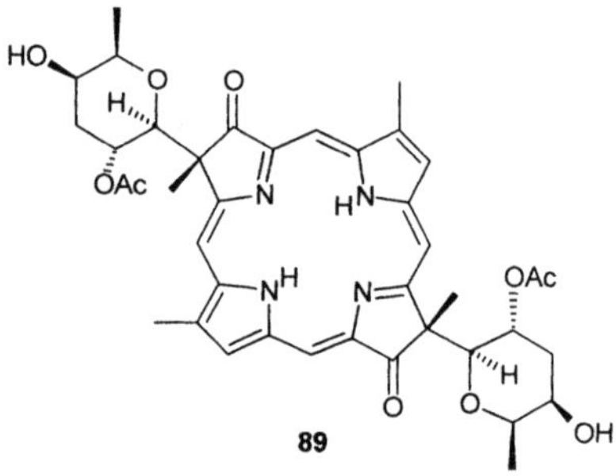

89

Tolyporphin A revised structure

Representatives of the bacteriochlorin class of compounds which are not involved in bacterial photosynthesis are the recently discovered tolyporphins (*18, 62, 63*). Tolyporphin A (**89**) is the first isolated and main member (*62*) of the unusual class of compounds found in the Pacific cyanobacterium *Tolypothrix nodosa*. The substitution pattern, especially the tetrahydropyrane residues of the bacteriochlorin, is unique among the naturally occurring porphyrinoids. For the constitution and relative configuration of tolyporphin A formula (**88**) was originally assigned based on structural investigations by classical spectroscopic methods, with ^{1}H NMR spectroscopy naturally playing the most important part (*62*). Formula picture (**88**) had to be revised to stereoisomeric structure (**89**) during synthetic studies directed to the total synthesis of tolyporphin A (*15, 16, 17*).

The studies revealed that compound (**88**) derived by total synthesis was not identical with naturally occurring tolyporphin A (*15, 16*). Repetition of NMR studies as well as total synthesis of compound (**89**) confirmed structure (**89**) for naturally occurring tolyporphin A (*16, 17*). Today a series of tolyporphins A–J (**89–98**) are known which differ in the residues attached to the geminally disubstituted positions of the saturated rings (*18, 63*). Tolyporphin K is a chlorin with the same substitution pattern at the reduced ring D as found for **92** (*63*). It was demonstrated in biological experiments that tolyporphin A (**89**) eliminates the multidrug resistance (MDR) in a vinblastine-resistant subline of the human ovarian adenocarcinoma line (*64*). Tolyporphin

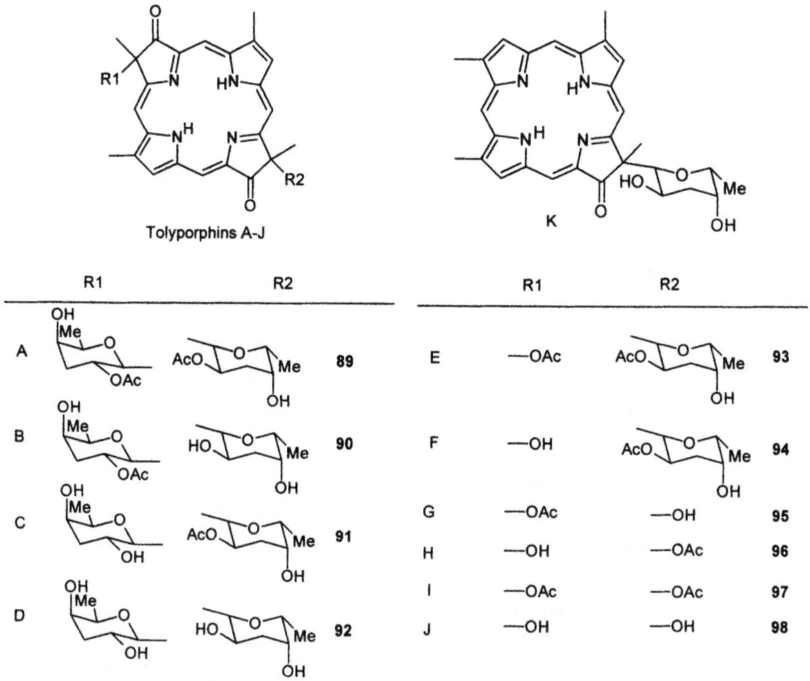

potentiates the cytotoxicity of adriamycin and vinblastine in these cell lines at doses as low as 1 μg/cm^3. The tolyporphins (**89–98**) show also very potent photosensitizing activity against tumor cells *in vitro* and *in vivo* (*64b*).

6.2. Aspects of Synthesis

Partial synthesis of bacteriochlorophylls starting from chlorophylls have been developed with respect to structural investigations in the bacteriochlorophyll series (*59*). The extremely high sensitivity of bacteriochlorins to various reaction conditions makes their chemistry very difficult. This might be one reason why methods for the total synthesis of bacteriochlorins had not, until very recently, been developed. Total synthesis of tolyporphin models (*65*), tolyporphin stereoisomer (**88**) (*15*) and tolyporphin A (**89**) (*17*) itself, which all contain the characteristic geminally dialkylated parts at the saturated rings, have been reported by Kishi using a synthetic approach closely related to Eschenmoser's syntheses of hexahydroporphyrins (see section 8.1.).

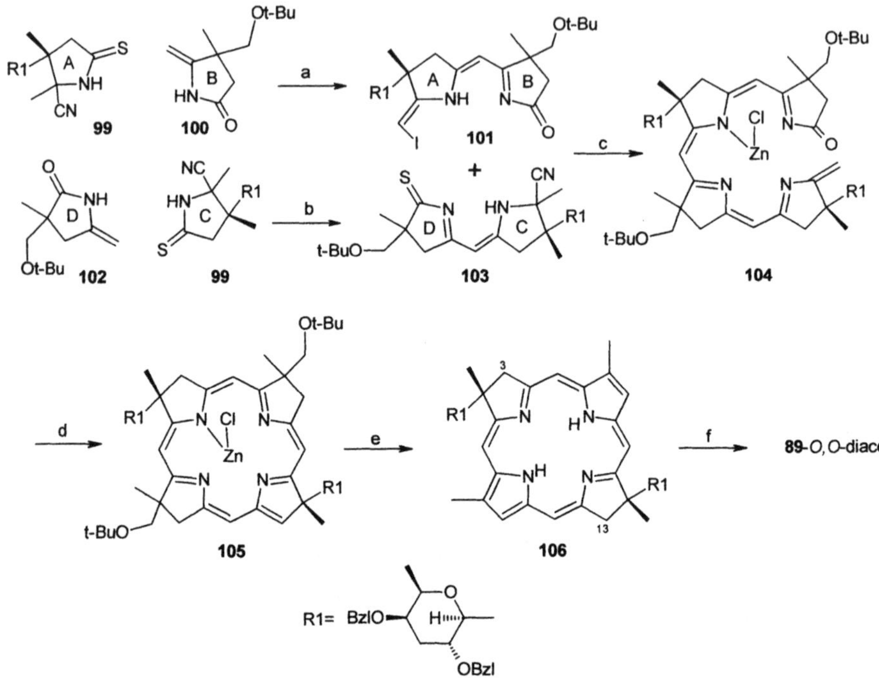

Fig. 7. **a 100**, NIS, *t*-BuOK, *t*-BuOH, benzene + **99**; $(EtO)_3P$, xylene; *t*-BuOK, *t*-BuOH; I_2, K_2CO_3, CH_2Cl_2; **b 102**, NIS, *t*-BuOK, *t*-BuOH, benzene + **99**; $(EtO)_3P$, xylene; Lawesson's reagent, toluene; **c** DBU, CH_3CN; $Ni(ClO_4)_2$, Ph_3P, CH_3CN; KCN, CH_3OH; *t*-BuOK, *t*-BuOH; $Zn(ClO_4)_2 \times 6\,H_2O$, CH_3OH; **d** CH_3OTf, pempidine, CH_3OH; **e** TFA, dimedone, Bu_3P, anisole; CH_3OH; *t*-BuOK, *t*-BuOH; 20% HCl; **f** $ZnCl_2$, EtSH, CH_2Cl_2; Ac_2O, Py; $CrO_3 \times$ DMP, CH_2Cl_2

Base mediated coupling of the constitutionally isomeric enamides (**100**) and (**102**) with the enantiomerically pure cyanothiolactam (**99**) in the presence of *N*-iodosuccinic imide, followed by sulfide contraction with triethylphosphite furnished the two bicyclic AB- and DC-building blocks (**101**) and (**103**).

In the case of the AB-component sulfide contraction was followed by base induced elimination of the cyano function and iodination to give **101**, whereas transformation of the lactam into the thiolactam function formed **103**. The sulfide contraction procedure was once again applied to couple the dimeric moieties under the influence of $Ni(ClO_4)_2$ as a template, followed by decomplexation, recomplexation with Zinc(II) and base induced decyanodation to yield the linear tetracyclic lactam (**104**). Upon treatment with MeOTf, the precorphin (**104**) underwent an enamine/iminoester cyclization to produce the Zn-corphin (**105**). Finally the corphin (**105**) was subjected to a retroaldol type elimination process of the *t*-butoxymethyl residues in the presence of trifuoroacetic acid, *t*-BuOK/*t*-BuOH and aqueous hydrochloric acid to furnish the bacteriochlorin (**106**). Debenzylation and acetylation at the pyrane moieties of **106** and oxidation at the 3- and 13-positions completed the total synthesis of tolyporphin A *O,O*-diacetate in all respects identical with the diacetate derived from tolyporphin A (**89**) originating from natural sources.

7. Isobacteriochlorins

7.1. Occurrence, Structure, and Biological Activity

Isobacteriochlorins (**8**) with two adjacent saturated pyrrole rings are the constitutional isomers of bacteriochlorins, in which two opposite rings are partially reduced. In 1973 the first naturally occurring isobacteriochlorin, iron-containing siroheme (**110**), was isolated by Siegel *et al.* (*66*) from a sulfite reductase of *Escherichia coli.* Later it was also discovered in sulfite and nitrite reductases of numerous bacteria and plants (*67*).

The complete structural elucidation of **110** by means of spectroscopic methods and biogenetic considerations was not accomplished until Battersby *et al.* (*68*) and Bykhovsky *et al.* (*69*) were able to isolate factor II (sirohydrochlorin) (**108**) from bacteria producing vitamin B_{12}. Factor II (**108**) is the metal-free isobacteriochlorin chromophore of siroheme. The reduced forms of factor II (**108**) and of the subsequently discovered factor III (**109**) were identified, together with factor I (**65**), as the first links in the biosynthetic chain from uroporphyrinogen III (**14**) to vitamin B_{12} (**4**) (*11*). The vitamin B_{12} biosynthesis will be dealt with in more detail in a later section (see section 9). Factor III (**109**) differs from factor II (**108**) by virtue of an additional methyl group in the methine position 20.

Heme d_1 (**107**), which was isolated by Timkovich *et al.* (*70*) and Chang *et al.* (*71*) occurs as one of two cofactors in the reductase cytochrome cd_1. Cytochrome cd_1 participates in the reduction of nitrite to nitrous oxide (N_2O) in chemoautotrophic bacteria, such as *Pseudomonas aeruginosa, Paracoccus denitrificans*, and *Thiobacillus denitrificans* (*13*). From recent investigations it seems very likely that cytochrome cd_1 mediates the nitrite reduction to nitric oxide (NO) and that a second enzyme produces N_2O from NO (*13*). A structure was

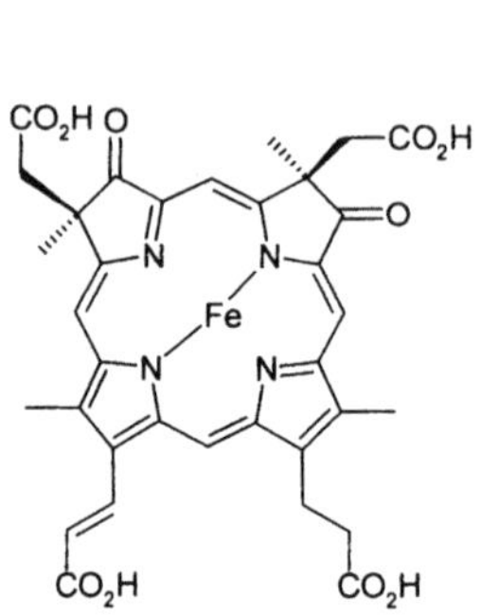

107

Heme d_1

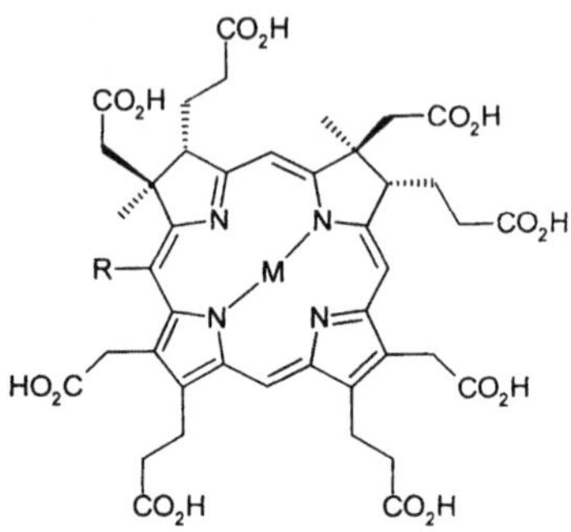

R= H M = 2H Factor II **108**
R= CH_3 M = 2H Factor III **109**
R= H M = Fe Siroheme **110**

originally proposed for heme d_1 on the basis of UV/Vis, MS, IR, and NMR spectroscopic data on the metal free, esterified macrocyclic framework (*70*). The original structural formula was revised on the basis of synthetic studies (*71*) and a new interpretation of spectroscopic data (*72*), leading to a dioxoisobacteriochlorin structural formula with unknown configuration and rather uncertain position of the acrylic acid double bond. In a less selective partial synthesis (*73*) the dioxoisobacteriochlorin chromophore of heme d_1 was obtained from protoporphyrin IX (**26**). The configurations of stereoisomeric intermediates in the synthesis were tentatively assigned on the basis of their different chromatographic behavior. The proposed *cis* intermediate then lead to heme d_1 with *cis* configuration. In another partial synthesis (see section 7.3) the constitution and the relative configuration of the final porphyrin d_1 was established by a crystal structure analysis (*74*). The synthetic porphyrin d_1 with *cis* configuration and 17-acrylic acid side chain corresponds to heme d_1 (**107**). This finding has unambiguously confirmed the constitution and relative configuration of heme d_1 (**107**).

The absolute configuration of heme d_1 then arose independently from the extension of the partial synthesis to the preparation of the enantiomer of (**107**) and from a total synthesis of heme d_1 (**107**) (*75*) using configurationally defined chiral building blocks (see section 7.3).

Both cofactors, siroheme (**110**) and heme d_1 (**107**), play an important role in the sulfur and nitrogen metabolisms of numerous organisms. It can be assumed that the global cycles of nitrogen and sulfur are decisively influenced by these organisms.

7.2. Biosynthesis

As mentioned in 7.1. the reduced forms of sirohydrochlorins (**108**) and (**109**) are both intermediates in the biosynthesis of vitamin B_{12} (*11, 12*) which will be discussed in section 9. So called precorrin-2 the reduced form of factor II (**108**) is biosynthesized from uroporphyrinogen III (**14**) *via* precorrin-1 [reduced form of factor I (**65**)] by a twofold methylation with *S*-adenosylmethionine (SAM). Precorrin-2 is then methylated with SAM to yield precorrin-3 which forms factor III (**109**) by aerial oxidation on isolation from natural sources. From precorrin-2 siroheme (**110**) is also formed in many bacteria for instance in *Escherichia coli* by a multifunctional enzyme named siroheme synthase. The enzyme acts as an oxidation agent under anaerobic conditions for precorrin-2 by using NAD^+ and is also able to insert Fe^{2+} into sirohydrochlorin (**108**) (*76*). Biosynthetic investigations on heme d_1 (**107**) have demonstrated that the carbon skeleton of the macrocycle originates from δ-aminolevulinic acid (**18**) (*77*) and that methyl groups at C-2 and C-7 are introduced by *S*-adenosylmethionine (SAM) (*78*). It is not yet clear how the propionic acid side chains at C-3 and C-8 can be removed and how oxygen is introduced. Another interesting question which has to be answered is whether methylation occurs after the acid side chains have been modified or whether methylation takes place directly from uroporphyrinogen III (**14**). In that case precorrin-2 or factor II (**108**) would be biosynthetic intermediates.

7.3. Aspects of Synthesis

All of the naturally occurring isobacteriochlorins contain the geminally dialkylated structural parts in the saturated pyrrole rings, which require as in the chlorin and bacteriochlorin series special approaches for their synthesis. Until the discovery of siroheme (**110**) and sirohydrochlorins (**108, 109**) this structural element was only known from the vitamin B_{12} structure. Using the synthetic potential which was invented during numerous syntheses (*79*) of model corrins and vitamin B_{12}, Eschenmoser developed the first synthesis of model isobacteriochlorins (*80*). Based on these synthetic experiences and in the light of investigations into vitamin B_{12} biosynthesis, Battersby has performed a total synthesis of the naturally occurring sirohydrochlorin (**108**) (*81*).

The geminally dialkylated building blocks **111** and **116** for rings A and B of the macrotetracycle are obtained from cobyrinic acid heptamethylester, a vitamin B_{12} derivative, by ozonolytic degradation and

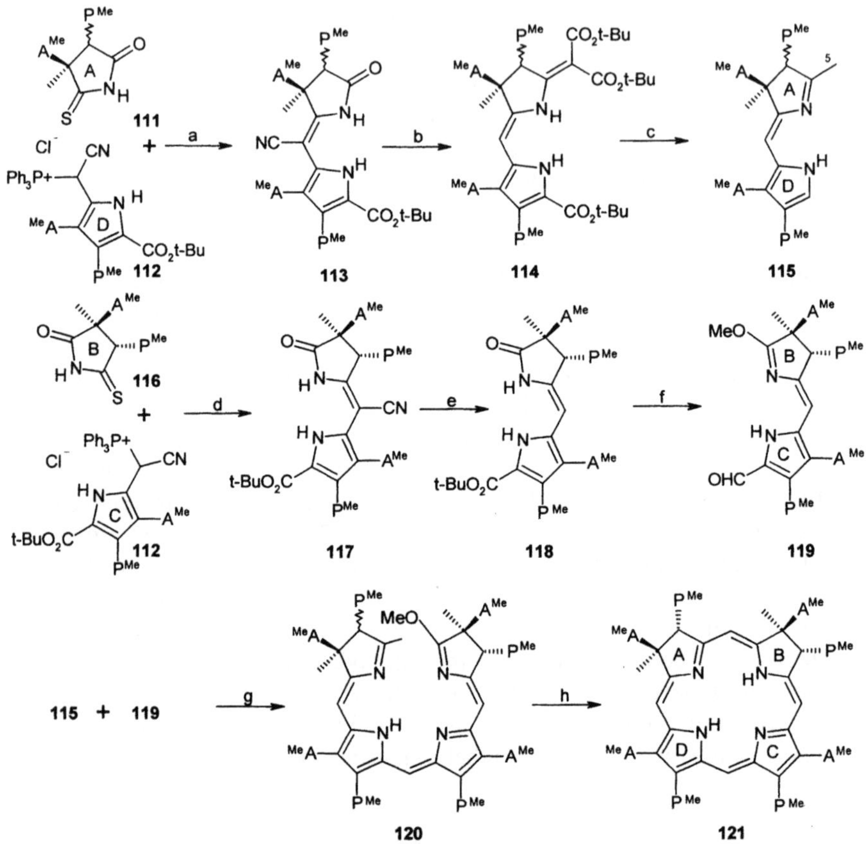

Fig. 8. **a** KO*t*-Bu, toluene; **b** H_2, Raney-Ni, CH_3OH, H_2O, MsOH; MsCl, DMAP; 1,2-bis(ethylamino)ethane, anisole; Lawesson's reagent; $BrCH(CO_2t\text{-}Bu)_2$, DBU; PPh_3, DBU, toluene; **c** TFA, labile **115** was condensed immediately with **119**; **d** KO*t*-Bu, toluene; **e** H_2, Raney-Ni, CH_3OH, H_2O, MsOH; MsCl, DMAP; 1,2-bis(ethylamino)ethane, anisole; **f** TFA, $HC(OCH_3)_3$; $(CH_3)_3O^+BF_4^-$; *i*-Pr_2NEt; **g** TFA, CH_3OH, h: hν (*vis*), TFA, *i*-Pr_2NEt, THF, CH_3OH

subsequent introduction of sulfur. Thio-Wittig reactions of the A-thiosuccinimide (**111**) and the B-thiosuccinimide (**116**) with the same pyrrole building block (**112**) yielded the AD-component (**113**) as well as the BC-dimer (**117**). The synthetic concept to construct the macrotetracycle from two molecular components in "north-south" direction has the advantage that one and the same pyrrole building block (**112**) can be used to form rings C and D and furthermore rings A and B also originate from a common precursor. Removal of the cyano group of **113**,

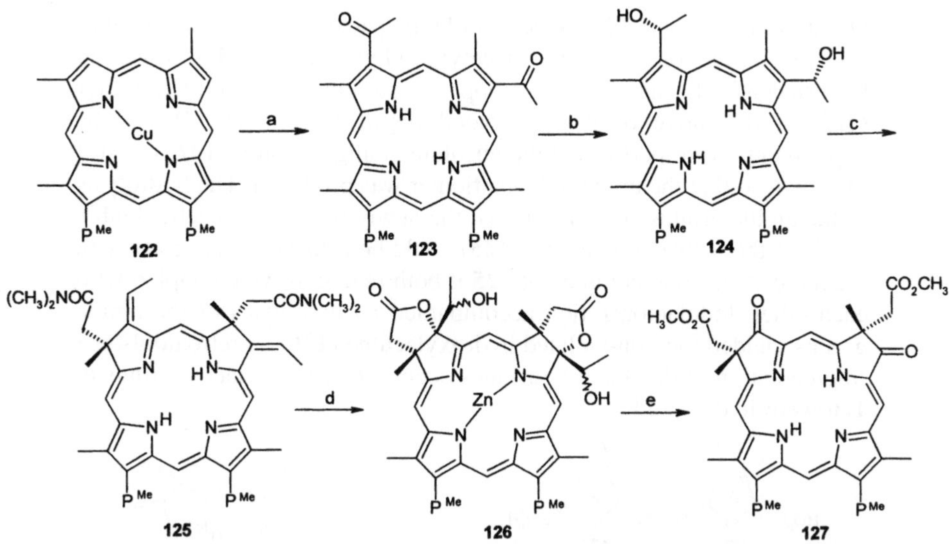

Fig. 9. **a** Ac_2O, $SnCl_4$, CH_2Cl_2; conc. H_2SO_4; CH_2N_2, Et_2O; **b** (S)-Tetrahydro-1-methyl-3,3-diphenyl-1*H*,3*H*-pyrrolo-[1,2-c][1,3,2]oxazaborol, $BH_3 \times SMe_2$, CH_2Cl_2; **c** $MeC(OMe)_2NMe_2$, *o*-xylene; **d** $Zn(acac)_2$, $CHCl_3$; I_2, THF, H_2O; $Na_2S_2O_3$, H_2O; **e** KOH, H_2O, CH_3OH; HCl, H_2O; CH_2N_2, Et_2O, CH_3OH

introduction of the carbon atom 5 by sulfide contraction, and subsequent cleavage of the *t*-butoxycarbonyl residues lead to the complete "western" AD-component (**115**). The BC-component (**119**) was formed along a similar reaction sequence which was completed by acid induced formylation with simultaneous cleavage of the *t*-butylester and methylation of the lactam function. Both parts of the target molecule, **115** and **119**, underwent acid-catalyzed condensation to yield the *seco*-isobacteriochlorin (**120**). Photochemical cyclization as in the synthesis of factor I (**65**) yielded sirohydrochlorin octamethylester (**121**) and after hydrolysis sirohydrochlorin (**108**).

Three different synthetic routes leading to heme d_1 (**107**) were developed in the laboratories of Battersby (*75*), Chang (*72, 73*), and Montforts (*74*). Chang's partial synthesis of heme d_1 is of less practical use due to its low selectivity and should therefore not be reviewed here. The partial synthesis performed by Montforts leads to the enantiomer of porphyrin d_1, the metal free porphyrin ligand of *ent*-**107** *via* the dioxoisobacteriochlorin intermediate (**127**).

The key step in this synthesis is a stereoselective two-fold amide acetal Claisen rearrangement of the enantiomerically pure hematoporphyrin dimethylester (**124**) which can be prepared by chirogenic enantioselective reduction from the diacetylporphyrin **123**. **123** is obtained by Friedel-Crafts acylation of copper deuteroporphyrin dimethylester (**122**), a derivative of the red blood pigment heme (**1**). When the experiments were performed the absolute configuration of **107** was still unknown so that the unnatural enantiomer was produced. The hydrolysis of the amide groups and cleavage of the newly formed exocyclic double bonds in the isobacteriochlorin (**125**) could be achieved very easily with the more stable zinc complex of **125** if both reactions were coupled with each other. Iodolactonization accompanied by hydrolysis of the amide groups yielded the non-isolated hydroxylactone (**126**). A retroaldol-type fragmentation of the 3 and 8 residues occurred when the lactone rings in **126** were hydrolyzed.

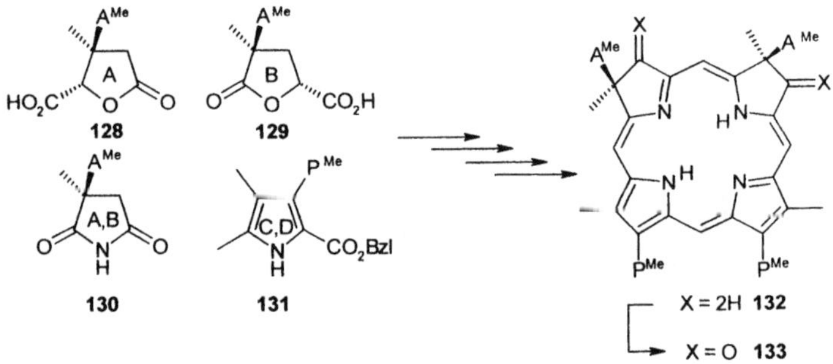

Autoxidation, followed by removal of zinc and reesterification yielded the dioxoisobacteriochlorin (**127**), which could be further transformed into unnatural porphyrin d_1. The total synthetic approach leading to porphyrin d_1 developed by Battersby (*75*) made use of chiral monocyclic building blocks (**128**–**130**) derived in multistep reaction sequences from *L*-glutamic acid. These building blocks were used together with pyrrole (**131**) to synthesize porphyrin d_1 along routes very similar to that leading to sirohydrochlorin (**108**), including the final photochemical cyclization. The isobacteriochlorin intermediate (**132**) obtained, was then oxidized to the dioxoisobacteriochlorin (**133**) which finally established the absolute configuration of natural heme d_1 (**107**).

8. Higher Saturated Hydroporphyrins

8.1. Occurrence, Structure, and Biological Activity

As already mentioned the investigation of vitamin B_{12} biosynthesis has brought several new hydroporphyrinoid structures to light which

References, pp. 42–51

are biosynthetic intermediates for the vitamin. It is intended to review these hydroporphyrinoid structures in section 9. in connection with the biosynthesis of vitamin B_{12} (**4**). The discovery (*13, 82*), structural elucidation (*13, 83*), chemistry (*7, 84*), and biology (*13, 85*) of factor F 430 (**134**) a dodecahydroporphyrin (**135**) in the laboratories of Wolfe, Thauer, Eschenmoser, Pfaltz, and Jaun has been an important source of additional impetus for the study of the chemistry of highly reduced porphyrins.

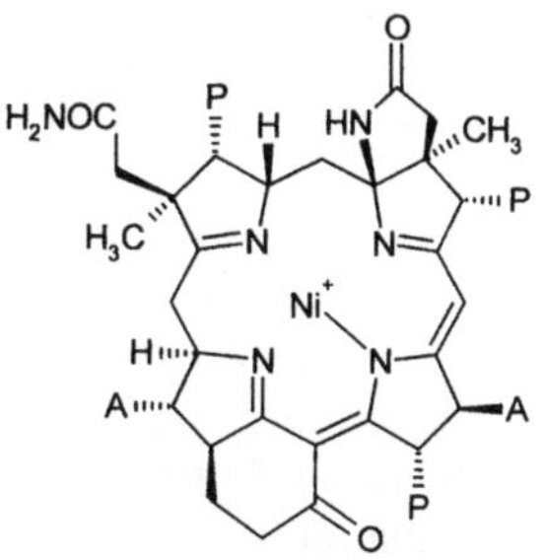

134

Factor F430

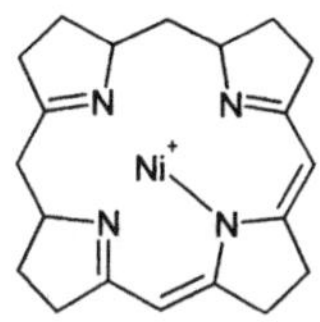

135

Dodecahydroporphyrin

The nickel-containing factor F 430 (**134**) provides an example of how nature exploits the reactivity of organometallic compounds, as is the case with vitamin B_{12}. Factor F 430 (**134**) plays a key role as cofactor for the coenzyme M reductase of primitive methanogenic bacteria in the formation of methane from 2-(methylthio)-ethanesulfonate (*86*). The structural elucidation of factor F 430 (**134**) is based on a combination of classical spectroscopic methods, chemical degradation, and biosynthetic studies with ^{13}C-labelled precursors (*83a,b*). These biosynthetic investigations will be addressed in section 8.2. Chemical degradation products obtained by ozonolysis of factor F 430 (**134**) allowed the determination of the absolute configuration by comparison with reference compounds derived from chlorophyll *a* (**2**) and vitamin B_{12} (**4**) (*83a,b*).

Detailed studies on the precise function of **134** are still under investigation, but have already demonstrated that the conformationally flexible dodecahydroporphyrin ligand of factor F 430 (**134**) influences the redox properties and reactivity of the central nickel atom depending on the conformation of the ligand (*7, 8, 83e*). During enzymatic transformation of urophyrinogen III (**14**) into vitamin B_{12} (**4**) the hexahydroporphyrin oxidation level of **14** is retained in the precorrin intermediates which occur along the biosynthetic pathway. Biosynthetic investigations have made different enzymes of vitamin B_{12} biosynthesis available by application of methods from molecular biology (*12a,d*). These enzymes are not very specific and accept porphyrinoid substrates other than those of the biosynthetic chain. This opens up the possibility of using these methylation enzymes for the synthesis of methylated hydroporphyrins.

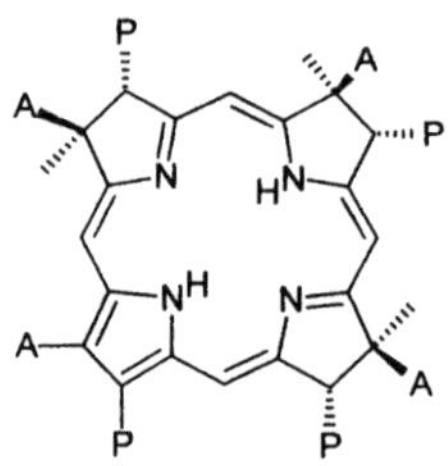

136

Trimethylpyrrocorphin

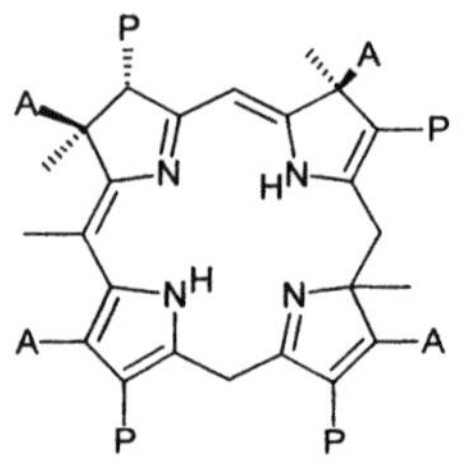

137

Compound-4

The uroporphyrinogen III methylase (M1) from *Escherichia coli*, which was over expressed and therefore available in sufficient amounts, methylates uroporphyrinogen III (**14**) in the presence of *S*-adenosyl methionine (SAM) on the biosynthetic pathway to siroheme (**110**) (*76*).

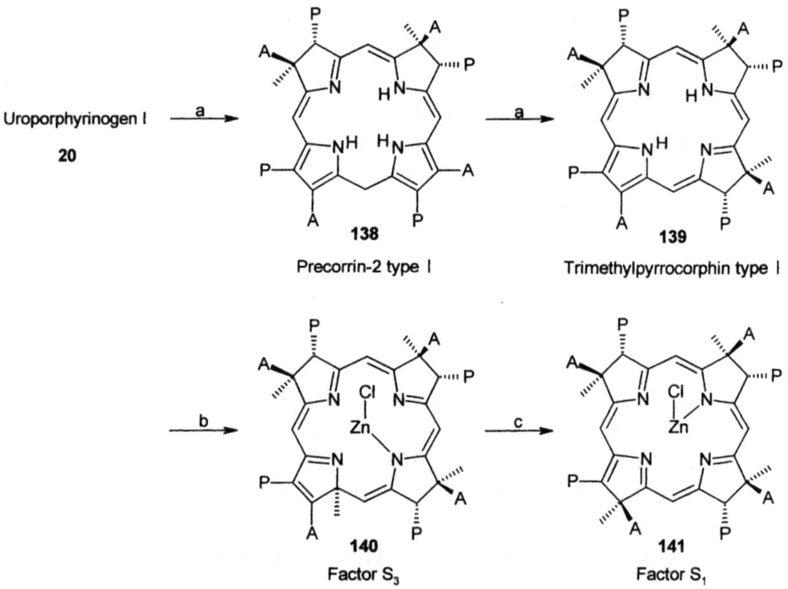

Fig. 10. **a** M-1, SAM or S-adenosyl methionine uroporphyrinogen III methyl transferase (SUMT), SAM; **b** cbiF, SAM; **c** [1,5]sigmatropic methyl shift

References, pp. 42–51

On prolonged incubation trimethylpyrrocorphine (**136**) is formed by over methylation (*12a,d*). The product of the methylation does not act as an intermediate in the biosynthesis of vitamin B_{12}. The so-called compound-4 (**137**) is also an over methylated product formed in the vitamin B_{12} biosynthesis, which is formed with an over expressed methyltransferase (cbi F) from *Salmonella typhimurum* (*87*). Compound-4 (**137**) is closely related to factor S_3 (**140**) (*88*).

Although uroporphyrinogen I (**20**), a constitutional isomer of uroporphyrinogen III (**14**), plays no direct role in porphyrin biosynthesis, this unnatural substrate is methylated in the presence of SAM by the methyl transferases of some bacteria.

A constitutional type I precorrin-2 (**138**) is obtained by methylation of uroporphyrinogen I (**20**) with the methylase M1 or *S*-adenosyl methionine uroporphyrinogen III methyl transferase (SUMT) from *Propionibacterium shermanii* and over expressed in *Paracoccus denitrificans* (*12a, 87*). The methyltransferase M1 is able to methylate the unnatural precorrin (**138**) once more to give the trimethylpyrrocorphin type I (**139**). The Zinc-containing factors S_3 (**140**) and S_1 (**141**) (*88*) produced by further methylation were also isolated from *Propionibacterium shermanii*. Metal-free factor S_3 is obtained by methylation of the trimethylpyrrocorphin type I (**139**) with the methyltransferase (cbi F) from *Salmonella typhimorum*. A [1,5] sigmatropic rearrangement of the 16-methyl group to position 17 converts factor S_3 (**140**) into factor S_1 (**141**) with its corphin chromophore.

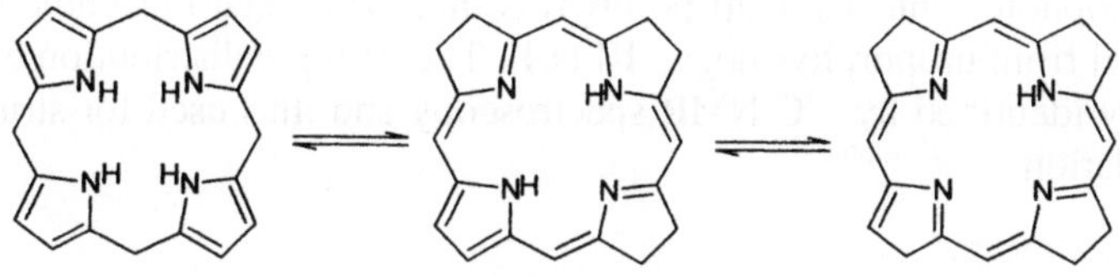

9 Porphyrinogen — **11** Pyrrocorphin — **10** Corphin

The enzymatic methylations correspond with biomimetic investigations on the structural types of pyrrocorphin (**11**) and porphyrinogen (**9**) which were performed by Eschenmoser *et al.* (*7, 13, 14*). From these experiments it was concluded that the tetrapyrrolic porphyrinogen structure (**9**) can be transformed by methylation reactions into the thermodynamically less stable, tautomeric, but similarly hexahydroporphyrinoid corphin type structure (**10**), in which the methyl groups fix the corphin chromophore.

8.2. Biosynthesis

From feeding experiments with isotopically labelled δ-aminolevulinic acid (**18**) and methionine (**142**) it became clear that the carbon skeleton of the naturally occurring highly saturated porphyrins are formed from uroporphyrinogen III (**14**) and that the additional methyl groups are transferred from *S*-adenosyl methione (*11, 12*).

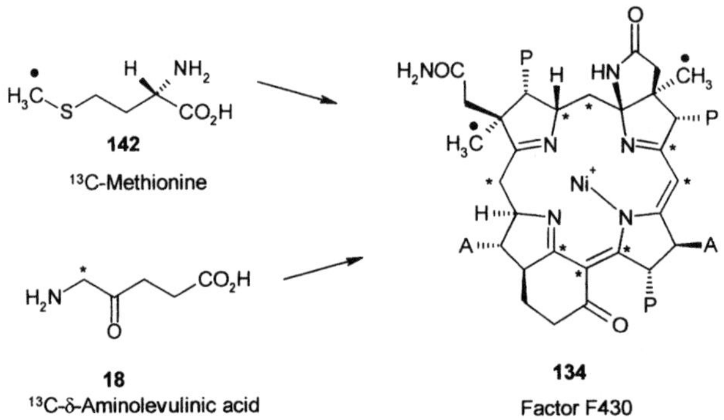

The structural elucidation of factor F 430 (**134**) is based to a large extent on biosynthetic investigations with δ-aminolevulinic acid (**18**) and methionine (**142**) (*83a,b*). The peripheral methyl groups in positions 2 and 7 stem from ^{13}C-methionine (**142**). δ-Aminolevulinic acid (**18**) which was ^{13}C-labelled in position 5, produces an isotope pattern in the chromophore typical for all porphyrins that have been biosynthetically derived from uroporphyrinogen III (**14**). The isotopic distribution can be readily identified by ^{13}C NMR spectroscopy and thus used for structure elucidation.

9. Corrins

Since the discovery of vitamin B_{12} (**4**) in 1948 (*89*), the fascination arising from its unique molecular structure and its unusual biochemical function has had a profound influence on the development of organic chemistry during the second half of the last century. In the 1950s the focus of vitamin B_{12} research was on its physiological role and on the determination of its structure. Following intensive but unsuccessful attempts to elucidate the constitution and configuration of vitamin B_{12} by chemical degradation, X-ray crystallography finally celebrated its first triumph by unlocking the unusual structure of the natural product (*90*).

In the 1960s the work of Eschenmoser (*91*) and Woodward (*92*) on the synthesis of vitamin B_{12} opened a new era of organic synthesis. The strategic planning of syntheses, the development of new synthetic methods directed to a specific target molecule, and the discovery of the Woodward-Hoffmann rules, which all came in the course of the synthesis of vitamin B_{12}, were some of the outstanding innovative ideas which, along with other important insights, had far reaching influences on organic synthesis following the synthetic investigations in the vitamin B_{12} field. As a logical consequence of the vitamin B_{12} synthetic work, the questions of the biosynthesis of vitamin B_{12} (*11, 12*) and the possibility of its prebiotic formation (*7, 91b*) began to be considered in the 1970s. In a remarkable example of chemical collaboration, using the latest spectroscopic methods and molecular-biological techniques, these questions were successfully answered in the laboratories of Arigoni (*44, 93*), Battersby (*11*), Scott (*12*), Thibaut (*11b, 94*), and Eschenmoser (*7, 91b*), so that today the biosynthesis of vitamin B_{12} is fully understood except for some very minor details.

Progress in the different phases of vitamin B_{12} research during the past 50 years was respectively reviewed by four conferences and conference reports (*5, 10, 95*) as well as by monographs (*6*) and reviews (*7, 11, 12, 91, 92*).

Progress of biological action in enzymatic methyl transfer and rearrangement reactions, medicinal aspects, structure and reactivity, and biosynthesis of vitamin B_{12} and B_{12}-coenzymes was very recently comprehensively presented at the 4th European Symposium on Vitamin B_{12} and B_{12}-Proteins and reviewed in an excellent monograph (*10*). Therefore it is intended with this contribution to address mainly results from research in the field of vitamin B_{12} biosynthesis. The procedure makes sense because of two reasons. First, many of the biosynthetic intermediates are closely related to hydroporphyrinoid structures discussed in previous sections and second, reactions of the biosynthetic pathway concern the chemistry of the hydroporphyrinoid and corrinoid frameworks involved, whereas the biochemical reactivity of vitamin B_{12} is mainly restricted to the central cobalt ion of the corrin macrocycle.

A milestone in vitamin B_{12} biosynthesis was the discovery of the chlorin factor I (**65**) by Müller (*43, 44*) and Arigoni (*44*) and of the isobacteriochlorin factor II (**108**) by Bykhovsky (*69*) and Battersby (*68, 69*). The reduced forms of factor I (**65**) and factor II (**108**) named as precorrin-1 and precorrin-2 were then identified as biosynthetic precursors of vitamin B_{12} (**4**).

Methylation of the hexahydroporphyrinoid uroporphyrinogen III (**14**) with methyl groups from *S*-adenosyl-methionine (SAM) at pyrrole ring

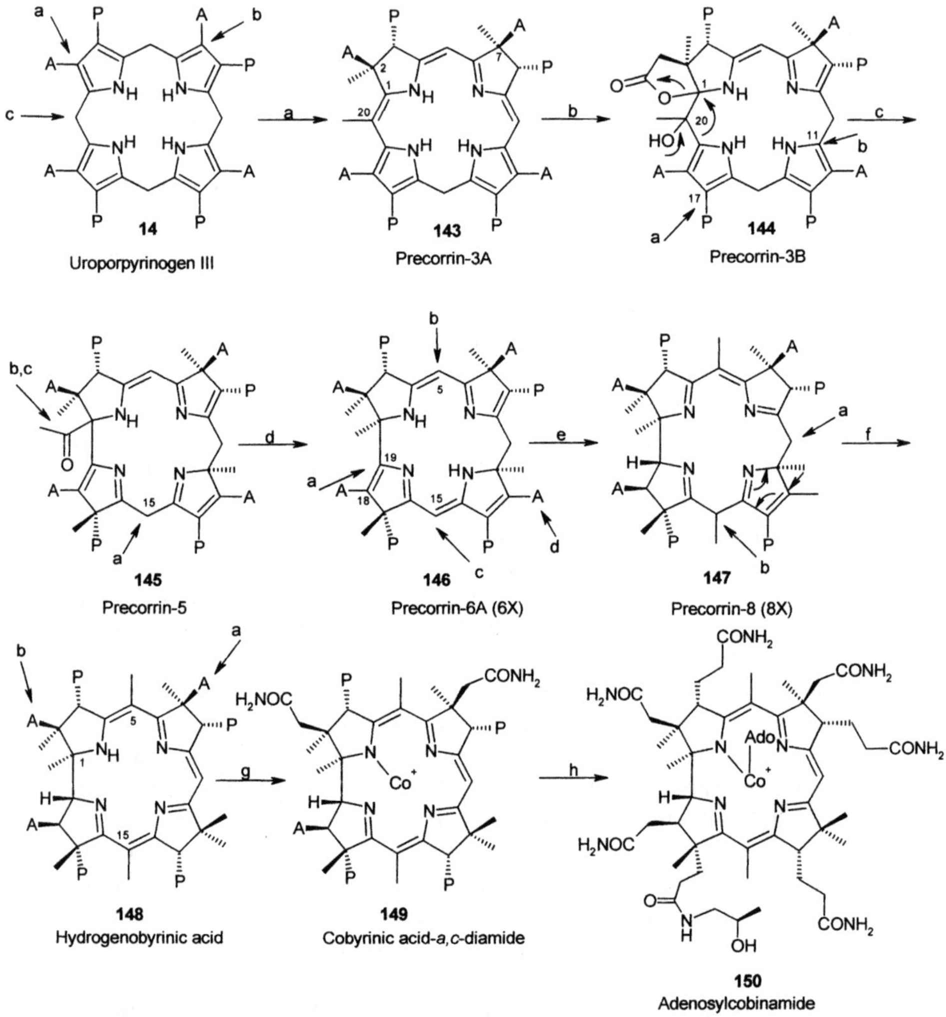

Fig. 11. **a** SUMT (CobA), SAM; SUMT (CobA), SAM; SP2MT (CobI), SAM; **b** Precorrin-3B synthase (CobG), O_2, Fe_4S_4; **c** SP3MT (CobJ), SAM; SP4MT (CobM), SAM; **d** SP5MT (CobF), SAM, H_2O; **e** Precorrin-6A reductase (CobK), NADPH; P8 synthase (CobL), 2 SAM; **f** HBY synthase (CobH); **g** HBY diamide synthase (CobB), 2 glutamine, 2 ATP, 2 Mg^{2+}; Cobaltchelatase (CobN, CobS,T), Co^{2+}, ATP, Mg^{2+}; **h** Cobalt(II) reductase (?), 0.5 NADH; Ado-transferase (CobO), ATP, Mg^{2+}; Cobyrinic acid synthase (CobQ), 4 glutamine, ATP, Mg^{2+}; Cobinamide synthase (CobC, D,protein α), 1-amino-2-propanol, ATP, Mg^{2+}

A and then ring B yielding the precorrins separates off the vitamin B_{12} branch from those directed towards the biosynthesis of heme and chlorophylls. The third methyl group is transferred from SAM into the 20 position of the macrotetracycle giving precorrin-3A (**143**). Another important finding was that this methyl group is lost again together with the *meso*-carbon, thus indicating the assistance of the crucial corrin forming ring contraction by this structural unit (*96*). In numerous feeding experiments with δ-aminolevulinic acid isotopically labelled at different positions and with genetically defined and modified microorganisms all biosynthetic intermediates of vitamin B_{12} biosynthesis could be identified. Beyond that, the experiments allowed a detailed insight into the mechanistic course of the enzymatically controlled transformations even at the level of the genes encoding for vitamin B_{12} biosynthesis. The investigations revealed a second splitting in the biosynthetic pathway, namely an aerobic pathway with late cobalt insertion (*11, 12, 94*) and an anaerobic way characterized by early cobalt incorporation (*11, 12*). Both pathways which also make use of different ring contraction sequences merge at the stage of the cobyrinic acid-*a*,*c*-diamide (**149**). The reactions steps which form the characteristic corrin framework and modify substituents are initiated or accompanied by further methylation whereby it was very early demonstrated by application of "chiral" methyl groups that the methyl groups were transferred from SAM with inversion of configuration (*93*).

The cloning of *Pseudomonas denitrificans* genes involved in vitamin B_{12} biosynthesis opened the way to the elucidation of the aerobic biosynthetic pathway (*97*). Twenty enzymes were purified and twenty-two *cob* genes were identified. In the aerobic pathway corrin ring formation is initiated by oxidation with molecular oxygen.

Precorrin-3A (**143**) formed by three-fold methylation (a, b, c) of uroporphyrinogen III (**14**) is oxidized to yield precorrin-3B (**144**) by simultaneous lactone ring formation with the 2 acetic acid residue. The oxygen of the 20 hydroxy function originates from molecular oxygen as demonstrated in labelling experiments (*98*). Precorrin-3B (**144**) undergoes a pinacol type rearrangement forming the direct linkage between rings A and B characteristic for the corrin macrocycle. Methylation at the 17 position (a) gives precorrin-4 which is further transformed by methylation (b) into precorrin-5 (**145**). After tautomerization at position 15 (a) of **145** the acyl substituent formed during the ring contraction is removed (b) in a retro-Claisen type fragmentation as acetic acid. Subsequent methylation at position 1 (c) of the thus formed intermediate gives precorrin-6A (**146**), which is reduced at the 18,19 double bond (a) with a NADPH dependent reductase producing precorrin-6B. Then methylations at the 5 meso-carbon (b) and 15 methine bridges (c) occur. The latter methylation process promotes the decarboxylation (d) of the 12 acetic acid side chain finally giving precorrin-8X (**147**).

An interesting [1.5] sigmatropic rearrangement of the 11 methyl group into the 12 position of ring C and tautomerizations (a, b) at the 10,15 methylene linkages complete the formation of the corrin ligand (**148**) of vitamin B_{12}.

Subsequent amidation (a, b) with amino transfer from glutamine in an ATP dependent reaction gives hydrogenobyrinic acid-*a*,*c*-diamide and at this stage cobalt (II) is inserted by an ATP dependent cobalt chelatase to produce cobyrinic acid-*a*,*c*-diamide (**149**) (*94, 97*). Reduction of Co(II) to Co(I) by a special reductase allows adenosylation of the central metal ion followed by amidation of the *b*, *d*, *e*, and *g* carboxylic acid side chains again with ATP dependent amino transfer from glutamine. The last amidation reaction takes place at the *f* propionic acid side chain in position 17 with 1-amino-2-propanol to give adenosylcobinamide (**150**) (*94, 97*) to which different nucleotides forming the α loop of B_{12} coenzymes can be attached (*10*).

A different but somewhat related pathway is operating in anaerobic *Propionibacterium shermanii (12)*. The main difference between the

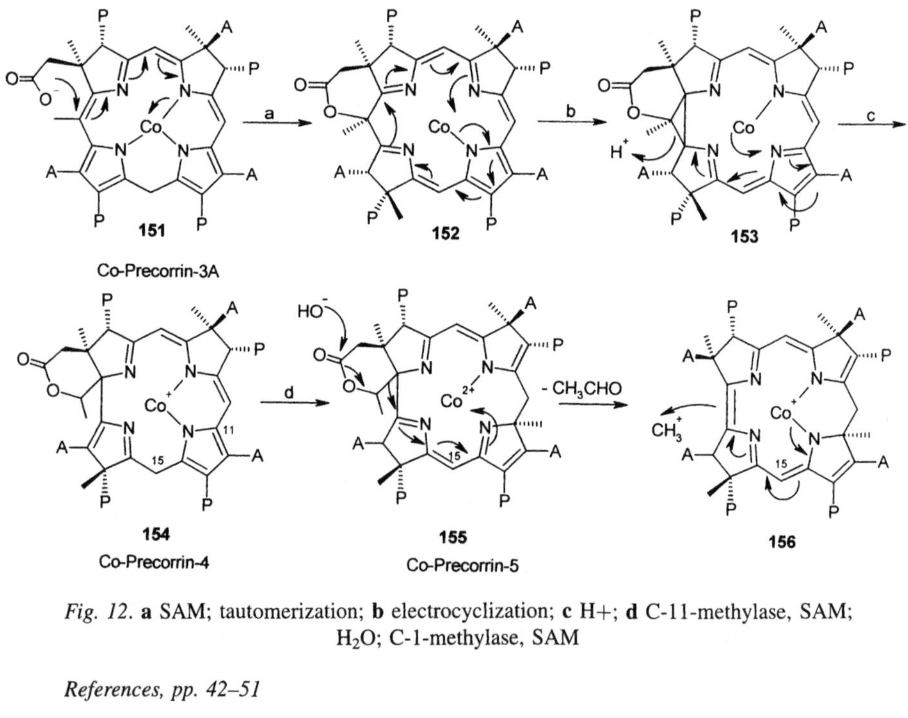

Fig. 12. **a** SAM; tautomerization; **b** electrocyclization; **c** H+; **d** C-11-methylase, SAM; H_2O; C-1-methylase, SAM

References, pp. 42–51

aerobic and the anaerobic pathways comes from the very early insertion of cobalt in anaerobic biosynthesis. In that case cobalt could then promote the ring contraction in the absence of molecular oxygen by acting as an internal oxidant. Cobalt(III) insertion occurs at the precorrin-3 stage giving Co-precorrin-3A (**151**) which undergoes lactone ring formation at the 20 methine bridge with the acetic acid side chain of ring A. Lactone formation is possible by participation of the central cobalt ion which experiences an internal reduction to cobalt(I). The lactonization is followed by methylation with SAM and tautomerization to yield intermediate (**152**). Electrocyclization of **152** brings out the ring contraction under formation of the first corrinoid intermediate (**153**). Protonation of the cyclopropane ring of this intermediate (**153**), whereby cobalt(I) acts as a base and is reoxidized to the +III state, gives Co-precorrin-4 (**154**) which is closely related to precorrin-3B of the aerobic pathway. After methylation at the 11 position giving Co-precorrin-5 (**155**) a retro-aldol type extrusion of the former 20 methine unit as acetaldehyde occurs. The intermediate (**156**) thus formed undergoes methylation with SAM to yield Co-precorrin-6X whose ligand is structurally identical with precorrin-6A (**146**) of the aerobic path.

The pathways of vitamin B_{12} biosynthesis merge at the stage of cobyrinic acid-*a,c*-diamide (**149**), which is formed by cobalt insertion in the aerobic route. Though there are differences in both routes of vitamin B_{12} biosynthesis nature has chosen a common strategy to form the corrin macrocycle. First fully conjugated intermediates were avoided by introducing quaternary carbon atoms at C-1 and C-11, thus the macrotetracycle is divided in two parts. All intermediates are pale yellow or uncoloured compounds and sensitive to metal ions. The last step of corrin formation is the liquidation of the C-11 quaternary center by methyl migration from C-11 to C-12 which unlocks the non-conjugated systems and forms the pink coloured corrins. Another common feature of both pathways is the requirement of a NADPH dependent reduction of the 18,19 double bond, because in previous reaction steps oxidation by molecular oxygen or with cobalt is performed.

However, the early cobalt incorporation during the anaerobic pathway could not be essential for ring contraction since metal free corrinoids have been found in anaerobic bacteria (*94*) like *Chromatium vinosum* and *Rhodospirillum rubrum* in the absence of cobalt. Moreover, metal free corrinoids are produced by strictly anaerobic methanogenic bacteria in the absence of cobalt. Thus further biosynthetic explorations are necessary to discover the primitive anaerobic corrin formation in Archaebacteria.

Whereas the main focus of vitamin B_{12} research during the past 25 years was coined by investigation in the fields of corrin biosynthesis and biological and physiological function of the vitamin, only minor activities in the field of synthesis and chemical reactivity of the corrin ligand happened. Attempts directed towards the total synthesis of the vitamin B_{12} came from the laboratories of Stevens (*99*), Jacobi (*100*), and Mulzer (*101*). Chemical reactions including degradation of the corrin macrocycle was reviewed by Bonnett (*102*) a pioneer in corrin chemistry. Interesting demethylation reactions at the 5- and 15-meso positions of the corrin ligand of vitamin B_{12} were observed during attempts to remove the central cobalt ion from the macrocycle (*103*).

Acknowledgements

We are indebted to those who helped to collect and to screen the literature describing the explorations in the wide field of chemistry of naturally occurring cyclic tetrapyrroles, namely Benjamin Gerlach, Frank Höper, Olaf Kutzki, Maria Stapelbroek-Möllmann, Francesc Pont, Klaus Rischka, and Michael Wedel.

References

1. Dolphin D (ed) (1978–1979) The Porphyrins Vols 1–7. Academic Press, New York
2. Smith KM (ed) (1975) Porphyrins and Metalloporphyrins. Elsevier, Amsterdam
3. Vernon LP, Seely GR (eds) (1966) The Chlorophylls. Academic Press, New York
4. Stoll A, Wiedemann E (1952) Chlorophyll. Fortschr Chem Forsch **2**: 538
5. Zagalak B, Friedrich W (eds) (1979) Vitamin B_{12}. de Gruyter, Berlin
6. Dolphin D (ed) (1982) Vitamin B_{12} Vols I, II. Wiley, New York
7. Eschenmoser A (1988) Vitamin B_{12}: Experimente zur Frage nach dem Ursprung seiner molekularen Struktur. Angew Chem **100**: 5
8. Kräutler B (1987) The Porphinoids – Versatile Biological Catalyst Molecules. Chimia **41**: 277
9. Scheer H (ed) (1991) Chlorophylls. CRC Press, Boca Raton
10. Kräutler B, Arigoni D, Golding BT (eds) (1998) Vitamin B_{12} and B_{12}-Proteins. Wiley – VCH, Weinheim
11. a) Battersby AR (1994) How Nature Builds the Pigment of Life: The Conquest of Vitamin B_{12}. Science **264**: 1551; b) Battersby AR, Blanche F, Cameron B, Crouzet J, Debussche L, Thibaut D, Vuilhorgne M, Leeper FJ (1995) Vitamin B_{12}: Wie das Problem seiner Biosynthese gelöst wurde. Angew Chem **107**: 421
12. a) Scott AI (1993) Wie die Natur Vitamin B_{12} synthetisiert – ein Überblick über die letzten vier Milliarden Jahre. Angew Chem **105**: 1281; b) Scott AI (1998) How Nature Synthesizes B_{12} Without Oxygen. In: Kräutler B, Arigoni D, Golding BT (eds) Vitamin B_{12} and B_{12}-Proteins. Wiley – VCH, Weinheim, p 81; c) Scott AI, Santander

PJ (1991) The biosynthesis of vitamin B_{12}. In: Jordan PM (ed) Biosynthesis of Tetrapyrroles. Elsevier, Amsterdam, p 101; d) Scott AI (1994) The Discovery of Nature's Pathway To Vitamin B_{12}. A 25 Year Odyssey. Tetrahedron **50**: 13315

13. Montforts FP, Gerlach B, Höper F (1994) Discovery and Synthesis of Less Common Natural Hydroporphyrins. Chem Rev **94**: 327

14. Montforts FP, Glasenapp-Breiling M (1998) The Synthesis of Chlorins, Bacteriochlorins, Isobacteriochlorins and Higher Reduced Porphyrins. In: Gribble GW, Gilchrist TL (eds) Progress in Heterocyclic Chemistry Vol. 10 p 1

15. Minehan TG, Kishi Y (1999) Totalsynthese des (+)-Tolyporphin-A-*O,O*-diacetats mit der vorgeschlagenen Struktur. Angew Chem **111**: 972

16. Minehan TG, Cook-Blumberg L, Kishi Y, Prinsep MR, Moore RE (1999) Revision der Struktur von Tolyporphin A. Angew Chem **111**: 975

17. Wang W, Kishi Y (1999) Synthesis and Structure of Tolyporphin A *O,O*-Diacetate. Org Lett **1**: 1129

18. Prinsep MR, Patterson GML, Larsen LK, Smith CD (1995) Further Tolyporphins from the Blue-Green Alga *Tolypothrix nodosa*. Tetrahedron **51**: 10523

19. Montforts FP, Glasenapp-Breiling M, Kusch D (1998) Porphyrins and Related Compounds. In: Schaumann E (ed) Houben-Weyl-Methods of Organic Chemistry Vol E9 d. Thieme, Stuttgard/New York, p 577

20. Bonnett R (1978) Nomenclature. In: Dolphin D (ed) The Porphyrins Vol 1. Academic Press, New York, p 1

21. a) Jordan PM (1991) The biosynthesis of 5-aminolaevulinic acid and its transformation into uroporphyrinogen III. In: Jordan PM (ed) Biosynthesis of Tetrapyrroles. Elsevier, Amsterdam, p 1; b) Neier R (1996) Chemical Synthesis of Porphobilinogen and Studies of its Biosynthesis. Advances in Nitrogen Heterocycles **2**: 35; c) Tietze LF, Schulz G (1993) Investigations Towards the Biosynthesis of Porphyrins and Corrins. Angew Chem **105**: 1090

22. Lever ABP, Gray HB (eds) (1989) Iron Porphyrins Vols 1,2,4,5. VCH, New York

23. DiNello RK, Chang CK (1978) Isolation and Modification of Natural Porphyrins. In: Dolphin D (ed) The Porphyrins Vol 1. Academic Press, New York, p 290

24. Wu W, Chang CK, Vartosis C, Babcock GT, Puustinen A, Wikström M (1992) Structure of Heme *o* Prosthetic Group from the Terminal Quinol Oxidase of *Escherichia coli*. J Am Chem Soc **114**: 1182

25. a) Kusch D, Montforts FP (1995) Enantioselective Synthesis of Hematoporphyrin Stereoisomers. Tetrahedron Asymmetry **6**: 867; b) Kojo S, Fukunishi K, Tsukamato I (1989) Erhalt der Konfiguration bei der nicht – Darwinistischen Evolution von Cytochrom c. Angew Chem **101**: 64; c) Dickerson RE, Timkovich R, Boyer PD (1975) The Enzymes. Vol II. Academic Press, New York

26. a) D'Ambrosio M, Guerriero A, Debitus C, Ribes O, de Forges BR, Pietra F (1989) Corallistin A, a Second Example of a Free Porphyrin from Living Organism. Helv Chim Acta **72**: 1451; b) D'Ambrosio M, Guerriero A, Debitus C, Ribes O, Pietra F (1993) On the Novel Free Porphyrins Corallistin B, C, D and E. Helv Chim Acta **76**: 1489

27. a) Scheer H (1991) Structure and Occurrence of Chlorophylls. In: Scheer H (ed) Chlorophylls. CRC Press, Boca Raton, p 3 b) Fookes CJR, Jeffry S (1989) The structure of chlorophyll c_3. A novel marine photosynthetic pigment. J Chem Soc Chem Commun 1827

28. a) Baker EW, Palmer SE (1978) Geochemistry of Porphyrins. In: Dolphin D (ed) The Porphyrins Vol 1. Academic Press, New York, p 485; b) Callot HJ, Ocampo R (2000) Geochemistry of Porphyrins. In: Kadish MK, Smith KM, Guilard R (eds) The

Porphyrin Handbook Vol 1. Academic Press, San Diego, San Francisco, New York, Boston, London, Sydney, Toronto; c) Ocampo R, Bauder C, Callot HJ, Albrecht P (1992) Porphyrins from Messel oil shale. Geochimica et Cosmochimica Acta **56**: 745

29. Akhtar M (1991) Mechanism and stereochemistry of enzymes involved in the conversion of uroporphyrinogen III into haem. In: Jordan PM (ed) Biosynthesis of Tetrapyrroles. Elsevier, Amsterdam, p 67

30. Sessler JL, Weghorn SJ (1997) Expanded Contracted and Isomeric Porphyrins. Pergamon, Oxford

31. Kadish MK, Smith KM, Guilard R (eds) (2000) The Porphyrin Handbook Vol 1,2. Academic Press, San Diego, San Francisco, New York, Boston, London, Sydney, Toronto

32. a) Deisenhofer J, Norris JR (1993) The Photosynthetic Reaction Center. Academic Press, San Diego; b) Häder DP (1999) Photosynthese. Thieme, Stuttgart, New York; c) Jordan P, Fromme P, Witt HT, Klukas O, Saenger W, Krauß N (2001) Three-dimensional structure of cyanobacterial photosystem I at 2.4 Å resolution. Nature **411**: 909

33. a) van den Bergh P, Cornaz P (1985) Lokalisierung und Therapie von Tumoren mit Porphyrinen. Nachr Chem Tech Lab **33**: 582; b) Moan J, Berg K (1992) Photochemotherapy of Cancer: Experimental Research. Photochem Photobiol **55**: 931; c) Bonnett R (1999) Photodynamic Therapy in Historical Perspective. Rev Contemp Pharmacother **10**: 1; d) Bonnet R (2000) Chemical aspects of photodynamic therapy. Gordon and Breach Science Publisher, Singapore

34. a) Douglas RH, Partridge JC, Dulai KS, Hunt DM, Mullineaux CW, Hynninen PH (1999) Enhanced retinal longwave sensitivity using a chlorophyll-derived photosensitiser in *Malacosteus niger*, a deep-sea fish with far red bioluminescence. Vision Research **39**: 2817; b) Douglas RH, Partridge JC, Dulai K, Hunt D, Mullineaux CW, Tauber A, Hynninen PH (1998) Dragon fish see using chlorophyll. Nature **393**: 423

35. a) Rinehart KL, Kishore V, Bible KC, Sakai R, Sullins DW, Li KM (1988) Didemnins and Tunichlorin: Novel Natural Products from the Marine Tunicate *Trididemnum Solidum*. J Nat Prod **51**:1; b) Bible KC, Buytendorp M, Zierath PD, Rinehart KL (1988) Tunichlorin: A nickel chlorin isolated from the Caribbean tunicate *Trididemnum solidum*. Proc Natl Acad Sci **85**: 4582; c) Pettit GR, Kantoci D, Doubek DL, Tucker BE (1993) Isolation of the Nickel-Chlorin Chelate Tunichlorin from the South Pacific Ocean Sea Hare *Dolabella Auricularia*. J Nat Prod **56**: 1981; d) Sings HL, Bible KC, Rinehart KL (1996) Acyl tunichlorins: A new class of nickel chlorins isolated from the Caribbean tunicate *Tridemnum solidum*. Proc Natl Acad Sci **93**: 10560

36. Karuso P, Bergquist PR, Buckleton JS, Cambie RC, Clark GR, Rickard CEF (1986) 13^2, 17^3-Cyclopheophorbide Enol, the first porphyrin isolated from a sponge. Tetrahedron Lett **27**: 2177

37. Falk H, Hoornaert G, Isenring HP, Eschenmoser A (1975) Über Enolderivate der Chlorophyllreihe. Darstellung von 13^2, 17^3-Cyclophäophorbid-enolen. Helv Chim Acta **58**: 2347

38. a) Watanabe N, Yamamoto K, Ihshikawa H, Yagi A (1993) New Chlorophyll-A Related Compounds Isolated as Antioxidants from Marine Bivalves. J Nat Prod **56**: 305; b) Ma L, Dolphin D (1995) Asymmetric Hydroxylation of Chlorophyll Derivatives: A Facile Entry to Both Diastereomers of Chlorophyllone *a*. Tetrahedron Asym **6**: 313; c) Ma L, Dolphin D (1996) Synthesis and spectral studies of natural chlorins having antioxidative activity. Pure Appl Chem **68**: 765

39. a) Sakata K, Yamamoto K, Ishikawa H, Yagi A, Etoh H, Ina K (1990) Chlorophyllone-A, A new Pheophorbide-A related Compound isolated from *Ruditapes Philippinarum* as an Antioxidative Compound. Tetrahedron Lett **31**: 1165; b) Yamamoto K, Sakata K, Watanabe N, Yagi A, Brinen LS, Clardy J (1992) Chlorophyllonic Acid *a* Methylester, a New Chlorophyll *a* related Compound isolated as an Antioxidant from short-necked Clam, *Ruditapes Philippinarum*. Tetrahedron Lett **33**: 2587

40. Lederer E (1939) Sur l'isolement et la constitution chimique de la bonelline, pigment vert de *Bonellia viridis*. C R Acad Sci **209**: 528

41. a) Pelter A, Ballantine JA, Ferrito V, Jaccarini V, Psaila AF, Schembri PJ (1976) Bonnelin, a most Unusual Chlorin. J Chem Soc Chem Commun **1976**: 999; b) Pelter A, Ballantine JA, Murray-Rust P, Ferrito V, Psaila AF (1978) The Structure of Anhydrobonellin and Bonellin, the Physiologically Active Pigment from the Marine Echiuroid *Bonellia Viridis*. Tetrahedron Lett **21**: 1881; c) Pelter A, Abela-Medici A, Ballantine JA, Ferrito V, Ford S, Jaccarini V, Psaila AF (1978) The Structure of Amino-Acid Conjugates of Bonellin Derived from the Marine Echiuroid *Bonellia Viridis*. Tetrahedron Lett **23**: 2017; d) Agius L, Ballantine JA, Ferrito V, Jaccarini V, Murray-Rust P, Pelter A, Psaila AF, Schembri PJ (1979) The Structure and Physiological Activity of Bonellin – A Unique Chlorin Derived from *Bonellia Viridis*. Pure Appl Chem **51**: 1847; e) Ballantine JA, Psaila AF, Pelter A, Murray-Rust P, Ferrito V, Schembri PJ (1980) The Structure of Bonellin and its Derivatives. Unique Physiologically Active Chlorins from the Marine Echurian *Bonellia Viridis*. J Chem Soc Perkin Trans I **1980**: 1080

42. Montforts FP, Müller CM, Lincke A (1990) Bonellin: Aufklärung der absoluten Konfiguration. Liebigs Ann Chem **1990**: 415

43. Deeg R, Kriemler HP, Bergmann KH, Müller G (1977) Neuartige, methylierte Hydroporphyrine und deren Bedeutung bei der Cobyrinsäure Bildung. Hoppe-Seyler's Z Physiol Chem **358**: 339

44. Arigoni D, Imfeld M (1979) Factor I ex *Clostridium tetanomorphum*: Proof of Structure and Relationship to Vitamin B_{12} Biosynthesis. In: Zagalak B, Friedrich W (eds) Vitamin B_{12}. de Gruyter, Berlin, New York, p 315

45. Obinger C, Maj M, Nicholls P, Loewen P (1997) Activity, Peroxide Compound Formation, and Heme *d* Synthesis in *Escherichia coli* HPII Catalase. Arch Biochem Biophys **342**: 58

46. Murshudov GN, Grebenko AI, Barynin V, Dauter Z, Wilson KS, Vainshtein BK, Melik-Adamyan W, Bravo J, Ferran JM, Ferrer JC, Switala J, Loewen PC, Fita I (1996) Structure of the Heme *d* of *Penicillium vitale* and *Escherichia coli* Catalases. J Biol Chem **271**: 8863

47. a) Lemberg R, Barret J (1973) Cytochromes. Academic Press, New York, p 8–14 and 233–240; b) Stolzenberg AM, Strauss SH, Kolm RH (1981) Iron (II, III)-Chlorin and Isobacteriochlorin Complexes. Models of the Heme Prosthetic Groups in Nitrite and Sulfite Reductases: Means of Formation and Spectroscopic and Redox Properties. J Am Chem Soc **103**: 4763; c) Shimada H, Orii Y (1978) The pH-Dependent Reactions of *Pseudomonas aeruginosa* Nitrite Reductase with Nitric Oxide and Nitrite. J Biochem **84**: 1553; d) Orii Y, Shimada H, Nozawa T, Hatano M (1977) The Interaction Between the Heme *c* and Heme *d* Moieties of *Pseudomonas* Nitrite Reductase as Reveals by Magnetic and Natural Circular Dichroism Studies. J Biochem **76**: 983; e) Shimada H, Orii Y (1976) Oxidation-Reduction Behavior of the Heme *c* and Heme *d* Moieties of *Pseudomonas aeruginosa* Nitrite Reductase and the Formation of an Oxygenated Intermediate at Heme d_1. J Biochem **80**: 135

48. Kita K, Konishi K, Anraku Y (1984) Terminal Oxidases of *Escherichia coli* Aerobic Respiratory Chain. J Biol Chem **259**: 3375
49. Leeper FJ (1991) Intermediate Steps in the Biosynthesis of Chlorophylls. In: Scheer H (ed) Chlorophylls. CRC Press, Boca Raton, p 407
50. Griffiths WT (1991) Protochlorophyllide Photoreaction. In: Scheer H (ed) Chlorophylls. CRC Press, Boca Raton, p 433
51. Rüdiger W, Schoch S (1991) The Last Steps of Chlorophyll Biosynthesis. In: Scheer H (ed) Chlorophylls. CRC Press, Boca Raton, p 451
52. a) Woodward RB (1960) Totalsynthese des Chlorophylls. Angew Chem **72**: 651; b) Woodward RB, Ayer WA, Beaton JM, Bickelhaupt F, Bonnett R, Buchschacher P, Closs GL, Dutler H, Hannah J, Hauck FP, Ito S, Langemann A, Le Goff E, Leimgruber W, Lwowski W, Sauer J, Valenta Z, Volz H (1960) The Total Synthesis of Chlorophyll. J Am Chem Soc **82**: 3800; c) Woodward RB, Ayer WA, Beaton JM, Bickelhaupt F, Bonnett R, Buchschacher P, Closs GL, Dutler H, Hannah J, Hauck FP, Ito S, Langemann A, Le Goff E, Leimgruber W, Lwowski W, Sauer J, Valenta Z, Volz H (1990) The Total Synthesis of Chlorophyll *a*. Tetrahedron **46**: 7599
53. a) Snow RJ, Fookes CJR, Battersby AR (1981) Synthetic Routes to C-Methylated Chlorins. J Chem Soc Chem Commun 524; b) Harrison PJ, Fookes CJR, Battersby AR (1981) Synthesis of the isobacteriochlorin macrocycle: a photochemical approach. J Chem Soc Chem Commun 797; c) Battersby AR, Dutton CJ, Fookes CJR, Turner PD (1983) Synthesis of the Chlorin Macrocycle by a Photochemical Approach. J Chem Soc Chem Commun 1235; d) Arnott DM, Battersby AR, Harrison PJ, Henderson GB, Sheng ZC (1984) Synthesis of 20-methyl and 20-cyano isobacteriochlorins: the Wittig-photochemical approach. J Chem Soc Chem Commun 525; e) Battersby AR, Fookes CJR, Snow RJ (1984) Synthetic studies relevant to biosynthetic research on vitamin B_{12}. Part 2. Syntheses of C-methylated chlorins via lactams. J Chem Soc Perkin Trans 1, 2733; f) Battersby AR, Dutton CJ, Fookes CJR, Turner SPD (1988) Synthetic studies relevant to biosynthetic research on vitamin B_{12}. Part 6. Synthesis of chlorins by a photochemical approach. J Chem Soc Perkin Trans 1, 1557 g) Battersby AR, Reiter LA (1984) Synthetic studies relevant to biosynthetic research on vitamin B_{12}. Part 3. An approach to isobacteriochlorins via nitrones. J Chem Soc Perkin Trans 1, 2742; h) Battersby AR, Fookes CJR, Snow RJ (1984) Synthetic studies relevant to biosynthetic research on vitamin B_{12}. Part 1. Syntheses of C-methylated chlorins based on 1-pyrrolines (3,4-dihydropyrroles). J Chem Soc Perkin Trans 1, 2725
54. a) Montforts FP (1981) Ein gezielter Aufbau des Chlorinsystems. Angew Chem **93**: 795; b) Montforts FP, Schwartz UM (1985) Ein gezielter Aufbau des Chlorinsystems. Liebigs Ann Chem 1228
55. a) Block MH, Zimmermann SC, Henderson GB, Turner SPD, Westwood SW, Leeper FJ, Battersby AR (1985) Syntheses relevant to vitamin B_{12} biosynthesis: synthesis of sirohydrochlorin and of its octamethyl ester. J Chem Soc Chem Commun 1061; b) Brunt RD, Leeper FJ, Grgurina I, Battersby AR (1989) Biosynthesis of vitamin B_{12}: synthesis of (+/–) [5-^{13}C] factor-1 ester: determination of the oxidation state of precorrin-1. J Chem Soc Chem Commun 428
56. Battersby AR, Dutton CJ, Fookes CJR (1988) Synthetic studies relevant to biosynthetic research on vitamin B_{12}. Part 7. Synthesis of (+/–)-bonellin dimethyl ester. J Chem Soc Perkin Trans 1, 1569
57. a) Montforts FP, Schwartz UM (1985) Totalsynthese von (+/–) Bonellindimethylester. Angew Chem **97**: 767; b) Montforts FP, Schwartz UM (1991) Totalsynthese von (+/–) Bonellin-dimethylester. Liebigs Ann Chem 709

58. a) Roth M, Dubs P, Götschi E, Eschenmoser A (1971) Sulfidkontraktion via alkylative Kupplung: eine Methode zur Darstellung von β-Dicarbonylderivaten. Helv Chim Acta **60**: 3039; b) Eschenmoser A, Wintner CE (1977) Natural product synthesis and vitamin B_{12}. Science **196**: 1410

59. Hynninen PH (1991) Chemistry of Chlorophylls: Modifications. Scheer H (ed) Chlorophylls. CRC Press, Boca Raton, p 145

60. Sotiriou C, Chang CKJ (1988) Synthesis of the Heme Prosthetic Group of Bacterial Terminal Oxidase. J Am Chem Soc **110**: 2264

61. a) Michalski TJ, Hunt JE, Bowman MK, Smith U, Bardeen K, Gest H, Norris JR, Katz JJ (1987) Bacteriopheophytin *g* – properties and some speculations on a possible primary role for Bacteriochlorophyll *g* and *b* in the biosynthesis of chlorophylls. Proc Natl Acad Sci USA **84**: 2570; b) Oelze J (1985) Analysis in bacteriochlorophylls. Meth Microbiol **18**: 257; c) Smith KM (1991) The structure and biosynthesis of bacteriochlorophylls. In: Jordan PM (ed) Biosynthesis of Tetrapyrroles. Elsevier, Amsterdam, p 237

62. Prinsep MR, Caplan FR, Moore RE, Patterson GML, Smith CD (1992) Tolyporphin, a Novel Multidrug Resistance Reversing Agent from the Blue Green Alga *Tolypothrix nodosa*. J Am Chem Soc **114**: 385

63. Prinsep MR, Patterson GML, Larsen LK, Smith CD (1998) Tolyporphins J and K, Two Further Porphinoid Metabolites from the Cyanobacterium *Tolypothrix nodosa*. J Nat Prod **61**: 1133

64. a) Smith CD, Prinsep MR, Caplan FR, Moore RE, Patterson GML (1994) Reversal of Multiple Drug Resistance by Tolyporphin, a Novel Cyanobacterial Natural Product. Oncol Res **6**: 211; b) Morliére P, Maziére JC, Santus R, Smith CD, Prinsep MR, Stobbe CC, Fenning MC, Golberg JL, Chapman JD (1998) Tolyporphin: A Natural Product from Cyanobacteria with Potent Photosensitizing Activity against Tumor Cells *in Vitro* and *in Vivo*. Cancer Res **58**: 3571

65. Minehan TG, Kishi Y (1997) Extension of the Eschenmoser Sulfide Contraction/ Iminoester Cyclization Method to the Synthesis of Tolyporphin Chromophore. Tetrahedron Lett **38**: 6811

66. a) Siegel LM, Murphy MJ, Kamin H (1973) Reduced Nicotinamide Adenine Dinucleotide Phosphat-Sulfite Reductase of Enterobacteria. I. The *Escherichia coli* Hemoflavoprotein: Molecular Parameters and Prosthetic Groups. J Biol Chem **248**: 251; b) Siegel LM (1978) Structure and Function of Siroheme and Siroheme Enzymes. In: Singer TP, Ondarza RN (eds) Mechanisms of Oxidizing Enzymes (Dev. Biochem. 1) Elsevier/North Holland, New York, p 201; c) Siegel LM, Murphy MJ, Kamin H (1978) Siroheme: Methods of Isolation and Characterisation. Methods Enzymol **52**: 436

67. a) Lee JP, LeGall J, Peck HD jr (1973) Isolation of Assimilatory-Dissimilatory-Type Sulfite Reductase from *Desulfovibrio vulgaris*. J Bacteriol **115**: 529; b) Murphy MJ, Siegel LM, Tove SR, Kamin H (1974) Siroheme: A New Prosthetic Group Participating in Six-Electron Reduction Reactions Catalyzed by Both Sulfite and Nitrite Reductases. Proc Natl Acad Sci USA **71**: 612; c) Aparicio PJ, Knaff DB, Malkin R (1975) The Role of an Iron-Sulfur Center and Siroheme in Spinach Nitrite Reductase. Arch Biochem Biophys **169**: 102; d) Vega JM, Garrett RH, Siegel LM (1975) Siroheme: A Prosthetic Group of the *Neurospora crassa* Assimilatory Nitrite Reductase. J Biol Chem **250**: 7980; e) Porra RJ, Skyring GW (1976) The Isobacteriochlorin-Type Prosthetic Groups of Various Microbiol Sulphite Reductases. In: Doss M (ed) Porphyrins in Human Diseases. Karger, Basel, p 459; f) Stoller ML, Malkin R, Knaff DB (1977) Oxidation-Reduction Properties of Photosynthetic Nitrite Reductase. FEBS Letters **81**: 271; g) Seki Y, Sogawa N, Ishimoto M (1981) Siroheme as an Active Catalyst in Sulfite Reduction. J Biochem **90**: 1487

68. Battersby AR, Jones K, McDonald E, Robinson JA, Morris HR (1977) The structures and chemistry of isobacteriochlorins from *Desulphovibrio gigas*. Tetrahedron Lett **25**: 2213

69. a) Battersby AR, McDonald E, Morris HR, Thompson M, Williams DC, Bykhovsky VY, Zaitseva NI, Bukin VN (1977) Biosynthesis of vitamin B_{12}: structural studies on the corriphyrins from *Propionibacterium shermanii* and the link with sirohydrochlorin. Tetrahedron Lett **25**: 2217; b) Battersby AR, McDonald E, Thompson M, Bykhovsky VY (1978) Biosynthesis of vitamin B_{12}: proof of A-B structure for sirohydrochlorin by its specific incorporation into cobyrinic acid. J Chem Soc Chem Commun **3**: 150

70. a) Timkovich R, Cork MS, Taylor PV (1984) Proposed Structure for the Noncovalently Associated Heme Prosthetic Group of Dissimilatory Nitrite Reductases. J Biol Chem **259**: 1577; b) Timkovich R, Cork MS, Taylor PV (1984) Proposed Structure for the Noncovalently Associated Heme Prosthetic Group of Dissimilatory Nitrite Reductases. J Biol Chem **259**: 15089

71. a) Chang CK (1985) On the Structure of Heme d_1. J Biol Chem **260**: 9520; b) Chang CK (1994) Heme d_1 and other heme cofactors from bacteria. Ciba Found Symp **180**: 228; c) Chang CK (1993) Studies on pigment of life: novel porphyrinoids from bacteria. Youji Huaxue **13**: 171

72. a) Chang CK, Wu W (1986) The Porphinedione Structure of Heme d_1. J Biol Chem **261**: 8593; b) Chang CK, Timkovich R, Wu W (1986) Evidence That Heme d_1 Is a 1,3-Porphindione. Biochemistry **25**: 8447

73. a) Wu W, Chang CK (1987) Structure of "Dioneheme". Total Synthesis of the Green Heme Prosthetic Group in Cytochrome cd_1 Dissimilatory Nitrite Reductase. J Am Chem Soc **109**: 3149; b) Chang CK, Sotiriou C (1987) A Novel Method of Functionalizing the Ethyl Chain of Octaethylporphyrin. J Org Chem **52**: 926

74. a) Montforts FP, Romanowski F, Bats JW (1992) A Convenient Synthesis Of The Nitrite Reducing Cofactor Heme d_1 From Hematoporphyrin. Tetrahedron Lett **33**: 765; b) Montforts FP, Romanowski F, Bats JW (1989) Eine einfache Synthese von Dioxoisobakteriochlorinen aus Hämatoporphyrin. Angew Chem **101**: 471; c) Kusch D, Töllner E, Lincke A, Montforts FP (1995) Eine einfache chirogen enatioselektive Synthese von Chlorinen und Isobakteriochlorinen. Angew Chem **107**: 874; d) Kusch D, Montforts FP (1995) Enantioselective Synthesis of Hematoporphyrin Stereoisomers. Tetrahedron Asym **6**: 867; e) Romanowski F, Mai G, Kusch D, Montforts FP, Bats JW (1996) Stereoselektive Synthese des nitritreduzierenden Cofaktors Häm d_1 ausgehend von Hämatoporphyrin. Helv Chim Acta **79**: 1572

75. a) Micklefield J, Mackman RL, Aucken CJ, Beckmann M, Block MH, Leeper FJ, Battersby AR (1993) A Novel Stereoselective Synthesis of the Macrocycle of Haem d_1 that Establishes its Absolute Configuration as 2*R*,7*R*. J Chem Soc Chem Commun **3**: 275; b) Aucken CJ, Leeper FJ, Battersby AR (1997) Haem d_1: stereoselective synthesis of the reduced form of its parent macrocycle using the original coupling strategy. J Chem Soc Perkin Trans 1, 2099; c) Mackman RL, Micklefield J, Block MH, Leeper FJ, Battersby AR (1997) Haem d_1: development of a new coupling procedure leading to the synthesis of isobacteriochlorins. J Chem Soc Perkin Trans 1, 2111; d) Micklefield J, Beckmann M, Mackman RL, Block MH, Leeper FJ, Battersby AR (1997) Haem d_1: stereoselective synthesis of the macrocyle to establish its absolute configuration as 2*R*,7*R*. J Chem Soc Perkin Trans 1, 2123

76. Spencer JB, Stolowich NJ, Roessner CA, Scott AI (1993) The *Escherichia coli cys*G gene encodes the multifunctional protein, siroheme synthase. FEBS **335**: 57

77. a) Matthews JC, Timkovich R (1993) Biosynthetic Origins of the Carbon Skelton of Heme d_1. Bioorg Chem **21**: 71; b) Palmedo G, Seither P, Koerner H, Matthews J, Burkhalter RS, Timkovich R, Zumft WG (1995) Resolution of the nitD locus for heme d_1 synthesis of cytochrome cd_1 from *Pseudomonas stutzeri*. Eur J Biochem **232**: 737; c) Kawasaki S, Arai H, Igarashi Y, Kodama T (1995) Sequencing and characterization of the downstream region of the genes encoding nitrite reductase and cytochrome *c*-551 (nirSM) from *Pseudomonas aeruginosa*: identification of the gene necessary for biosynthesis of heme d_1. Gene **167**: 87

78. Yap-Bondoc F, Bondoc LL, Timkovich R, Baker DC, Hebbler A (1990) C-methylation Occurs during the Biosynthesis of Heme d_1. J Biol Chem **265**: 13498

79. a) Eschenmoser A (1974) Organische Naturstoffsyn these heute. Vitamin B_{12} als Beispiel. Naturwissenschaften **61**: 513; b) Eschenmoser A (1970) Roads to corrins. Quart Rev Chem Soc **24**: 366; c) Eschenmoser A (1976) Post-B_{12} problems in corrin synthesis. Chem Soc Rev **5**: 377

80. Montforts FP, Ofner S, Rasetti V, Eschenmoser A, Woggon WD, Jones K, Battersby AR (1979) Ein synthetischer Zugang zum Strukturtyp der Isobakteriochlorine. Angew Chem **91**: 752

81. Battersby AR, Westwood SW (1989) Synthetic studies relevant to biosynthetic research on vitamin B_{12}. Part 5. Synthesis of (*RS*)-ring-B imide. J Chem Soc Perkin Trans 1, 1679

82. Diekert G, Jaenchen R, Thauer RK (1980) Biosynthetic evidence for a Nickel tetrapyrrole structure of Factor F430 from *Methanobacterium thermoautotrophicum*. FEBS Lett. **119**: 118

83. a) Pfaltz A, Jaun B, Fässler A, Eschenmoser A, Jaenchen R, Gilles HH, Diekert G, Thauer RK (1982) Factor F430 from methanogenic bacteria: structure of porphiniod ligand system. Helv Chim Acta **65**: 828; b) Fässler A, Kobelt A, Pfaltz A, Eschenmoser A, Baldon C, Battersby AR, Thauer RK (1985) Factor F430 from methanogenic bacteria. Absolute configuration. Helv Chim Acta **68**: 2287; c) Livingston DA, Pfaltz A, Schreiber J, Eschenmoser A, Ankel-Fuchs D, Moll J, Jaenchen R, Thauer RK (1984) Factor F430 from methanogenic bacteria: structure of the protein-free factor. Helv Chim Acta **67**: 334; d) Pfaltz A, Livingston DA, Jaun B, Diekert G, Thauer RK, Eschenmoser A (1985) Factor F430 from methanogenic bacteria: nature of the isolation artifacts of F430, a contribution to the chemistry of F430 and the conformational stereochemistry of ligand periphery of hydroporphinoid nickel(II) complexes. Helv Chim Acta **68**: 1338; e) Fässler A, Pfaltz A, Müller PM, Farooq S, Kratky C, Kräutler B, Eschenmoser A (1982) Preparation and properties of some hydrocorphinoid nickel(II) complexes. Helv Chim Acta **65**: 812; e) Jaun B (1990) Coenzyme F430 from Methanogenic Bacteria: Oxidation of F430 Pentamethyl Ester to Ni(II) Form. Helv Chim Acta **73**: 2209

84. a) Kratky C, Fässler A, Pfaltz A, Kräutler B, Jaun B, Eschenmoser A (1984) Chemistry of corphinoids: structural properties of corphinoid nickel(II) complexes related to coenzyme F430. J Chem Soc Chem Commun 1368; b) Eschenmoser A (1986) Chemistry of cophinoids. Ann NY Acad Sci **471**: 108; c) Jaun B, Pfaltz A (1986) Coenzyme F430 from Methanogenic Bacteria: Reversible One-electron Reduction of F430 Pentamethyl Ester to Nickel(I) Form. J Chem Soc Chem Commun 1327; d) Färber G, Keller W, Kratky C, Jaun B, Pfaltz A, Spinner C, Kobelt A, Eschenmoser A (1991) Coenzyme F430 from methanogenic bacteria: complete assignment of configuration based on an X-ray analysis of 12,13-diepi-F430 pentamethyl ester and on NMR spectroscopy. Helv Chim Acta **74**: 697; e) Hamilton CL, Scott RA, Johnson MK (1989) The Magnetic and Electronic Properties of *Metha-*

nobacterium thermoautotrophicum (Strain Δ H) Methyl Coenzyme M Reductase and its Nickel Tetrapyrrole Cofactor F430. J Biol Chem **264**: 11605; f) Furenlid LR, Renner MW, Smith KM, Fajer J (1990) Structural Consequences of Nickel versus Macrocycle Reductions in F430 Models: EXAFS Studies of Ni(I) Anion and Ni(II) π Anion Radicals. J Am Chem Soc **112**: 1634; g) Crumbliss AL, McLachlan KL (1990) Electrochemistry of Heat-extracted Methanogenic Bacterial Cofactor F430. Inorg Chim Acta **170**: 161; h) Friedman HC, Klein A, Thauer RK (1991) Biochemistry of coenzyme F430, a nickel porphinoid involved in methanogenesis. In: Jordan PM (ed) Biosynthesis of Tetrapyrroles. Elsevier, Amsterdam, p 139

85. a) Diekert G, Weber B, Thauer RK (1980) Nickel dependence of factor F430 content of methanogenic bacteria. J Bacteriol **148**: 273; b) Ellefson WL, Wolfe RS (1981) Component C of the methylreductase system of Methanobacterium. J Biol Chem **256**: 4259; c) Holliger C, Schraa G, Stupperich E, Stams AJM, Zehnder AJB (1992) Evidence for the Involvement of Corrinoids and Factor F430 in the Reductive Dechlorination of 1,2-Dichloroethane by *Methanosarcina barkeri*. J Bacteriol **174**: 4427; d) Krone UE, Laufer K, Thauer RK (1989) Coenzyme F430 as a Possible Catalyst for the Reductive Dehalogenation of Chlorinated C_1 Hydrocarbons in Methanogenic Bacteria. Biochemistry **28**: 10061

86. a) Fässler A, Pfaltz A, Kräutler B, Eschenmoser A (1984) Chemistry of corphinoids: synthesis of a nickel(II) complex containing the chromophore system of coenzyme F430. J Chem Soc Chem Commun 1365; b) Kratky C, Waditschatka R, Angst C, Johansen JE, Plaquevent JC, Schreiber J, Eschenmoser A (1985) The saddle conformation of hydroporphinoid nickel(II) complexes: structure, origin, and stereochemical consequences. Helv Chim Acta **68**: 1312

87. Roessner CA, Warren MJ, Santander PJ, Atshaves BP, Ozaki SI, Stolowich NJ, Iida K, Scott AI (1992) Expression of 9 *Salmonella typhimurium* enzymes for cobinamide synthesis. FEBS **301**: 73

88. a) Müller G, Schmiedl J, Schneider E, Sedlmeier R, Wörner G, Scott AI, Williams HJ, Santander PJ, Stolowich NJ, Fagerness PE, Mackenzie NE (1986) Structure of Factor S3, a Metabolite of *Propionibacterium shermanii* Derived from Uroporphyrinogen I. J Am Chem Soc **108**: 7875; b) Müller G, Schmiedl J, Savidis L, Wirth G, Scott AI, Santander PJ, Williams HJ, Stolowich NJ, Kriemler HP (1987) Factor S1, a Natural Corphin from *Propionibacterium shermanii*. J Am Chem Soc **109**: 6902

89. Folkers K (1979) Introductory remarks. Historical perspectives of the isolation of crystalline vitamin B_{12}. In: Zagalak B, Friedrich W (eds) Vitamin B_{12}. de Gruyter, Berlin

90. Hodgkin DC (1979) New and old problems in the structure analysis of vitamin B_{12}. In: Zagalak B, Friedrich W (eds) Vitamin B_{12}. de Gruyter, Berlin

91. a) Eschenmoser A, Wintner CE (1977) Natural Product Synthesis and Vitamin B_{12}. Sience **196**: 1410; b) Eschenmoser A (1982) Über organische Naturstoffsynthese: Von der Synthese des Vitamin B_{12} zur Frage nach dem Ursprung der Corrinstruktur. Nova Acta Leopoldina **55**: 5

92. a) Woodward RB (1979) Synthetic vitamin B_{12}. In: Zagalak B, Friedrich W (eds) Vitamin B_{12}. de Gruyter, Berlin; b) Woodward RB (1968) Recent Advances in the Chemistry of Natural Products. Pure Appl Chem **17**: 519; c) Woodward RB (1973) The Total Synthesis of Vitamin B_{12}. Pure Appl Chem **33**: 145

93. Imfeld M, Townsend CA, Arigoni D (1976) Intact Transfer of Methyl Groups in the Biosynthesis of Vitamin B_{12}. J Chem Soc Chem Commun 541

94. Thibaut D, Blanche F, Cameron B, Crouzet J, Debussche L, Rémy E, Vuilhorgne M (1998) Vitamin B_{12} Biosynthesis in *Pseudomonas denitrificans*. In: Kräutler B, Arigoni D, Golding BT (eds) Vitamin B_{12} and B_{12}-Proteins. Wiley – VCH, Weinheim

95. a) Heinrich HC (ed) (1957) Vitamin B_{12} and Intrinsic Factor (1st European Symposium) Hamburg. Enke Verlag, Stuttgart; b) Heinrich HC (ed) (1962) Vitamin B_{12} and Intrinsic Factor (2nd European Symposium) Hamburg. Enke Verlag, Stuttgart

96. a) Mombelli L, Nussbaumer C, Weber H, Müller G, Arigoni D (1981) Biosynthesis of vitamin B_{12}: nature of the volatile fragment generated during formation of the corrin ring system. Proc Natl Acad Sci USA **78**: 11; b) Nussbaumer C, Imfeld M, Woerner G, Müller G, Arigoni D (1981) Biosynthesis of vitamin B_{12}: mode of incorporation of factor III into cobyrinic acid. Proc Natl Acad Sci USA **78**: 9; c) Battersby AR, Bushell MJ, Jones C, Lewis NG, Pfenniger A (1981) Biosynthesis of vitamin B_{12}: identity of fragment extruded during ring contraction to the corrin macrocycle. Proc Natl Acad Sci USA **78**: 13; d) Rasseti V, Pfaltz A, Kratky C, Eschenmoser A (1981) Ring contraction of hydroporphinoid to corrinoid complexes. Proc Natl Acad Sci USA **78**: 16

97. Blanche F, Thibaut D, Debussche L, Hertle R, Zipfel F, Müller G (1993) Parallelen und maßgebliche Unterschiede bei Vitamin-B_{12}-Biosynthesen. Angew Chem **105**: 1704

98. a) Scott AI, Roessner CA, Stolowich NJ, Spencer JB, Min C, Ozaki SI (1993) Biosynthesis of vitamin B_{12}. FEBS **331**: 105; b) Spencer JB, Stolowich NJ, Santander PJ, Pichon C, Kajiwara M, Tokiwa S, Takatori K, Scott AI (1994) Mechanism of the Ring Contraction Step in Vitamin B_{12} Biosynthesis: The Origin and Subsequent Fate of the Oxygen Functionalities in Precorrin-3x. J Am Chem Soc **116**: 4991

99. a) Stevens RV, Beaulieu N, Chan WH, Daniewski A, Takeda T, Waldner A, Williard P, Zutter UJ (1986) Studies on the Synthesis of Vitamin B_{12}. J Am Chem Soc **108**: 1039; b) Stevens RV (1979) Recent Studies on the Synthesis of Vitamin B_{12}. In: Zagalak B, Friedrich W (eds) Vitamin B_{12}. de Gruyter, Berlin, p 119

100. Jacobi PA, Brielmann HL, Hauck SI (1996) Toward the Synthesis of Biologically Important Chlorins, Isobacteriochlorins, and Corrins. Cyclic Enamides from Acteylenic Amides. J Org Chem **61**: 5013

101. a) Mulzer J, Riether D (1999) Synthesis of the C-ring fragment of cobyric acid. Tetrahedron Lett **40**: 6197; b) Mulzer J, List B, Bats JW (1997) Stereocontrolled Synthesis of a Nonracemic Vitamin B_{12} A-B Semicorrin. J Am Chem Soc **119**: 5512

102. Bonnett R (1982) Reactions of the Corrin Macrocycle. Dolphin D (ed) Vitamin B_{12} Vol I. Wiley, New York, p 201

103. a) Lewis NJ, Nussberger R, Kräutler B, Eschenmoser A (1983) 5,15-Bisnorcobester: eine unvorhergesehene Bildungsweise. Angew Chem **95**: 744; b) Nussbaumer C, Arigoni D (1983) Einfacher Zugang zu 5-Nor-, 15-Nor- und 5,15-Bisnorcobester. Angew Chem **95**: 746

The Chemistry of Taxol and Related Taxoids

D. G. I. Kingston, P. G. Jagtap, H. Yuan, and L. Samala
Department of Chemistry, Virginia Polytechnic Institute and State University, Blacksburg, VA, U.S.A.

Contents

1. Introduction

The anticancer compound taxol (**1.1.1**) was originally isolated in the late 1960's from the bark of the Western Yew, *Taxus brevifolia*, and its structure was published in 1971 in a paper that has been cited well over 1000 times since its publication (*1*). Taxol was originally isolated on the basis of its cytotoxicity and antileukemic activity, but it was not actively developed as a drug for several years because of its scarcity (the initial yield from yew bark was 0.02%) and because of its lack of water-solubility. Fortunately additional testing at the National Cancer Institute revealed its activity against various human solid tumor xenografts in nude mice, and a program to develop taxol as a drug was initiated in 1977. A key discovery from the Horwitz laboratory in 1979, in a paper that has also been cited over 1000 times since its publication, showed that taxol had a unique mechanism of action, binding to polymerized tubulin and stabilizing microtubules to dissociation (*2*). These twin discoveries of taxol's activity and its mechanism of action provided the impetus for continued clinical trials in the face of some initial disappointments, and the first positive clinical results against ovarian cancer were announced in 1989 (*3*). Following these results taxol was also found to be effective against breast cancer (*4*) and was licensed to Bristol-Myers Squibb; its sales are estimated at $1.6 billion in 2000 (*5*). A more detailed account of the history of the development of taxol has been published elsewhere (*6, 7, 8*). Thus from being a scientific curiosity 25 years ago, taxol (**1.1.1**, also known as paclitaxel or Taxol^{TM}) (*9*) and its semisynthetic analog docetaxel (Taxotere) (**1.1.2**) have developed into two of the most widely used anticancer drugs today.

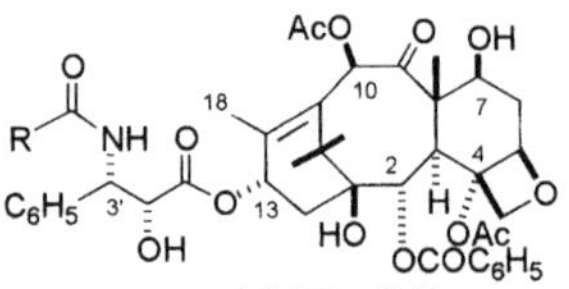

1.1.1 R = C_6H_5
1.1.2 R = OBu^t

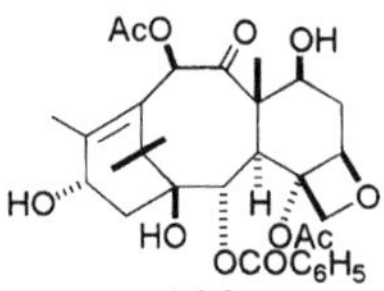

1.1.3

This increase in importance of the taxoid drugs has led to a surge of interest in taxol's chemistry and biology, as chemists have sought to understand its chemistry and to prepare improved analogs for a "second generation" taxoid drug, and as biologists have sought to understand its mechanism of action. The literature on the chemistry and biology of taxol has thus exploded, with the number of publications on taxol and related compounds mushrooming from a mere handful each year in the 1970's and early 1980's to over 100 in 1992, over 450 in 1995, and over 700 in 1999, the latest year for which complete figures are available. Publications on the clinical use of taxol and docetaxel now make up a large share of this total.

The last review of the taxane diterpenoids, or taxoids, in this series was published in 1993 by one of the present authors (*10*). This chapter thus covers the advances in the chemistry of the taxoids from mid 1992 to mid 2000. Because of the enormous volume of new work that has been published in the last seven years, the scope of this review is more restricted than the previous one, and its focus is primarily on the interesting and novel chemistry of taxol that has been discovered over the last eight years. The review does not include any discussion of the structures of the taxoids; readers interested in this aspect of the subject can obtain information from any of several recent reviews (*11–20*). The biosynthesis of taxol is likewise not covered; this subject has also been reviewed on several occasions (*19*, *21*, *22*, *23*, *24*). In addition, this review does not attempt to provide complete coverage of the many synthetic approaches that have been made to taxol and its analogs. The various methods that have been used to prepare the taxol side chain and to attach it to the baccatin III nucleus (**1.1.3**) are described, and the completed total syntheses of taxol are also summarized, but most other synthetic work in the taxol area is left to more specialized reviews. Finally, this review does not attempt to provide a complete coverage of all the medicinal chemistry that has been carried out on taxol over the past eight years. An excellent review by Georg (*25*) covered developments to 1995, and other excellent reviews by Ojima (*26*, *27*, *28*) and Ter Haar (*29*) bring the coverage of this subject to 2000.

2. A-Ring Chemistry

Taxoids which differ from taxol in the A ring, excluding those with different C-13 ester side chains, are relatively uncommon. The major avenues for preparing analogs by different functionalization of ester groups are lacking in the A ring, and the double bond is relatively unreactive and thus not readily amenable to derivative formation. In spite of this there have been a number of relatively recent papers describing reactions of the A-ring. These can be broadly characterized as reactions of the carbon–carbon double bond, reactions of the C-18 methyl group, and the synthesis of analogs with a C-14 hydroxyl group. In addition, a number of taxol analogs have been prepared with a five-membered A-ring (so-called A-*nor*taxoids or 11(15→1)abeotaxoids). These analogs will be discussed later, in Section 6.1

2.1. Modifications of the 11,12-Double Bond

The 11,12-double bond of the taxoids is rather unreactive to most reagents, because it is buried within the tetracyclic skeleton and is also tetrasubstituted. As examples of its general unreactivity, cephalomannine (**2.1.1**) can be brominated selectively on its side chain to give the dibromide **2.1.2** (*30*), and catalytic hydrogenation of taxol hydrogenates the aromatic rings but does *not* reduce the C-11,12 double bond (*31*)!

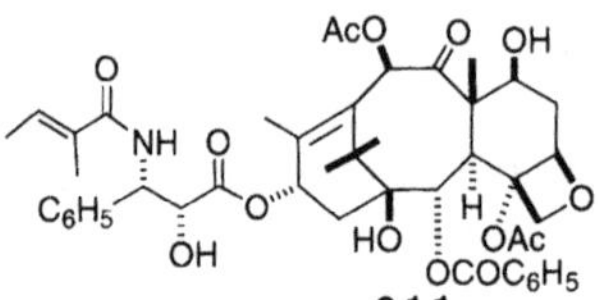
2.1.1

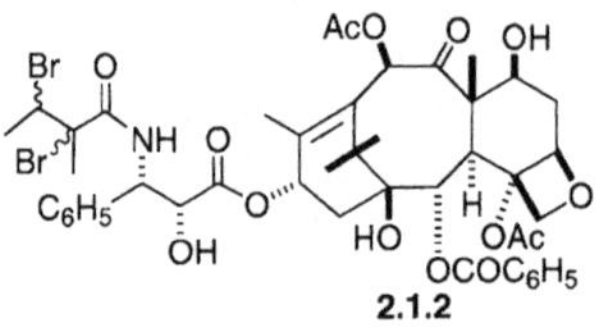
2.1.2

In spite of this unreactivity, some reactions at the C-11,12 double bond have been possible. Thus Georg and her collaborators investigated the epoxidation of baccatin III (*32*). Although baccatin III did not undergo epoxidation at the C-11,12 double bond, 10-deacetoxybaccatin III and 10-deacetoxytaxol did react with MCPBA; the product in the latter case was the taxol epoxide **2.1.3**, which was more active than taxol in a tubulin-assembly assay but was threefold less cytotoxic to B16 melanoma cells.

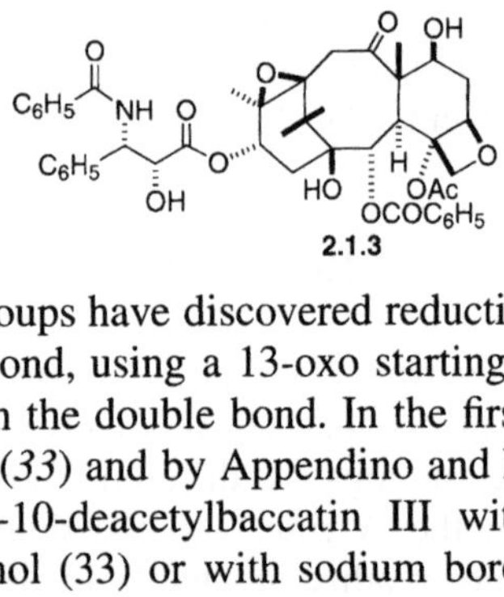
2.1.3

Three different groups have discovered reductive ways of modifying the C-11,12 double bond, using a 13-oxo starting material to allow for nucleophilic attack on the double bond. In the first approach, by Potier and his collaborators (*33*) and by Appendino and his collaborators (*34*), treatment of 13-oxo-10-deacetylbaccatin III with zinc and aqueous acetic acid in methanol (33) or with sodium borohydride and cuprous bromide (*34*) gave the 12,13-dihydro derivative **2.1.4** as well as a ring-B seco derivative **2.1.5** in the first case. If reduction was carried out with zinc in aqueous ammonia on the 7-triethylsilyl derivative, however, reduction gave the dihydro derivative cleanly, and this could be converted to the 13-hydroxyderivative **2.1.6**; the structure of this compound was confirmed by X-ray crystallography. Regrettably, attempts to attach the docetaxel side chain to compound **2.1.6** proved unsuccessful, but compound **2.1.7** was found to be approximately four times less active in a microtubule disassembly assay than 10-deacetylbaccatin III.

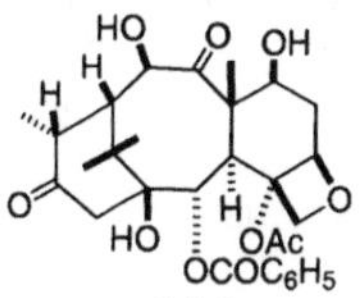

2.1.4

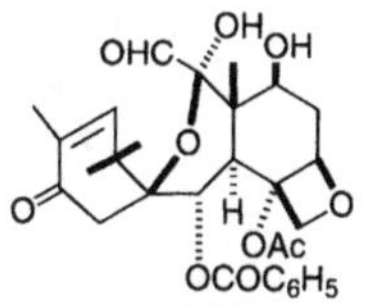

2.1.5

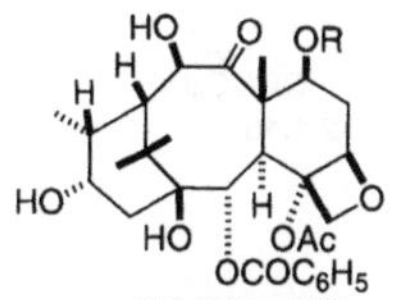

2.1.6 R = SiEt3
2.1.7 R = H

A later publication from the Potier group described a synthesis of the side chain derivative of **2.1.7**. The secret to attaching the side chain lay in removing the C-4 acetate group; when this was done the 4-deacetyl derivative **2.1.8** could be acylated at C-13 and then derivatized to give the docetaxel analog **2.1.9** (*35*). Disappointingly the C-4 acetate could not be reinstalled, so the fact that compound **2.1.9** was inactive in the microtubule disassembly assay does not prove that the dihydro modification is deleterious to activity.

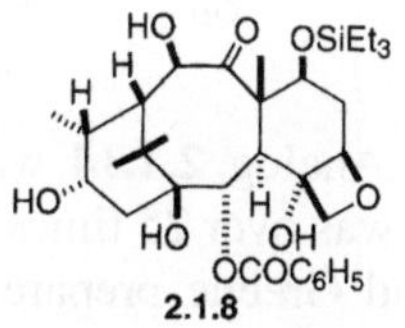

2.1.8

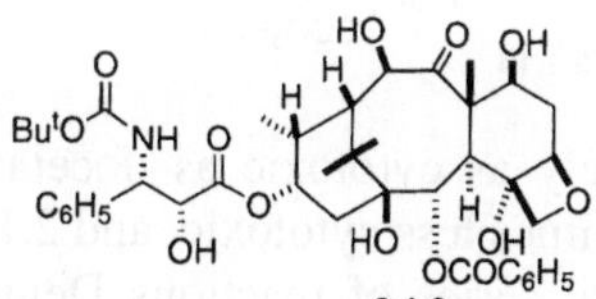

2.1.9

Reduction of the double bond of 7-triethylsilyl-13-oxo-baccatin III with zinc and acetic acid was also achieved by the Upjohn group (*36*). In this case, however, reduction was carried out on 7-triethylsilyl-13-oxo-baccatin III (**2.1.10**) rather than the 10-deacetyl analog used by Potier. Initial reduction surprisingly yielded the stable enol **2.1.11**. This enol was stable enough to be acylated at C-13 with protected side chain to give the isotaxol analog **2.1.12** after deprotection. Alternatively the enol **2.1.11** could be oxygenated at C-12 by treatment with oxygen followed by tetrabutylammonium borohydride to give the 12-hydroxy-11,12-dihydrobaccatin **2.1.13**. Acylation of **2.1.13** with the protected side chain gave the 12-hydroxy-11,12-dihydrotaxol analog **2.1.14**. A final series of transformations from **2.1.11** began with ketonization of the enol by treatment with silica gel to give the ketone **2.1.15**. This ketone was then reduced to the C-13 alcohol **2.1.16** and esterified to give the corresponding dihydrotaxol analog **2.1.17**.

The cytotoxicities of the analogs **2.1.12**, **2.1.14**, and **2.1.17** were determined against mouse L1210 cells, and **2.1.12** was found to be

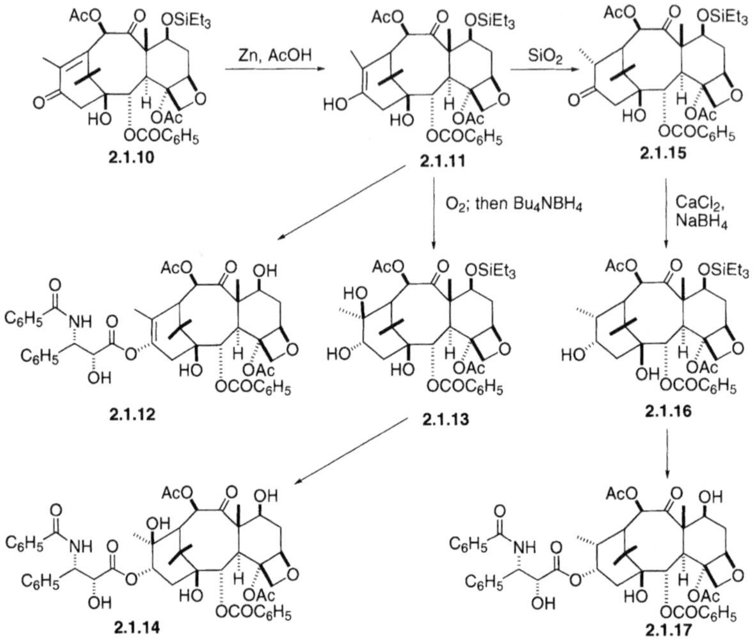

approximately as cytotoxic as docetaxel. Analog **2.1.14** was approximately ten times less cytotoxic, and **2.1.17** was over 25 times less active.

In a final series of reactions Denis and Greene prepared the taxol analog **2.1.19** by reduction of the naturally occurring taxoid taxagifine (**2.1.18**) to its dihydro derivative, followed by esterification with protected side chain and deprotection. The resulting compound, perhaps unsurprisingly because of its lack of the oxetane ring and the 2-benzoyl group, was inactive in both tubulin-assembly and cytotoxicity assays (*37*).

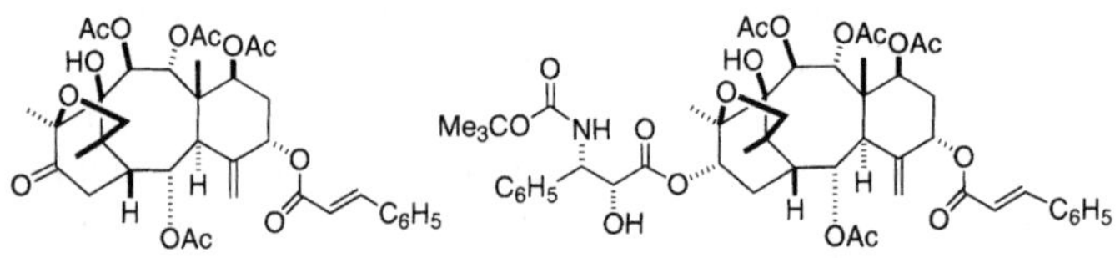

2.2. Modifications at C-13

A number of modified taxols where the major modification is in the nature of the attachment at the C-13 position have been made. Most of these start with baccatin III or 10-deacetylbaccatin III; the former compound has been available from taxol for many years by a reductive process using tetrabutylammonium borohydride as a selective reductant (*38*). Two additional methods to effect selective deacylation at C-13 have appeared in recent years. In one approach, 10-deacetyltaxol was protected as its 7-triethylsilyl ether and then treated with methanolic ammonia; desilylation gave 10-deacetylbaccatin III (*39*). In a second approach, treatment of 13-acetyl-7-(triethylsilyl)baccatin III (prepared by silylation and acetylation of 10-deacetylbaccatin III) with tetrabutylammonium borohydride in the presence of lithium chloride effected a selective deacetylation at C-13 (*40*).

Selective oxidation of 10-deacetylbaccatin III at C-13 can be accomplished by treatment with Collins' reagent (CrO_3-pyridine) at low temperature; the resulting 13-oxo-10-deacetylbaccatin III is stable (*41*, *42*).

Some more significant changes at C-13 have also been made. Thus Georg and her coworkers prepared 13-epitaxol from baccatin III (*43*). The key to their strategy was the stereoselective reduction of 4-deacetyl-13-oxo-7-(triethylsilyl)baccatin III (**2.2.1**) with tetramethylammonium triacetoxyborohydride, which gave a 13-epibaccatin derivative which was converted to the taxol analog **2.2.2**. This product was essentially inactive in a tubulin-assembly assay.

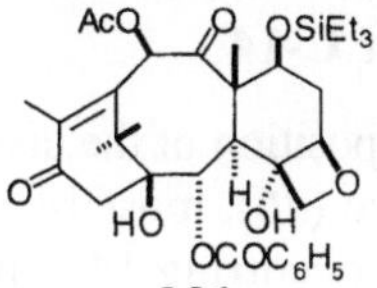

2.2.1

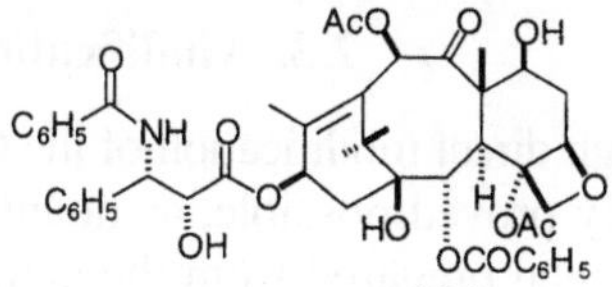

2.2.2

Two different groups have prepared 13-azataxol analogs. In the first approach, by Chen and his coworkers, the protected 13-azidobaccatin III **2.2.3** was prepared by treatment of 1-(dimethylsilyl)-7-(triethylsilyl)-baccatin III with DEAD/$Ph_3P/(BnO)_2P(O)N_3$ in THF. Treatment of **2.2.3** with protected β-lactam side chain gave the azataxol **2.2.4**. The analog **2.2.4** was over 200 times less active than taxol in a tubulin-assembly assay, and over 25 times less active in the HCT-116 cytotoxicity assay (*44*).

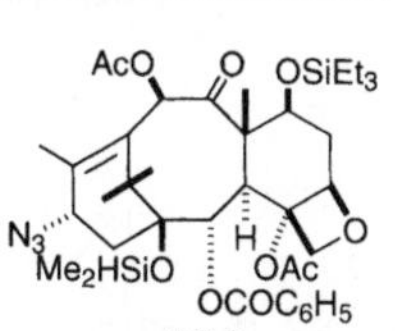

2.2.3

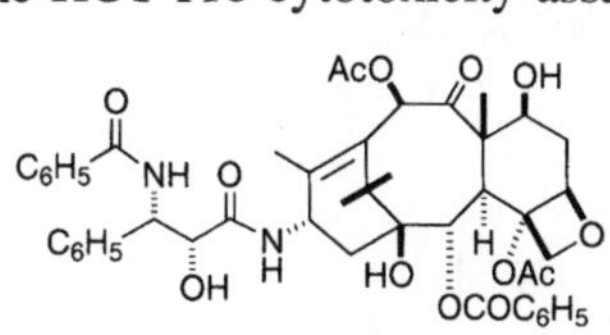

2.2.4

The second approach, by Menichincheri and her associates, used the oxime **2.2.5**, prepared from baccatin III, as their key intermediate (*45*). Acylation with protected side chain gave an unstable acyloxime which could not be isolated, but reduction with Raney nickel gave the 10-deacetoxyimine **2.2.6**. Acylation of **2.2.6** with protected side chain and deprotection gave the 10-deacetoxy-13,14-dehydro-13-azataxol **2.2.7**, together with its imide tautomer.

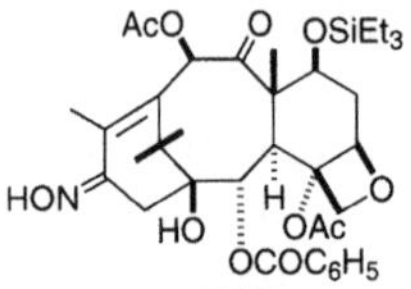
2.2.5

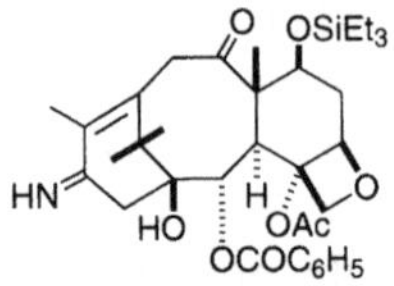
2.2.6

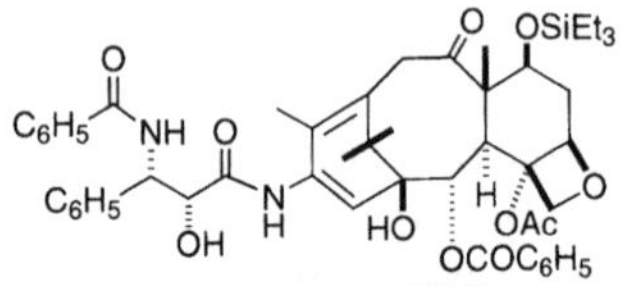
2.2.7

The azataxol **2.2.7** and other related compounds prepared by these workers were all inactive in tubulin-assembly and cytotoxicity bioassays. The 13-azabaccatins **2.2.8** and **2.2.9** were also prepared by these workers by methylation or oxidation of the imine **2.2.6** (*46*).

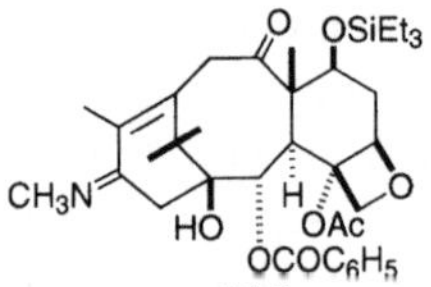
2.2.8

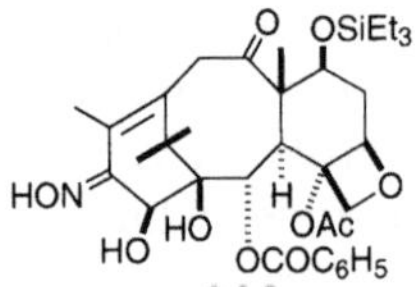
2.2.9

2.3. Modifications at C-14

Although direct modification of the C-14 position of the taxoid skeleton has recently proved possible, as noted above (*46*), most 14-hydroxytaxoids have been prepared from the naturally occurring 14-β-hydroxy-10-deacetylbaccatin III (**2.3.1**) (*47*). Thus Ojima and his collaborators prepared the protected 14β-docetaxel analog **2.3.2** by protection of the C-10 and C-7 hydroxyl groups as their Troc derivatives, and then of

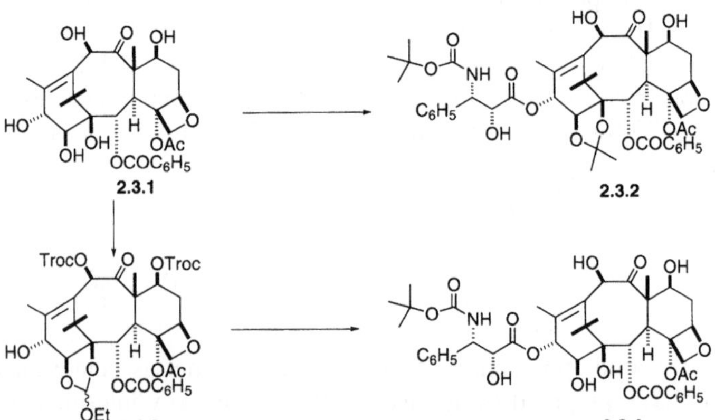
2.3.1 **2.3.2** **2.3.3** **2.3.4**

the C-1 and C-14 hydroxyl groups as their cyclic ketal. Acylation of the protected derivative with the side chain as its β-lactam derivative, followed by deprotection, gave **2.3.2**. The free C-14 hydroxyl analog **2.3.4** was prepared by protection of the Troc-protected compound as its orthoester **2.3.3**, acylation at C-13 with β-lactam, and deprotection (*48*).

A second series of transformations by Ojima and Appendino from 14β-hydroxy-10-deacetylbaccatin III (**2.3.1**) gave taxol analogs with the side chain at C-14 or with the side chain at C-13 and carbonate protection at C-14. Thus protection of **2.3.1** as its 2,7-di-Troc derivative by treatment with 4 equivalents of trichloroethoxy carbonyl chloride (Troc Cl) followed by esterification with the side chain β-lactam and deprotection, gave the analog **2.3.5** with the side chain at C-14β instead of the normal C-13α position. Alternatively, treatment of **2.3.1** with 6 equivalents of Troc Cl gave the cyclic carbonate **2.3.6**. Esterification of this derivative at C-13 with the β-lactam protected side chain, followed by deprotection, gave the analogs **2.3.7** and **2.3.8**, depending on which β-lactam derivative was used (*49*).

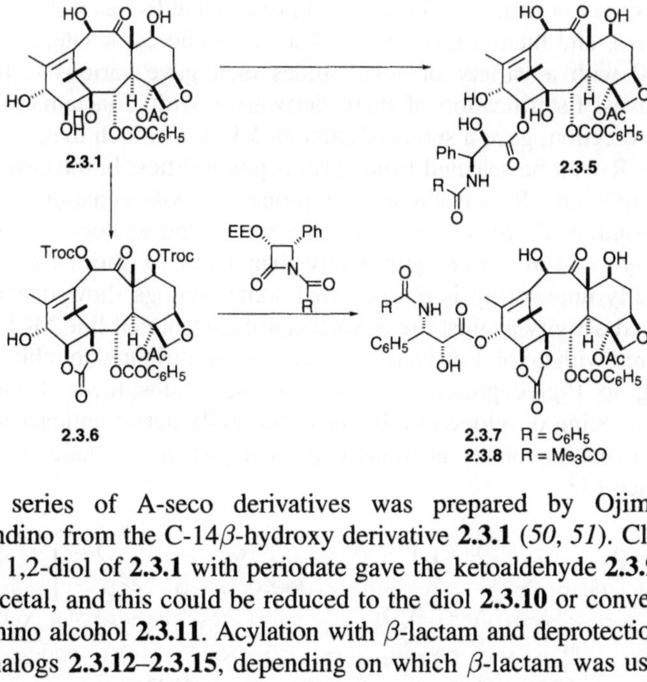

A series of A-seco derivatives was prepared by Ojima and Appendino from the C-14β-hydroxy derivative **2.3.1** (*50, 51*). Cleavage of the 1,2-diol of **2.3.1** with periodate gave the ketoaldehyde **2.3.9** as its hemiacetal, and this could be reduced to the diol **2.3.10** or converted to the amino alcohol **2.3.11**. Acylation with β-lactam and deprotection gave the analogs **2.3.12–2.3.15**, depending on which β-lactam was used and on whether the amino alcohol or the diol was esterified. Analogs **2.3.12** and **2.3.13** were only 15–20 times less cytotoxic than taxol, but analogs **2.3.14** and **2.3.15** were over 100 times less cytotoxic than taxol.

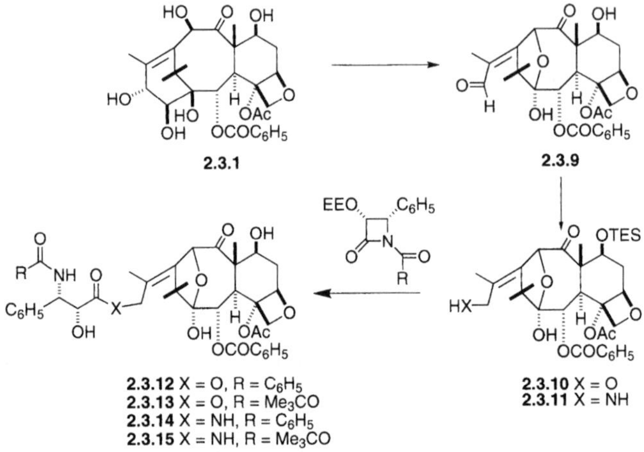

Additional protected 14β-hydroxy analogs were prepared by Ojima following a slightly modified methodology (*52*). In this methodology the 14β-hydroxy derivative **2.3.1** was first protected at C-7 as its triethylsilyl derivative, and then at C-1 and C-14 as its cyclic carbonate. Acylation at C-10 with a variety of acyl halides then gave various C-10 acyl derivatives. Esterification of these derivatives with β-lactam, followed by deprotection, gave a series of analogs **2.3.16** in which the groups R_1, R_2, and R_3 can be selected from various possibilities. In the case where $R_2 = {}^tBuO$- and R_3 = isobutenyl or propenyl, hydrogenation gave the corresponding dihydro derivatives. Several of the analogs prepared in this way showed better cytotoxicity than taxol to various cell lines. Especially noteworthy is the fact that some analogs showed markedly improved activity against the doxorubicin-resistant cell line MCF7-R as compared with taxol; this increase in activity is attributed to inhibition of binding to P-glycoprotein by the C-10 acyl substituent. Compound **2.3.17** is being developed by Bayer as an orally active anticancer drug, and information on its pharmacology and pharmacokinetics has been published (*53*).

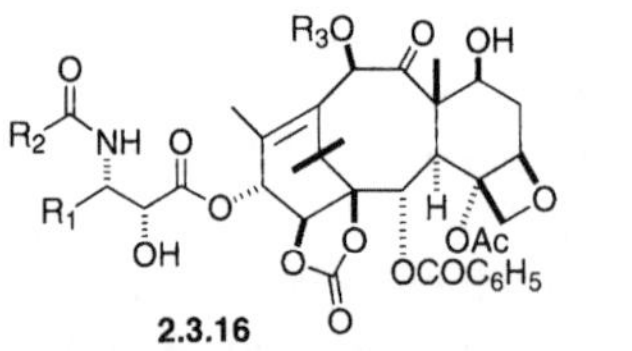

2.3.16

R_1 = 2-Furyl, $Me_2C{=}CH$, Me_2CCH_2, Ph
R_2 = ${}^tBuO$- and n-C_5H_{11}
R_3 = H, CH_3CO, EtCO, Me_2NCO, MeOCO

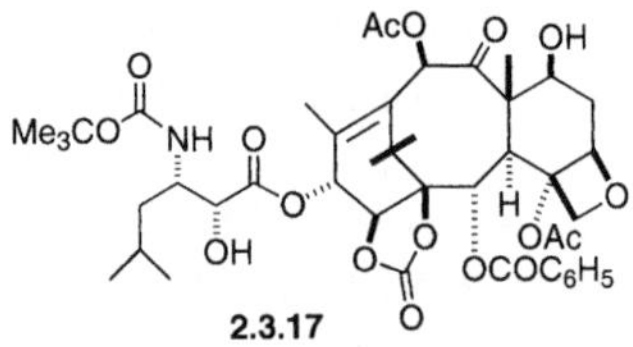

2.3.17

A final interesting finding in this area is that oxidation of **2.3.8** with cupric acetate gave the diketo derivative **2.3.18**, and treatment of this with hydrazine gave the pyrazoline **2.3.19**, which is claimed to be an antitumor agent (*54*).

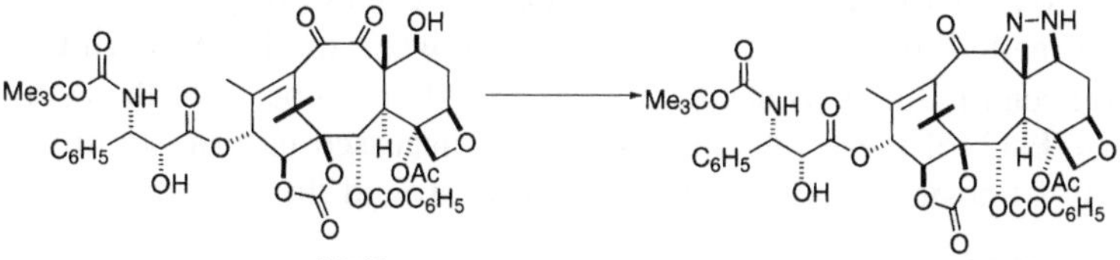

2.3.18 **2.3.19**

2.4. Modifications at C-18

Modification of the C-18 methyl group was achieved by Soga and his collaborators (*55*). The key step was allylic bromination of 7,13-di(triethylsilyl)baccatin III to the bromide **2.4.1**. Treatment of **2.4.1** with a variety of nucleophiles (Me_2CuLi, NaN_3, Bu_4NOAc, KCN) yielded the corresponding C-18 substituted analogs, and these were converted by standard chemistry to the 10-acetyldocetaxel analogs **2.4.2**–**2.4.5**. All of the analogs prepared were less cytotoxic than taxol, with the smallest decrease in activity being observed for the smaller methyl and cyano substituents. This result indicates that the locus of the C-18 binding site on tubulin is not very forgiving of steric bulk.

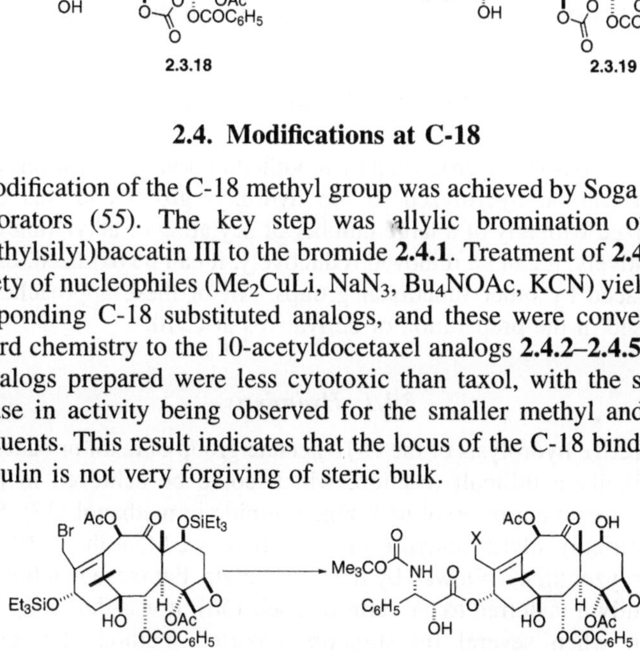

2.4.1

2.4.2 X = CH_3
2.4.3 X = N_3
2.4.4 X = OAc
2.4.5 X = CN

Treatment of 2′-(*t*-butyldimethylsilyl)-7-*epi*-10-deacetyltaxol with triflic anhydride at 0°C gave the dehydrated product **2.4.6** and the oxidized product **2.4.7**. No bioactivity data was reported for these compounds (*56*).

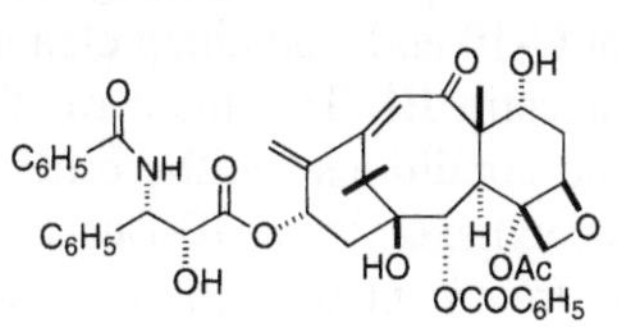

2.4.6

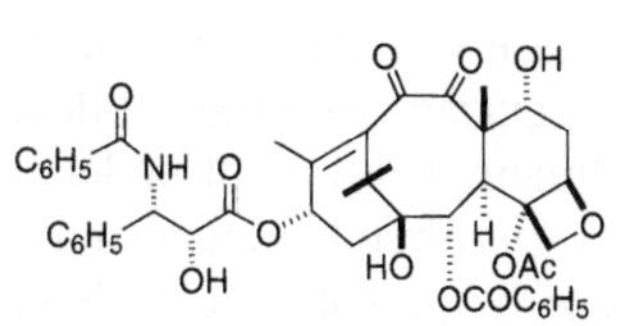

2.4.7

3. B-Ring Chemistry

The eight-membered B-ring of Taxol carries substituents at C-9 and C-10 on its "Northern" face and at C-2 on its "Southern" face, and numerous modifications have been made at each of these positions. The C-19 methyl group, on the bridgehead between the B and C rings, has also been modified, as has the C-1 hydroxyl group between the A and B rings. Each of these substituents will be discussed in turn, starting with the C-10 position and moving clockwise around the ring.

3.1. Modifications at C-10

The C-10 acetoxy group can be modified in principle in several ways. Thus it can be hydrolyzed to the hydroxyl group and this can be derivatized with any of a large number of acylating or alkylating agents. Alternatively, it can be removed reductively, it can be oxidized, or it can be replaced by other functional groups. All of these approaches have been used in the preparation of derivatives at C-10.

3.1.1. Hydrolysis

Selective hydrolysis of the C-10 acetate group of taxol or baccatin III was initially a difficult reaction, which could be achieved in modest yield by treatment of taxol with zinc bromide in methanol (*57*). Studies with 7-(triethylsilyl)hexahydrobaccatin III showed that the C-10 acetate is hydrolyzed first, followed by the C-4 acetate, but these reactions were not readily transferred to baccatin III itself (*58*). It was thus a significant advance when several investigators reported methods for selective hydrolysis of the C-10 acetate group.

Treatment of taxol with hydrazine in aqueous ethanol was shown by Johnson and Kelly and their colleagues to give selective deacylation at C-10; thus 10-acetyldocetaxel was converted in 86% yield to docetaxel under these conditions. The C-7 hydroxy group is required for this reaction; when it is absent or masked by a protecting group the reaction is much less effective (*59*). Similar results were also reported by Georg, although in this case simultaneous deacylation at C-10 and side-chain cleavage of taxol occurred, yielding 10-deacetylbaccatin III. Treatment of 2′-(*tert*-butyldimethylsilyl)taxol under the same conditions, on the other hand, gave 2′-(*tert*-butyldimethylsilyl)-10-deacetyltaxol (*60*). 10-Deacetyltaxol can also be prepared in high yield by treatment of taxol with sodium bicarbonate and hydrogen peroxide. The reaction works well with taxol itself, but gives poor yields or fails when the 2′- and 7-positions are protected as their triethylsilyl ethers (*61*). Recently it has been shown that scandium trifluoromethane sulfonate can catalyze the hydrolysis of the C-10 acetyloxy group, although the yields here (72%) are lower than those with the bicarbonate/peroxide method (*62*).

3.1.2. Acylation

The direct acylation of 10-deacetylbaccatin III (10-DAB) to baccatin III is a reaction of potential commercial importance, since 10-DAB is available from yew needles (*63*, *64*), and various methods have been developed to achieve this conversion. Thus treatment of 10-DAB with THF, lithium chloride, a tertiary amine, and acetyl chloride gave baccatin III (*65*). Two groups have reported that treatment of 10-DAB with acetic anhydride and a Lewis acid catalyst such as $ZnCl_2$, $CeCl_3$, or $La(OTF)_3$ gave good yields of baccatin III (*66*, *67*, *68*). The use of an adduct generated *in situ* from triflic anhydride and dimethylformamide gave 10-formyl-10-deacetylbaccatin III in good yield, and treatment of 10-DAB with the iminium salt formed from n-ethylacetamide and mesyl chloride, followed by hydrolysis with aqueous sodium acetate, gave baccatin III in about 85% yield (*69*). Finally, 10-DAB can be acylated using a lipase from *Pseudomonas cepacia*, but both the 7- and 10-positions were acylated and the reaction required the use of chloroacetic anhydride, so it is not a substitute for the chemical reactions described above (*70*).

A number of investigators have made C-10 acyl analogs of taxol. Thus Bourzat and his collaborators prepared various 10-*O*-carbamoyl analogs of taxol such as **3.1.2.2** by treatment of 7,10-di-troctaxol (**3.1.2.1**) with various amines, followed by deprotection of the unreacted C-7 Troc group (*71*).

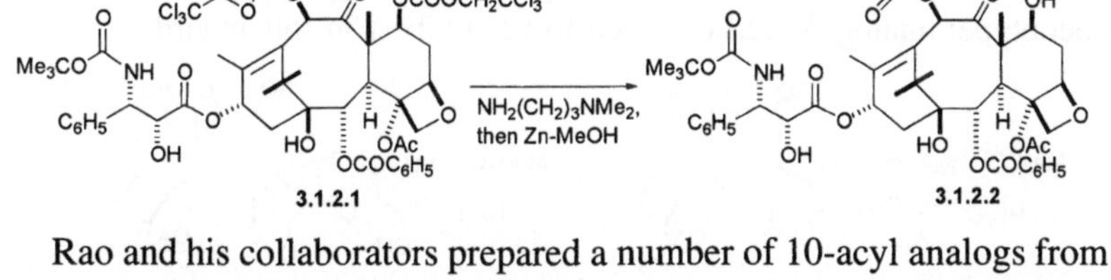

3.1.2.1 **3.1.2.2**

Rao and his collaborators prepared a number of 10-acyl analogs from 10-deacetyltaxol by protecting the 2′ and 7 positions as their chloroacetyl derivatives **3.1.2.3**, acylating at C-10 and deprotecting with thiourea to give the 10-acyl analogs **3.1.2.4** (*72*).

A different route was used by Kant and his coworkers (*73*) and by Ojima and his collaborators (*74*). Both groups selectively acylated 7-

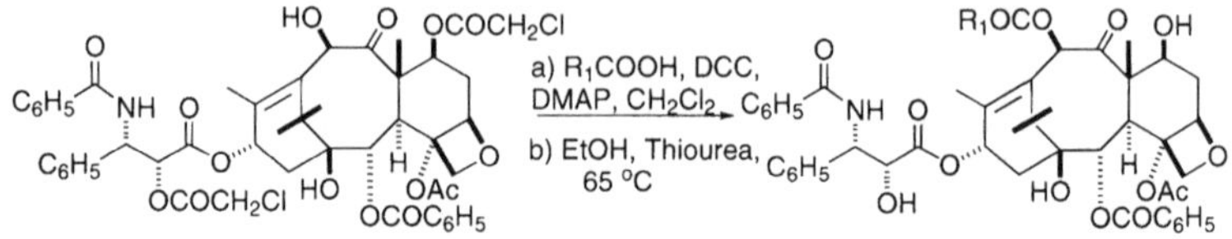

3.1.2.3 **3.1.2.4**

triethylsilyl-10-deacetyl baccatin III using strong base (n-BuLi or LHMDS) and an electrophile to get 10-acyl derivatives **3.1.2.5**, and then coupled these with various β-lactams to give 10-acyl analogs **3.1.2.6**. This approach is obviously more versatile than the Rao approach, in that analogs can be prepared which differ both at C-10 and on the side chain. Selective acetylation of the C-10 hydroxyl group of 2′-triethylsilyl-10-deacetyl-7-*epi*taxol has been achieved; this reaction is simplified by the congested location of 7-*epi*hydroxyl group (*75*).

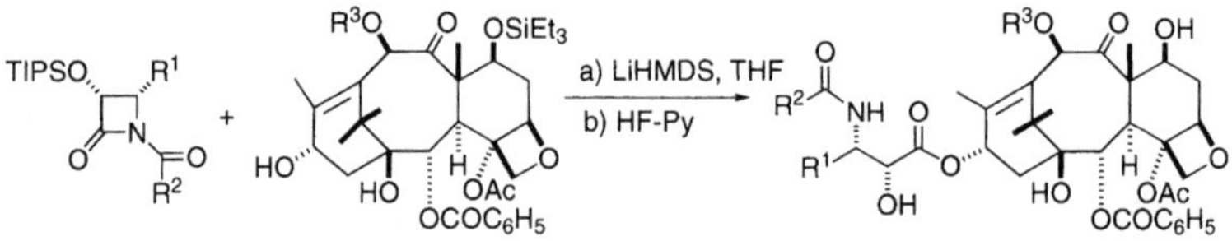

3.1.2.5 **3.1.2.6**

Taxoids with modified C-10 acyl substituents and modified side-chain substituents have been prepared by Ojima and his coworkers. In one recent study 3′-difluoromethyl-10-acyl analogs were prepared in both the normal series (**3.1.2.7–3.1.2.11**) and the 14β-hydroxy series (**3.1.2.12–3.1.2.17**) (*76*). The difluoromethyl side-chain was prepared by reaction of a formyl azetidinone precursor with DAST, and the resulting difluoromethyl-β-lactam was coupled with appropriate 10-acyl analogs of baccatin III or 14β-hydroxybaccatin III. Three of the analogs (**3.1.2.7**, **3.1.2.8**, and **3.1.2.9**) proved to be slightly more cytotoxic than docetaxel to normal LCC6-WT cells, but most interestingly several analogs were significantly more active than taxol or docetaxel against the drug-resistant LCC6-MDR phenotype; analog **3.1.2.11** proved to be the best in this regard.

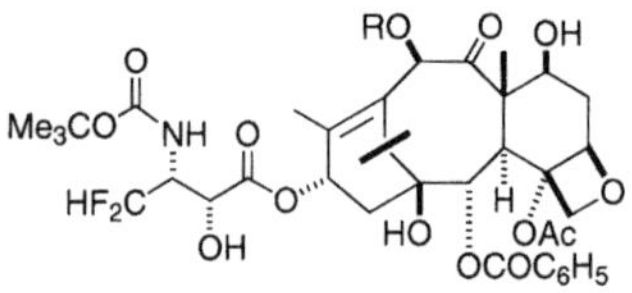

3.1.2.7 R = H
3.1.2.8 R = Ac
3.1.2.9 R = Me$_2$NCO
3.1.2.10 R = EtCO
3.1.2.11 R = Me$_3$CCH$_2$CO

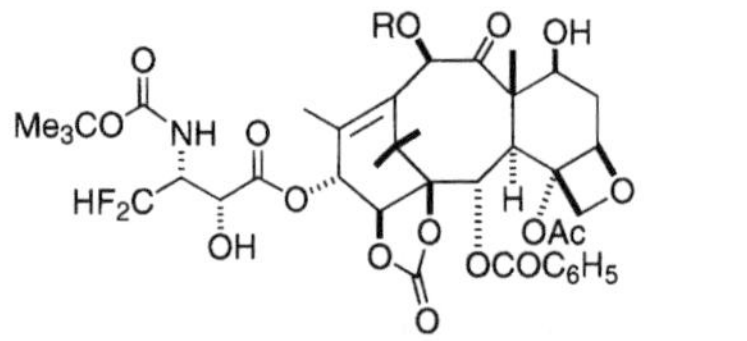

3.1.2.12 R = H
3.1.2.13 R = Ac
3.1.2.14 R = Me$_2$NCO
3.1.2.15 R = EtCO
3.1.2.16 R = Me$_3$CCH$_2$CO
3.1.2.17 R = cy-PrCO

3.1.3. Alkylation

C-10-Alkyl analogs have also been prepared. In a direct approach exemplified by Bouchard and his collaborators, a suitably protected 10-deacetylbaccatin III was alkylated with an alkyl iodide in the presence of sodium hydride to give a C-10 alkyl derivative. Appropriate protecting group manipulation and coupling with side chain then gave C-10 alkyl taxol analogs such as **3.1.3.1**. Alternatively, deprotection and methylation at C-7 followed by side chain attachment gave a 7,10-dialkyl analog **3.1.3.2** (*77*). A similar approach has been used by Hirofumi *et al.* in the preparation of 10-allyl analogs of taxol such as 10-allyl docetaxel (**3.1.3.1**, R = allyl) (*78*).

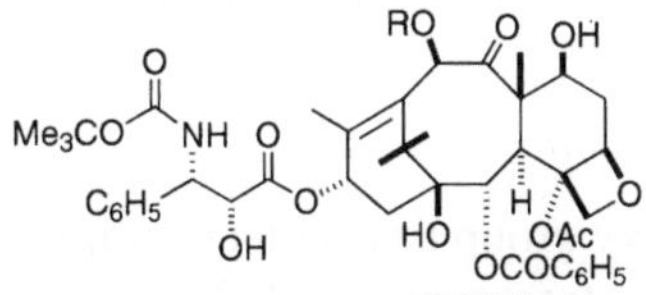

3.1.3.1

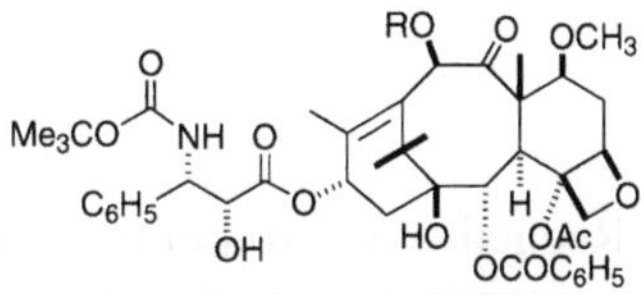

3.1.3.2

C-10 analogs have also been made by addition of a C-10 radical (generated by reaction of the C-10 xanthate **3.1.3.3** with an organostannane) to various $\alpha\beta$-unsaturated carbonyl and cyano derivatives. C-10β-alkylation was observed in all cases and was proved by a Noesy experiment. Addition of the side-chain and deprotection yielded the analogs **3.1.3.4** (*79*). A recent modification uses organosilanes instead of organostannanes to generate the radical with somewhat improved yields (*80*). The C-10 modified docetaxel analogs were tested against P388, PC-6 and PC-12 cell lines. The activities of analogs **3.1.3.4** with polar substituents R such as NH_2, NHCHO or COOH were significantly less than that of docetaxel, while the OH and CN analogs were only moderately less active. The COOEt and especially the COOMe analog showed fairly good activity, with the latter exhibiting more potent activity than docetaxel.

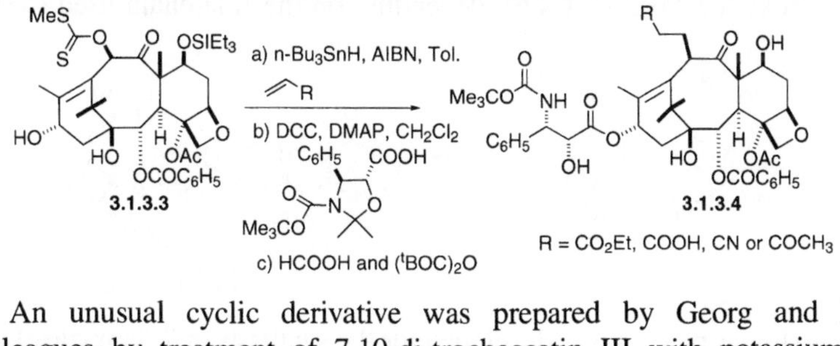

An unusual cyclic derivative was prepared by Georg and her colleagues by treatment of 7,10-di-trocbaccatin III with potassium *t*-butoxide. Enolate formation was followed by intramolecular attack of the enolate ion on the carbonyl group of the 10-Troc substituent, leading to the cyclic enol carbonate **3.1.3.5**. Appropriate chemistry was then used to convert this to the taxol analogs **3.1.3.6** and **3.1.3.7**, which were found to be reasonably competent at causing microtubule assembly but to be only weakly cytotoxic (*81*).

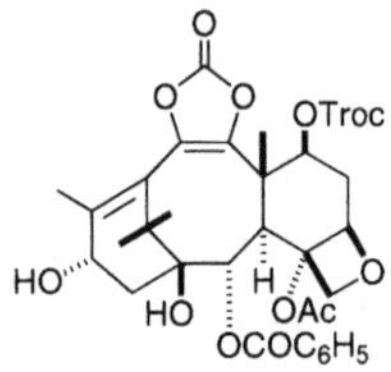

3.1.3.5

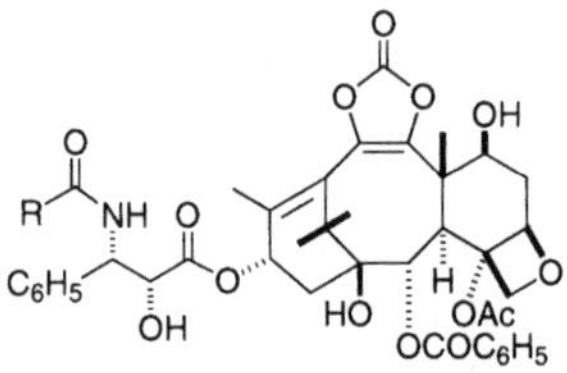

3.1.3.6 R = C_6H_5
3.1.3.7 R = $OCMe_3$

3.1.4. Reduction

Reductive removal of the C-10 acetoxy group has been accomplished in various ways. In the first approach, treatment of 2′,7-di-Troc-10-deacetyltaxol with Et_2NCF_2CHFCl (Yarovenko's reagent) unexpectedly gave the diene **3.1.4.1**, which could be hydrogenated and deprotected to give 10-deacetoxytaxol **3.1.4.2** (*82*).

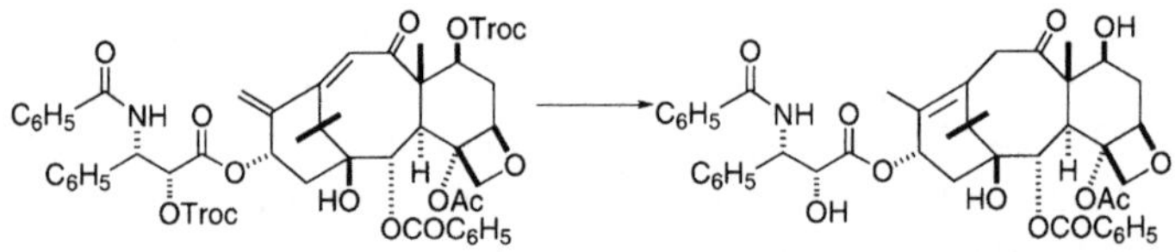

3.1.4.1 **3.1.4.2**

Alternate procedures were also developed using Barton's reagent. In the first of these, coupling of the xanthate **3.1.4.3** obtained from 10-deacetylbaccatin III with protected side chain, followed by free radical deoxygenation and deprotection, gave 10-deacetoxytaxol (**3.1.4.4**) or 10-dehydroxydocetaxel (**3.1.4.5**), depending on the side-chain used (*83*).

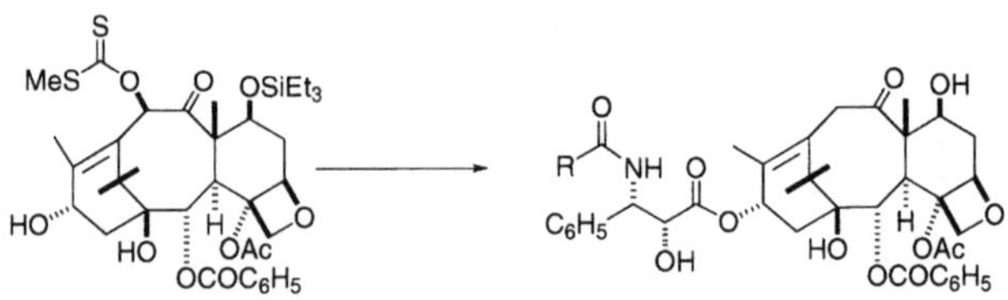

3.1.4.3

3.1.4.4 R = C_6H_5
3.1.4.5 R = $OCMe_3$

A similar process was developed for the synthesis of 7-deoxytaxol and 7,10-dideoxytaxol (*84*). In the latter case the rearranged baccatin **3.1.4.8** was formed as a by-product, and a reasonable mechanism through the intermediates **3.1.4.6** and **3.1.4.7** was proposed to account for its formation. A later publication from the same group reported the formation of additional rearrangement products formed via radical intermediates (*85*).

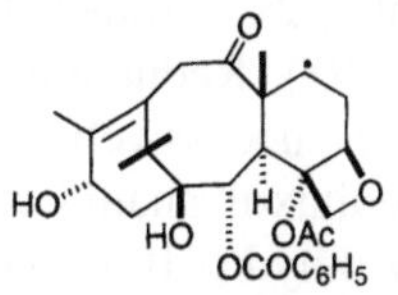

3.1.4.6

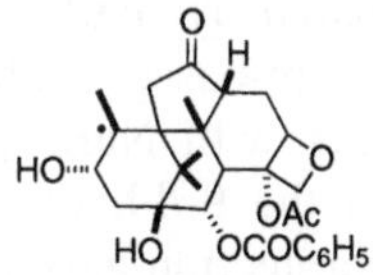

3.1.4.7

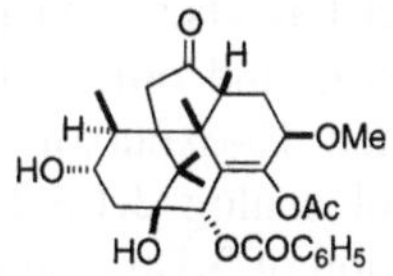

3.1.4.8

The lengthy procedures described above for deoxygenation at C-10 were rendered obsolete by independent reports from the Holton and Georg groups, who demonstrated that the 10-acetoxy group can readily be removed in a single step by treatment with samarium iodide (*86*, *87*). The reaction proceeds in high yield on both taxol and baccatin III, but is shut down by silylation at C-7. Under appropriate conditions reduction at C-9 also occurs, leading to 10-deacetoxy-9-dihydrobaccatin III (**3.1.4.9**) from baccatin III (*88*, *89*) and epimerization at C-7 can also occur (*90*). The use of D_2O instead of H_2O as a solvent gave the deuterated product **3.1.4.11**, indicating that protonation of the intermediate enol **3.1.4.10** occurs from the α-face (*91*).

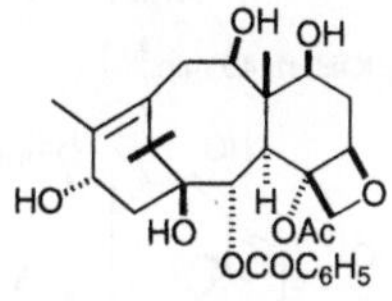

3.1.4.9

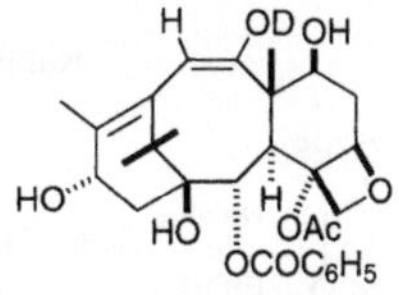

3.1.4.10

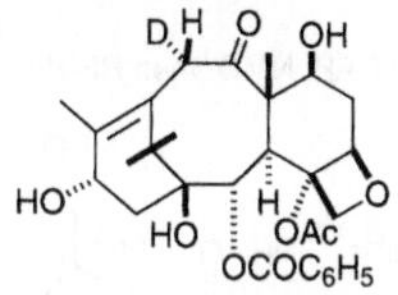

3.1.4.11

Electrochemical reduction of taxol and of docetaxel gave very different results (*92*). Thus reduction of taxol in the presence of ammonium chloride gave the 10-deacetoxy derivative in 45% yield, but the reduction of docetaxel under essentially the same conditions gave a mixture of 9α- and 9β-dihydrodocetaxels.

3.1.5. Oxidation and Epimerization

Direct oxidation of the C-10 hydroxyl group of 10-deacetylbaccatin III was studied by Appendino (*42*), who found that it could be oxidized with moderate selectivity as compared with the C-13 hydroxyl group by treatment with MnO_2 in acetonitrile. Treatment with $Cu(OAc)_2$ oxidized the C-10 hydroxyl group with complete selectivity, but also epimerized the C-7 hydroxyl group.

Taxol analogs with various oxidation levels at C-9 and C-10 have been prepared from taxol by Georg and her collaborators. Thus oxidation of 2′-(*tert*-butyldimethylsilyl)-7-(triethylsilyl)taxol with TPAP and NMO gave the 10-deacetyl-10-oxo product **3.1.5.1**. A similar product has also been prepared by oxidation of a 2′,7-protected taxol with PCC and other chromium-based oxidants (*93*). Reduction of the protected 10-oxotaxol **3.1.5.1** with $NaBH_4$ gave successively the 10-epitaxol analog **3.1.5.2** and the 9-dihydro-10-epitaxol analog **3.1.5.3**. Oxidation of **3.1.5.3** and deprotection gave the 9-dihydro-10-oxo analog **3.1.5.4**, while acetylation of **3.1.5.3** or **3.1.5.2** followed by deprotection gave the 9-dihydro-10-*epi* analog **3.1.5.5** and the 10-*epi* analog **3.1.5.6** respectively (*94*). The 10-*epi* analog **3.1.5.6** and the C-9,C-10 dihydroxy analog **3.1.5.5** showed increased activity in microtubule assembly assays and cytotoxicity relative to paclitaxel. Stereochemical inversion of the oxygen functionality at C-10 thus has a slightly positive effect on biological activity.

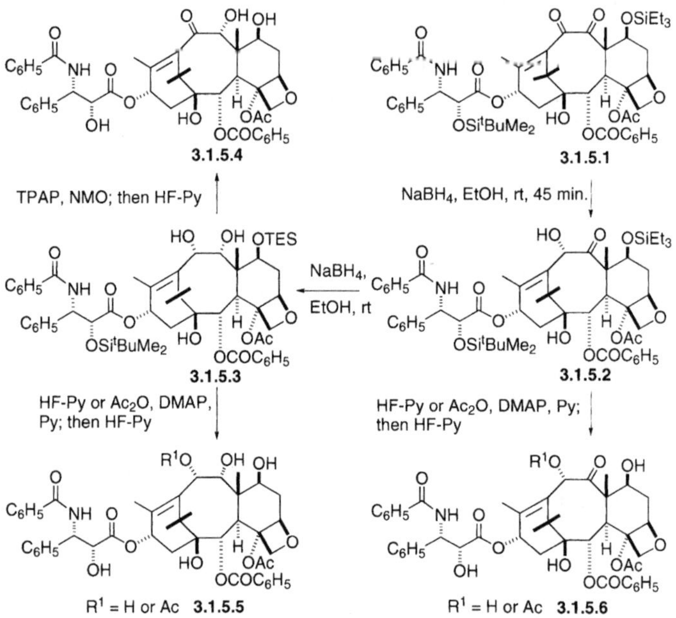

An interesting set of reactions of 10-oxo-10-deacetylbaccatin III was reported by Appendino (*95*), who treated it with various nitrogen nucleophiles. Pyrrolidine gave the product **3.1.5.7**, *O*-benzylhydroxylamine gave **3.1.5.8**, while hydrazine gave **3.1.5.9**. Simple reduction of the 10-oxo-10-deacetylbaccatin III with sodium borohydride and ceric chloride gave 10-*epi*-10-deacetylbaccatin III, the baccatin analog of **3.1.5.6** (*96*).

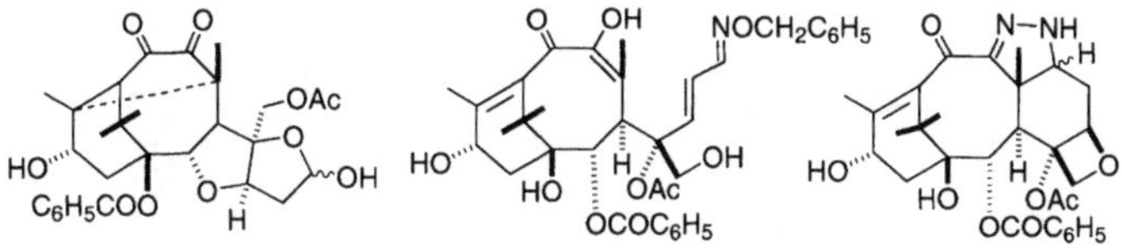

3.2. Modifications at C-9

Modifications at the C-9 position of taxol were initially facilitated by the isolation of 9-dihydrobaccatin (**3.2.1**) from *T. canadensis*, the Canadian yew (*97*). The availability of this compound was then used to advantage by Klein, who converted it into 9-dihydrotaxol. Protection of **3.2.1** as its 7,9-acetonide **3.2.2**, selective deacetylation at C-13 to give **3.2.3**, addition of the side chain to give **3.2.4**, and deprotection gave the (*9R*)-dihydrotaxol **3.2.5** (*98*).

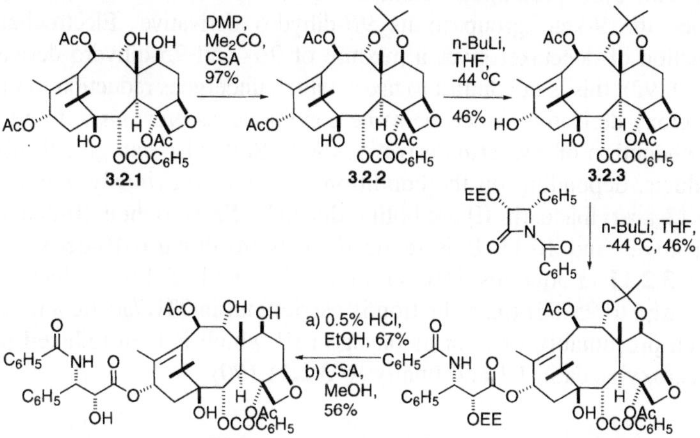

Subsequent work from the same group showed that protection as the acetonide was unnecessary as long as the C-7 hydroxyl group was protected as its triethylsilyl derivative **3.2.6**, and a variety of 3′-N-acyl analogs **3.2.7** (R^1 = various) (*99*) and 3′-modified analogs **3.2.7** (R^2 = various) (*100*) were prepared. The synthesis of additional analogs, including ether derivatives **3.2.8** and **3.2.9**, and their cytotoxic activities were reported by the same group in 1995 (*101*). Both ethers **3.2.8** and **3.2.9** were comparable to docetaxel in their cytotoxicity.

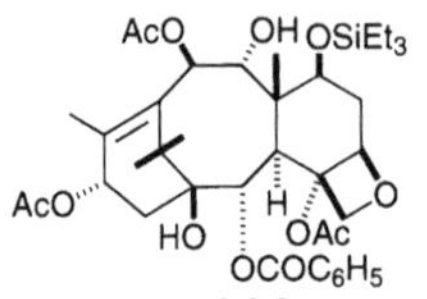

3.2.6

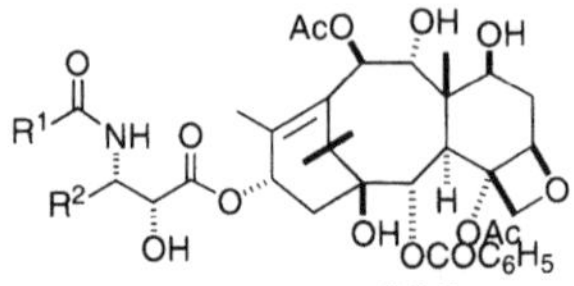

3.2.7

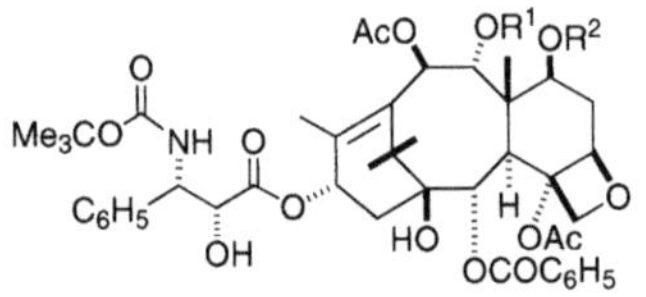

3.2.8 R^1 = H, R^2 = CH_3
3.2.9 R^1 = CH_3, R^2 = H

9-Deoxotaxol analogs have also been prepared by direct reductions of the C-9 carbonyl group. This group is relatively unreactive to sodium borohydride and similar reagents, and an important early observation on taxol was that the side chain could be removed reductively by treatment with Bu_4NBH_4 in CH_2Cl_2 without significant reduction of the 9-keto group (*38*). However, prolonged treatment of baccatin III with Bu_4NBH_4 will reduce the 9-keto group to its 9β-dihydro derivative. Electrochemical reduction of docetaxel gives a mixture of 9α- and 9β-dihydro derivatives **3.2.10** (*92*); this is in contrast to taxol, which undergoes reduction to the 10-deacetoxy derivative under the same conditions, as previously described.

Reduction of baccatin derivatives with SmI_2 also can give 9-dihydro products, depending on the conditions. Thus as previously noted, taxol and 13-acetylbaccatin III are both reduced by SmI_2 to their 10-deacetoxy derivatives and 10-DAB is reduced to its 9β-dihydro-10-deoxy derivative **3.2.11** in aqueous THF solution. The yield of this product can be increased to 95% if the reduction is carried out in THF/acetic anhydride, which presumably first forms baccatin III which is then reduced to 10-deacetoxybaccatin III and finally to **3.2.11** (*90*).

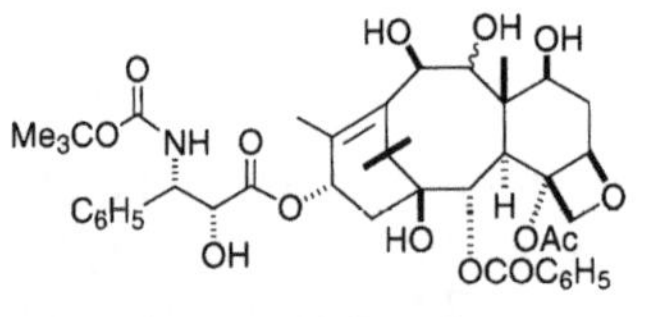

3.2.10

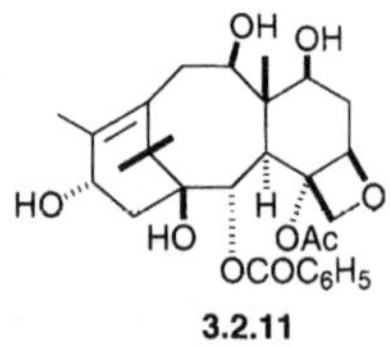

3.2.11

Reduction of taxol or docetaxel with SmI_2 under modified conditions has given both 9β-dihydro and 10-deoxy-9β-dihydro derivatives. Thus treatment of taxol with SmI_2 in H_2O/THF gave the deacetoxydihydro derivative **3.2.12**, and treatment of docetaxel under similar conditions (but for a shorter time) gave the derivatives **3.2.13** and **3.2.14** (*88*). The products **3.2.12–3.2.14** were all slightly less cytotoxic than their corresponding parent compounds, and also slightly less effective at causing microtubule polymerization.

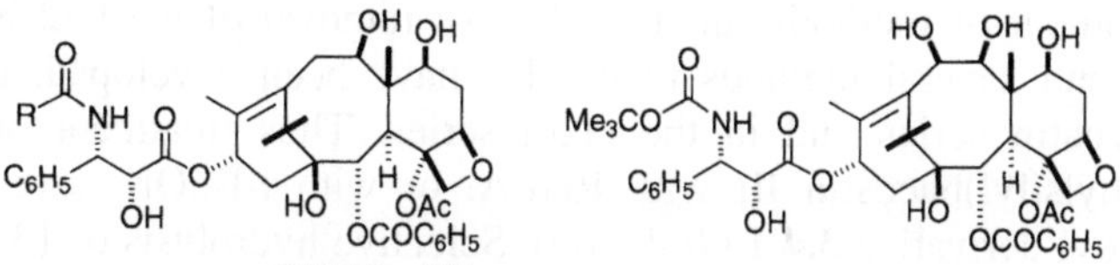

3.2.12 R = C_6H_5
3.2.13 R = $OCMe_3$
3.2.14

9-Deoxotaxol **3.2.16** has been prepared by deoxygenation of the 9-xanthate of (9R)-dihydrobaccatin **3.2.15** using the Barton procedure, followed by the usual side chain coupling and deprotection (*102*). It was found to be slightly more cytotoxic than taxol in four cell lines. 7,9-Dideoxy taxol and 7,9,10-trideoxytaxol were prepared by logical extensions of this method; the dideoxy analog was comparably cytotoxic to taxol, but the trideoxy analog was tenfold less active.

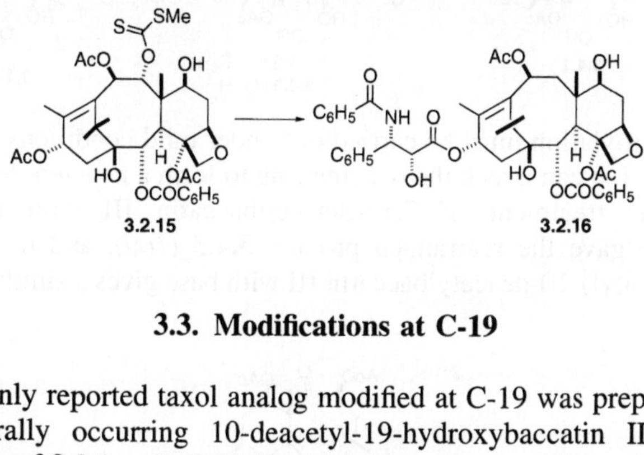

3.2.15 **3.2.16**

3.3. Modifications at C-19

The only reported taxol analog modified at C-19 was prepared from the naturally occurring 10-deacetyl-19-hydroxybaccatin III (**3.3.1**). Protection of **3.3.1** as its 7,10,19-tri(trichloroethoxycarbonate), followed by coupling with a protected side chain and appropriate deprotection gave the 19-hydroxy docetaxel analog **3.3.2** (*103*).

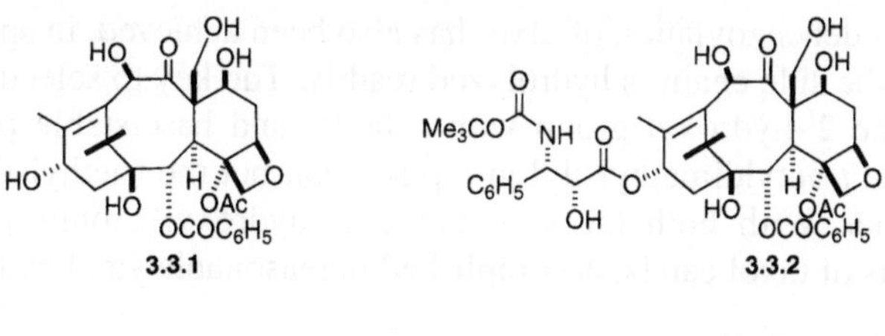

3.3.1 **3.3.2**

3.4. Modifications at C-2

Structural modification of taxol at the C-2 position has been fruitful in that several analogs with improved activity as compared with taxol have been discovered.

The first requirement for the synthesis of analogs at C-2 is the development of methods for the selective removal of the C-2 benzoyl group, and several methods to do this have been developed, both in the baccatin series and in the taxol series. Thus treatment of 7,13-di(triethylsilyl)baccatin III with Red-Al or with *t*-BuOK gave the 2-debenzoyl derivative **3.4.1** (*104*, *105*). Selective hydrolysis of 13-oxo-7-triethylsilyl-10-deacetylbaccatin **3.4.2** to its 2-debenzoyl derivative **3.4.3** could be achieved by treatment with K_2CO_3 in methanol (*106*), and selective debenzoylation of 7,10-di-(triethylsilyl)-13-trimethylsilyl-10-deacetylbaccatin III with Red-Al to give **3.4.4** has also been carried out (*107*, *108*).

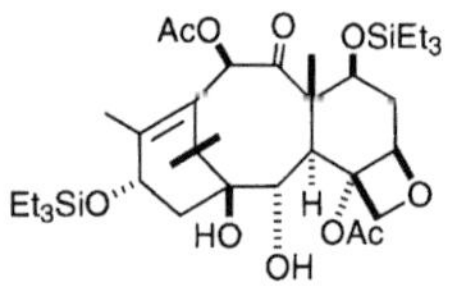

3.4.1

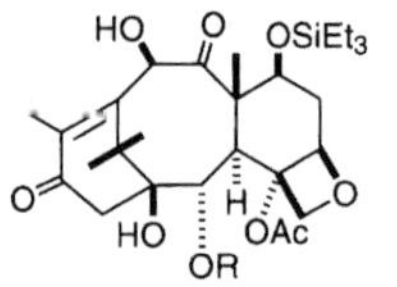

3.4.2 R = C_6H_5
3.4.3 R = H

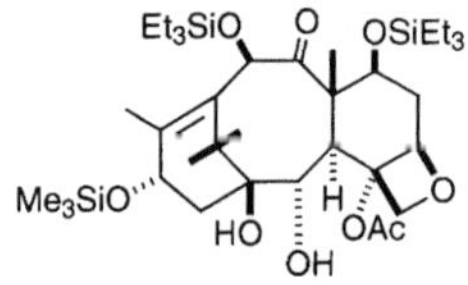

3.4.4

Debenzoylation must be carried out under mild conditions, since the 2-alkoxide ion can attack the oxetane ring to form a new tetrahydrofuran ring. Thus treatment of 7,13-diacetylbaccatin III with tributyltin methoxide gave the rearranged product **3.4.5** (*109*), and treatment of 7-(triethylsilyl)-10-deacetylbaccatin III with base gives a similar product (109).

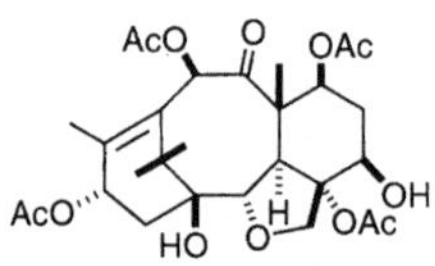

3.4.5

Direct debenzoylation of taxol has also been achieved, in spite of the fact that the side chain is hydrolyzed readily. The key to selectivity is to protect the 2′-hydroxyl group with a bulky and base-stable protecting group: the *t*-butyldimethylsilyl group is ideal, but the triethylsilyl group also works. With both the C-7 and C-2′ hydroxyl groups protected, hydrolysis of taxol can be accomplished in reasonable yield by treatment with NaOH under phase-transfer conditions (*110*), with potassium *t*-butoxide (*111*), or by the use of Triton B in CH_2Cl_2 at −78°C (*112*). Selective debenzoylation of a 2′,7-diprotected taxol has also been achieved by treatment with KOtBu (*113*) and selective debenzoylation of docetaxel has been accomplished by electrochemical reduction (*114*).

Conversion of protected 2-debenzoylbaccatins or 2-debenzoyltaxols to C-2 acyl analogs of taxol can be achieved in two different ways. The most direct procedure is simply acylation of the C-2 debenzoyl analog with a carboxylic acid in the presence of DCC and DMAP or comparable acylating agents (*110*). This method works with most acids, but it requires a large excess of acid and co-reactants to achieve a good yield. Direct acylation can also be achieved by treatment with BuLi in THF at −78°C, followed by an acyl chloride (114). The second general method, developed initially by Holton (*107*) and Nicolaou (*115*) makes use of the 1,2-carbonate such as **3.4.6** (*107*) or **3.4.7** (*115*). Treatment of these carbonates with an aryllithium derivative at low temperature gives the corresponding 2-acyl analog such as **3.4.8** in good-excellent yield. This route generally gives better yields than direct acylation, but it is limited to the formation of analogs with substituents that are compatible with the aryllithium intermediate.

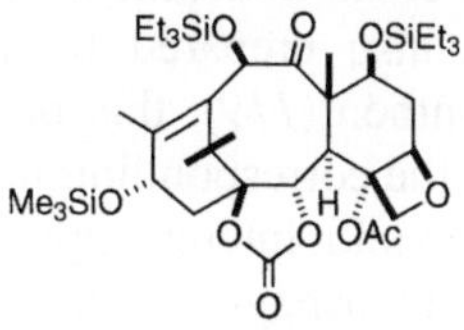

3.4.6

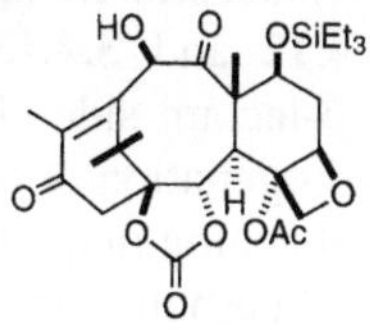

3.4.7

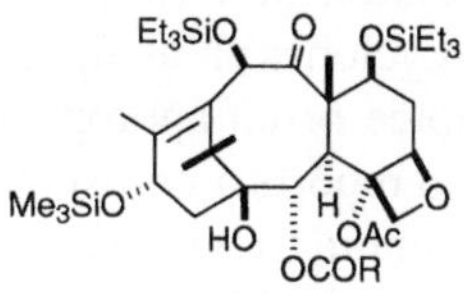

3.4.8

Debenzoylation of taxol as described above and direct reacylation was used to prepare a variety of 2-benzoyl analogs of taxol **3.4.9** (*110*).

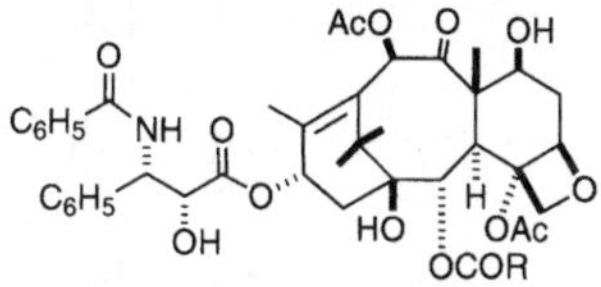

3.4.9

Interestingly most *meta*-substituted benzoyl groups gave analogs with improved activity, while all *para*-substituted derivatives were significantly less active than taxol. A full account of this work appeared in 1998 (*112*). Acylation of a protected 2-debenzoyltaxol derivative has also been used by Commerçon and his colleagues to give various 2-aroyl analogs (*114*).

Other workers have prepared 2-acyl analogs from 10-deacetyl baccatin III. Thus the 2-debenzoyl derivative **3.4.4** was converted to the 2-acyl analog **3.4.8** either by direct acylation or through the 1,2-carbonate **3.4.6**. Coupling with the side chain and deprotection then gave the 2-acyl analogs **3.4.10** (*107*, *104*, *116*). In the process developed by Nicolaou and his co-workers (*115*, *117*) the carbonate **3.4.7** was subjected to nucleophilic ring-opening followed by acetylation at C-10 to give the analogs **3.4.11**. Reduction of the C-13 carbonyl group and attachment of the side-chain by β-lactam chemistry gave taxol analogs **3.4.9**.

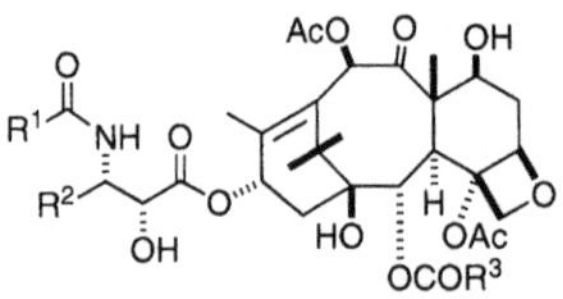

3.4.10

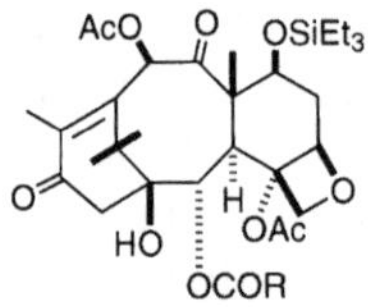

3.4.11

Various 2-*O*-heteroaroyl taxanes were prepared by a pathway similar to that of Chaudhary *et al.* (*110*) by Georg (*118*). The same group also prepared the 2-cyclohexyl analogs of taxol and docetaxel by hydrogenation of 7-(triethylsilyl)baccatin III and side chain attachment. The 3′-cyclohexyl analogs **3.4.12** and **3.4.13** were also prepared by the choice of an appropriate β-lactam side chain synthon (*119*); this paper also reports a one-step hydrogenation of taxol to the corresponding non-aromatic tricyclohexyltaxol analog in quantitative yield; interestingly the 11–12 double bond, the C-9 carbonyl group and the benzylic C–N bond remained intact. Similar analogs were prepared independently by similar routes by Ojima and his colleagues (*120*).

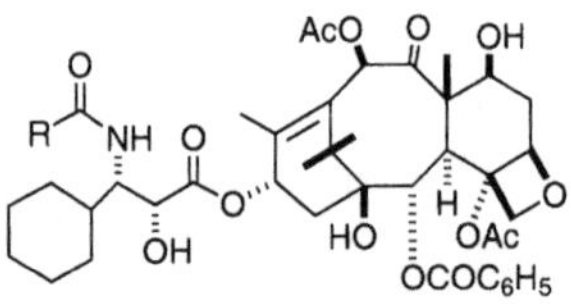

3.4.12 R = C_6H_5
3.4.13 R = Me_3CO

The effect of stereochemistry at the C-2 position was investigated by Chordia and Kingston, who oxidized a protected 2-debenzoyltaxol to the 2-oxo derivative **3.4.14**. Reduction with $NaBH_4$ followed by benzoylation and deprotection gave 2-epitaxol (**3.4.15**) which was inactive in both tubulin-assembly and HCT116 cytotoxicity assays (*121*).

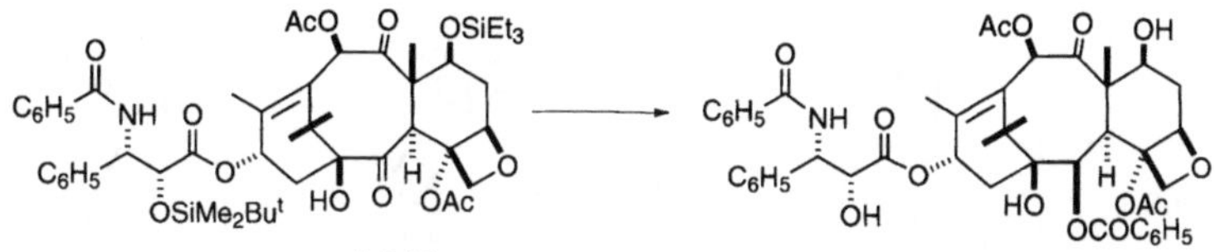

3.4.14 3.4.15

Removal of the C-2 substituent was accomplished by Chen and his colleagues, who prepared the xanthate **3.4.16** from the corresponding 2-debenzoylbaccatin III and then deoxygenated it by the Barton procedure. Side chain attachment gave 2-debenzoyloxytaxol **3.4.17** with 100-fold reduced cytotoxicity to the HCT116 cell line (*122*).

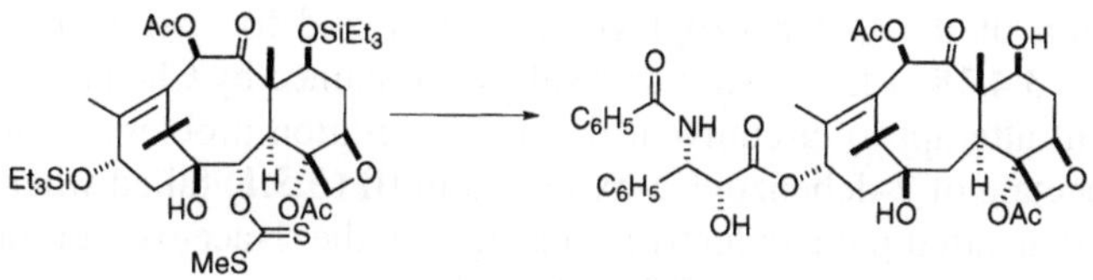

3.4.16 3.4.17

A 2-debenzoyl analog was also prepared by Kingston and his colleagues as the product of an attempt to deoxygenate taxol at C-1. Attempted formation of the 1-xanthate of 2′,7-di(triethylsilyl)taxol unexpectedly gave the 1-benzoyl-2-xanthyltaxol **3.4.18**; the structure of this compound was confirmed by an HMBC experiment. Barton deoxygenation and deprotection then gave 1-benzoyl-2-debenzoyloxytaxol **3.4.19**, which was over 50-fold less cytotoxic than taxol to P-388 leukemia cells (*123*).

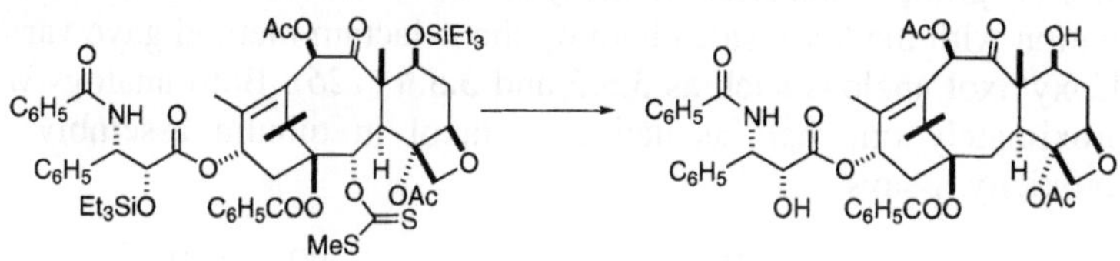

3.4.18 3.4.19

Careful studies by Chen revealed that deoxygenation of baccatin derivatives at C-2 can be accompanied by free-radical rearrangement reactions. As examples the cyclic 1,2-thiocarbonate **3.4.20** gave both a simple A-nor product **3.4.21** and the more complex rearrangement product **3.4.22** (124). Deoxygenation of the 10-acetoxy analog of **3.4.20** only gave A-nor product.

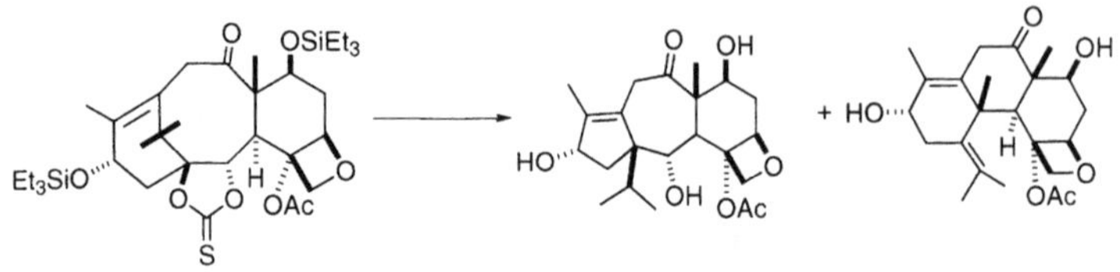

3.4.20 **3.4.21** **3.4.22**

3.5. Modifications at C-1

Modification at the C-1 position has proved to be a difficult task. As noted above deoxygenation of taxol by the Barton procedure resulted in migration of the C-2 benzoyl group to C-1 and formation of the C-2 xanthate **3.4.18** (123). A similar result was obtained by Chen *et al.* (*124*). A recent attempt to circumvent the acyl migration problem by forming the 1-acetate of 2-debenzoyl-2-oxobaccatin III (**3.5.1**) failed when it was found that attempted reductive cleavage of the 1-acetoxy group gave only rearranged products **3.5.2** and **3.5.3** (*125*).

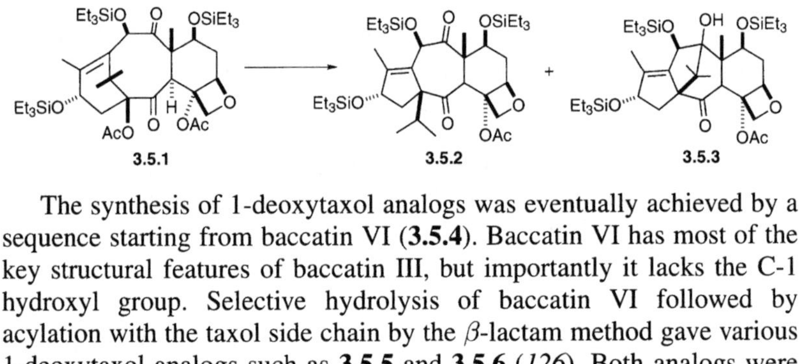

3.5.1 **3.5.2** **3.5.3**

The synthesis of 1-deoxytaxol analogs was eventually achieved by a sequence starting from baccatin VI (**3.5.4**). Baccatin VI has most of the key structural features of baccatin III, but importantly it lacks the C-1 hydroxyl group. Selective hydrolysis of baccatin VI followed by acylation with the taxol side chain by the β-lactam method gave various 1-deoxytaxol analogs such as **3.5.5** and **3.5.6** (*126*). Both analogs were approximately one half as active as taxol in tubulin assembly and cytotoxicity assays.

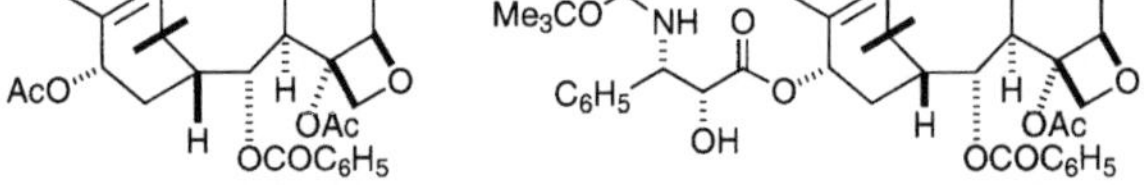

3.5.4

3.5.5 R = Ac
3.5.6 R = H

Recently 1-deoxytaxol itself (**3.5.7**) has been reported as a minor constituent from the isolation of taxol. It proved to have only moderate stability as compared with taxol, and it was approximately one third as active as taxol in the HCT116 and A2780 cell lines (*127*). The syntheses of 1-deoxybaccatin III and 1-deoxytaxol have also been claimed in a patent application (*128*). Finally, going in the opposite direction, an interesting hydroxylation of a 2-deacetoxytaxinine J derivative by dimethyldioxirane has been reported. Reaction of the 5-(triethylsilyl) derivative of 2-deacetoxytaxinine J with dimethyldioxirane gave the 1-hydroxy derivatives **3.5.8** and **3.5.9**, together with the normal unhydroxylated epoxides (*129*).

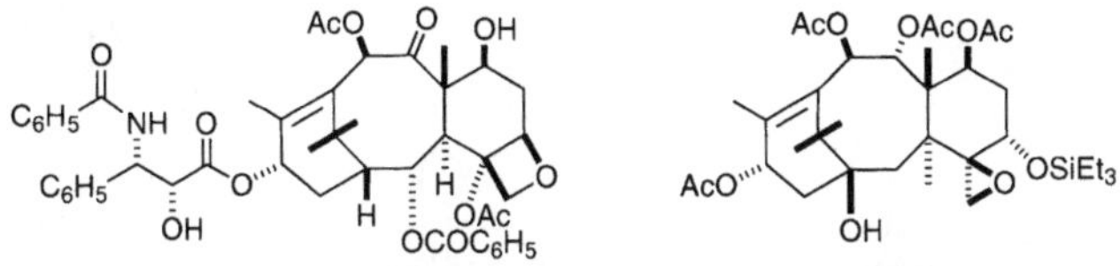

3.5.7

3.5.8
3.5.9 (epimeric at C-4)

3.6. Other B-Ring Analogs

An analog with a very different B-ring was prepared by putting the taxol side chain on a taxane with a 2(3→20) *abeo*taxane skeleton (**3.6.1**). The taxane **3.6.1** was protected as its triethylsilyl derivative, deacetylated, rebenzoylated at C-2, and then coupled with the β-lactam side chain to give the analog **3.6.2**. Compound **3.6.2** was only very weakly active, with IC_{50} values of 6 μg/mL in four different cell lines (*130*).

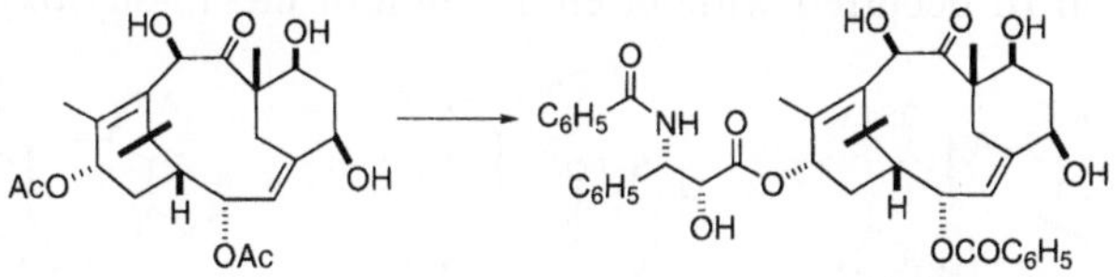

3.6.1

3.6.2

Analogs in which the B-ring has been cleaved have also been prepared. In the first such study, Khuong-Huu and her collaborators prepared products resulting from cleavage of the C-1/C-2, the C-9/C-10, and the C-11/C-12 bonds of 10-DAB (*131*). In the first of these reactions the protected tetraol **3.6.3** was cleaved with periodate to give the hemiacetal **3.6.4**. Similar treatment of the 10-deoxy analog of **3.6.3** gave the aldehyde **3.6.5**. Cleavage of the C-9/C-10 bond was effected by reduction of 13-acetyl-10-deacetyl-7-*epi*baccatin III with $BH_3.SMe_2$ to

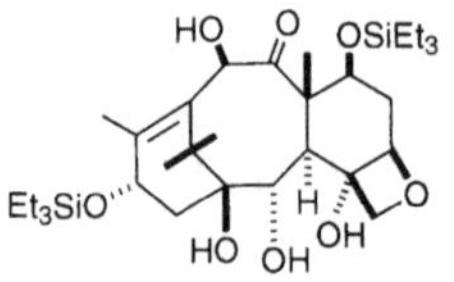

3.6.3

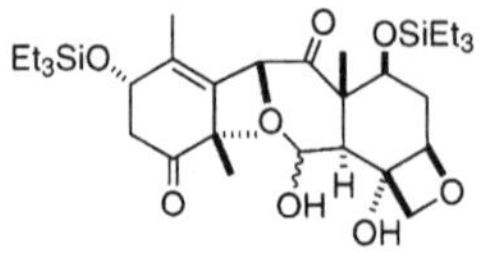

3.6.4

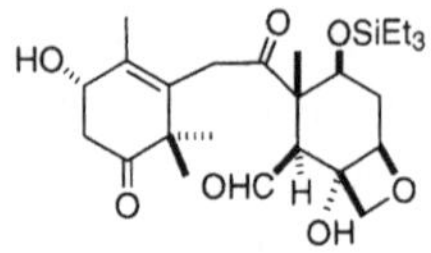

3.6.5

give the tetraol **3.6.6**, which was then cleaved with periodate to the hemiacetal **3.6.7**. Cleavage of the C11/C12 double bond was carried out on 13-acetyl-10-deacetoxy-7-methoxycarbonylbaccatin III. Oxidation of this compound with RuO_2 and periodate gave the hemiacetal **3.6.8**.

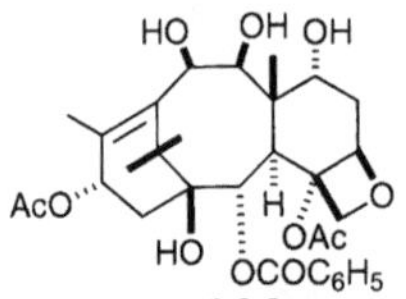

3.6.6

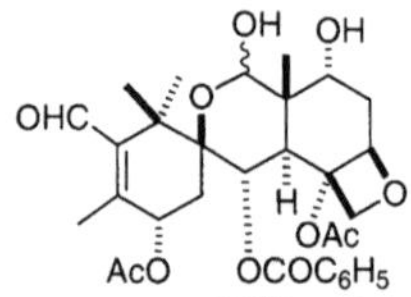

3.6.7

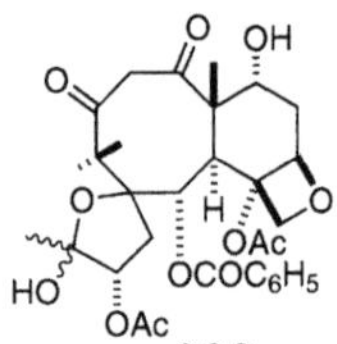

3.6.8

A second study of this reaction by Kingston and his collaborators gave somewhat different products (*132*). Treatment of the 13-chlorobaccatin III **3.6.9** with sodium azide in DMF/H_2O, in an attempt to make a 13-azido derivative, instead gave the B-seco product **3.6.10**. The mechanism for the formation of **3.6.10** probably involves neighboring group attack on the C-11/C-12 double bond by the C-10 acetate with subsequent trapping of the resulting acetoxonium ion by water and retroaldol-type cleavage. A final cleavage was effected by treatment of 2′-(*tert*-butyldimethylsilyl)-7-(triethylsilyl)-2-debenzoyltaxol with MnO_2; the 1,2 bond was cleaved with concomitant elimination of the side chain to give the ketoaldehyde **3.6.11**. A similar reaction on 7,13-di(triethylsilyl)baccatin III occurred without elimination of the 13-silyloxy group.

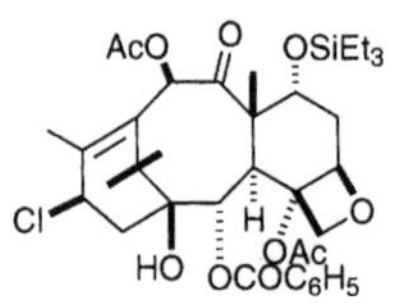

3.6.9

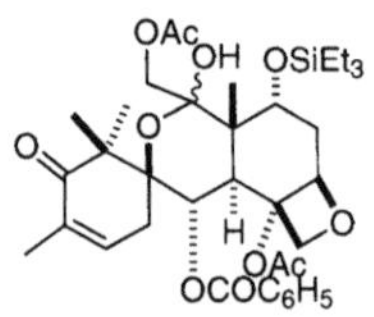

3.6.10

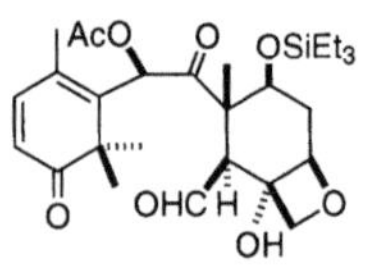

3.6.11

3.7. SAR of B-Ring Analogs

Extensive studies have been made of the effect of structural modifications to ring B on the activity of taxol. Most of the work has focussed on analogs at C-2, C-9, and C-10, although some 1-deoxy analogs have also been investigated.

This review, as noted earlier, is not designed to provide a complete summary of all the SAR work that has been performed, but some general trends and highlights can be observed as described below.

3.7.1. The C-1 Position

Deoxygenation at C-1 by itself has not been achieved, but various 1-deoxy-*9R*-dihydro analogs have been prepared as reported above (*126*). The bioactivity of these analogs can be compared with the corresponding (9*R*)-dihydrotaxol analogs prepared by Klein *et al.* (*133*); when this was done it was found that removal of the C-1 hydroxyl group caused a modest loss of activity as compared with the corresponding (*9R*)-dihydrotaxol analog. The recently discovered 1-deoxytaxol (*127*) is also less active than taxol as noted above. It thus appears that the C-1 hydroxyl group has a moderately positive effect on activity.

3.7.2. The C-2 Position

Modifications at the C-2 position have been made by various workers. Activity is retained with many different substituents on the benzene ring of 2-aroyl analogs, but interestingly only when the substituent is in the *ortho* or *meta* position; *para* substituted derivatives were uniformly inactive (*110*, *112*). The *m*-azidobenzoyl substituent in particular showed much higher cytotoxicity than taxol itself, and its tubulin-assembly activity has been studied in detail (*134*). A benzoyl group is not necessary for activity, and various heterocyclic groups at the C-2 position lead to active analogs (*118*). It has also been shown that certain non-aromatic groups such as 2-methylprop-l-enoyl can replace the benzoyl group at C-2 with maintenance or enhancement of activity (*135*). Studies of C-2 analogs have also been made by others (*116*, *136*).

Ojima has made extensive studies of taxol analogs with various substituents on the side chain and at the 2- or 10-positions. In one series he investigated the effects of substitution with alkyl or alkenyl groups at both the C-2 and the C-3′ positions (*135*). In a second series he found the best activity against drug-resistant LCC6 breast cancer cells with the two analogs **3.7.2.1** and **3.7.2.2**, each with a *m*-methoxybenzoyl group at the 2-position. These analogs were about three times as active as taxol against wild-type LCC6 cells, but over 300 times as active as taxol against drug-resistant LCC6 cells (*137*).

3.7.3. The C-9 Position

In a work reported earlier Klein and his colleagues prepared 17 9-dihydrotaxol analogs. The most potently cytotoxic compounds, with activities up to 25 times greater than those of taxol, were the ethers **3.2.8** and **3.2.9** previously described, but the simple 9-dihydro derivative of docetaxel **3.7.3.1** was also up to 10 times more active than taxol, having about the same activity as docetaxel itself (*101*).

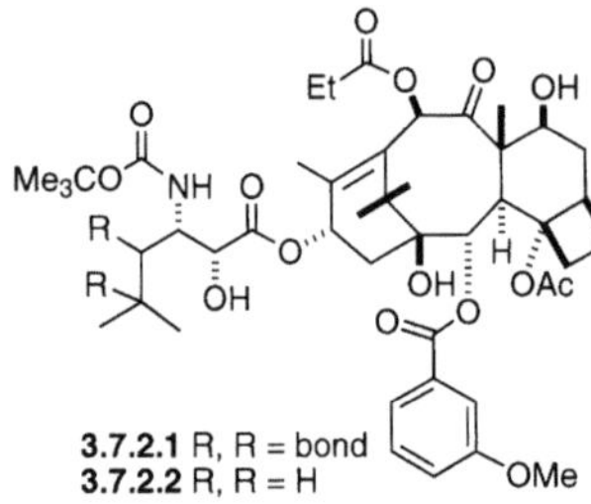

3.7.2.1 R, R = bond
3.7.2.2 R, R = H

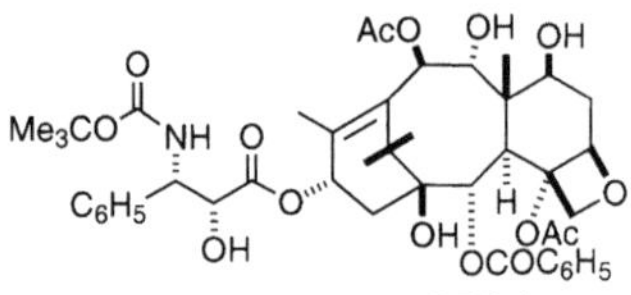

3.7.3.1

3.7.4. The C-10 Position

Soga and his colleagues prepared a series of water-soluble non-prodrug analogs of docetaxel with *sec*-aminoethyl groups at C-10. The 10-*O*-(2-morpholinoethyl) and 10-*O*-(2-thiomorpholinoethyl) derivatives **3.7.4.1** and **3.7.4.2** and their desoxydifluoro derivatives **3.7.4.3** and **3.7.4.4** showed slightly greater cytotoxicity than docetaxel in one or more cell lines (*138*).

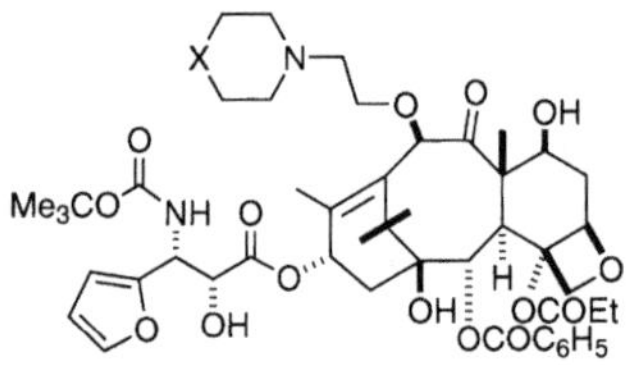

3.7.4.1 X = O
3.7.4.2 X = S

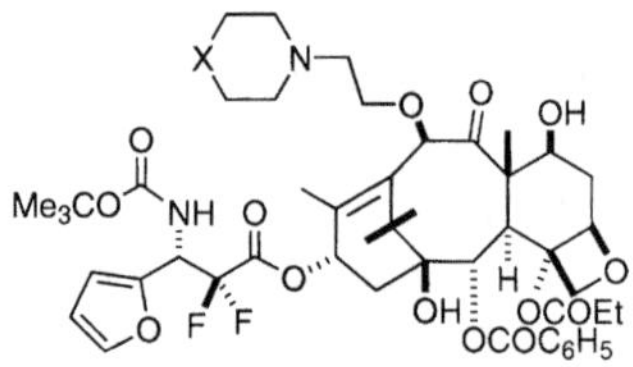

3.7.4.3 X = O
3.7.4.4 X = S

Ojima has also made a number of docetaxel analogs with 3′-(2-methyl-1-propenyl) or 3′-(2-methylpropyl) side chain groups and with varying C-10 substituents (*74*). He found the best activity against the drug-resistant MCF7-R breast cancer cell line, with cytotoxicities up to 100 times that of docetaxel, with analogs **3.7.4.5** and **3.7.4.6**. C-10 analogs have also been prepared by a parallel synthesis approach by Georg and her collaborators. Several analogs showed moderately improved activity against the MCF7-ADR adriamycin-resistant breast cancer cell line, but only analog **3.7.4.7** was more active than taxol against both MCF-7 and MCF-ADR cell lines (*139*).

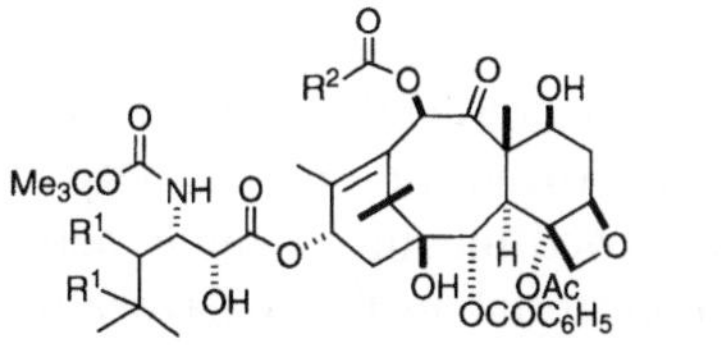

3.7.4.5 R^1, R^1 = bond, R^2 = Et, Cyclopropyl, Me_2N, or MeO
3.7.4.6 R^1, R^1 = H, R^2 = Et, Cyclopropyl, Me_2N, MeO, MeCH=CH, or Et_2N

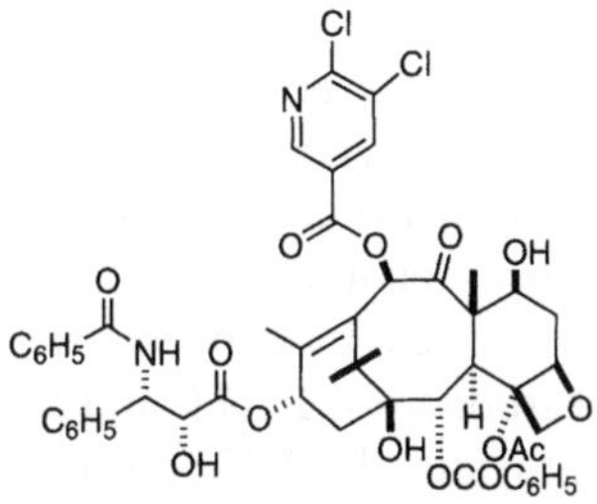

3.7.4.7

4. C-Ring Chemistry

The C-ring of taxol has two functional groups available for manipulation; a hydroxy group at C-7 and an acetyl group at C-4. The chemistry of the oxetane ring, which can be considered as a "functional group" at the C-5 position, is discussed in the following section.

4.1. The C-7 Hydroxyl Group

4.1.1. Acyl and Alkyl Derivatives

The C-7 hydroxyl group is in many ways the most accessible functional group on the taxane ring system, and is second in reactivity only to the C-2′ hydroxyl group in the side chain. For this reason its chemistry has been studied fairly extensively, and many simple derivatives have been prepared. Thus the acetyl derivative **4.1.1.1** (*140, 141*), benzoyl derivative **4.1.1.2** (*142*), glutaryl derivative **4.1.1.3** (*143*), amino acid derivatives **4.1.1.4–4.1.1.6** (*144*), chloroacetyl derivative **4.1.1.7** (*145*), and a docosahexaenoic acid derivative **4.1.1.8** (*146*) have all been prepared and found to have comparable activity to taxol.

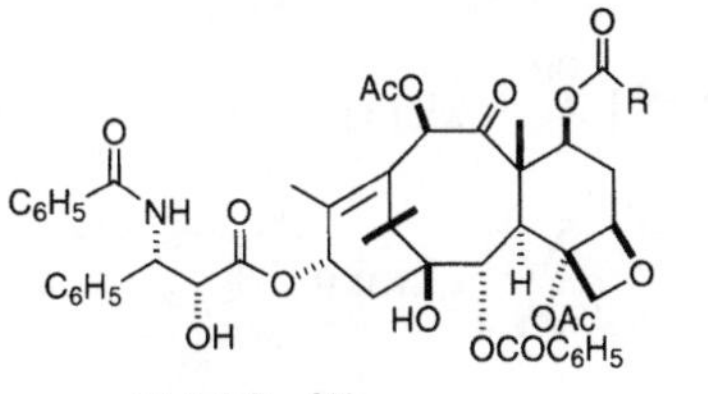

4.1.1.1 R = CH_3
4.1.1.2 R = C_6H_5
4.1.1.3 R = $CH_2CH_2CH_2COOH$
4.1.1.4 R = $CH_2CH(NH_2)C_6H_5$
4.1.1.5 R = CH_2NMe_2
4.1.1.6 R = $CH(NH_2)CH_3$
4.1.1.7 R = CH_2Cl
4.1.1.8 R = $CH_2(CH_2CH{=}CH)_6CH_2CH_3$

Acylation at the C-7 position has in some cases given derivatives with interesting new biological properties. Thus baccatin III derivatives **4.1.1.9** and **4.1.1.10** were found to be highly effective reversal agents for multidrug resistance in cancer cells; several related baccatin III analogs were also prepared (*147*). Even simple analogs at C-7 can have improved properties, as evidenced by the improved *in vivo* activity of the ethylcarbonate **4.1.1.11** and the improved tubulin-assembly activity of the mesylate **4.1.1.12** (*148*). Combinatorial methods have also been applied to the synthesis of C-7 acyl derivatives, and a library of 26 such analogs was prepared. None of the analogs were more active than taxol against the MCF7 breast cancer cell line, but three of them (the acetic, valeric, and methylthioacetic derivatives) showed slightly increased activity against the MCF-7-R drug-resistant cell line and also showed slightly increased tubulin-assembly activity (*149*).

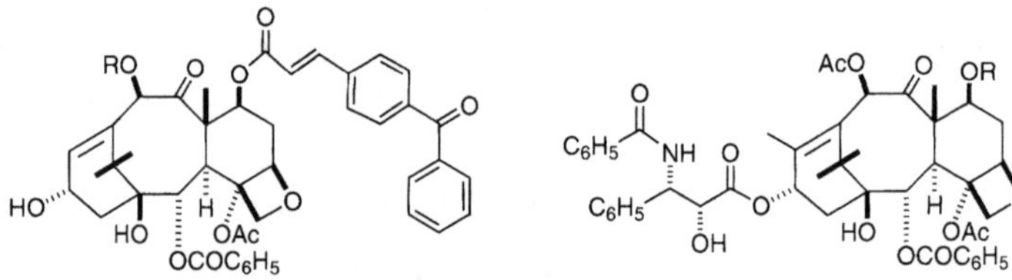

4.1.1.9 R = H
4.1.1.10 R = Ac

4.1.1.11 R = COOEt
4.1.1.12 R = SO_2Me

Acyl derivatives at C-7 have been used as probes with the potential for the study of the interaction of taxol and the microtubule. The results stemming from the use of some of these probes will be discussed in section 11. As examples the 7-azidobenzoyl analog **4.1.1.13**, (*150*), the diazirine **4.1.1.14** (*151*), various aminobenzoyl derivatives such as **4.1.1.15** (*152*) and the benzophenone derivative **4.1.1.16** (*153*) have all been prepared.

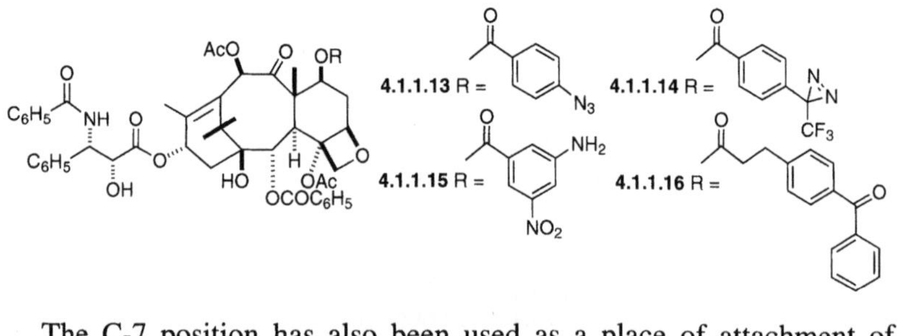

The C-7 position has also been used as a place of attachment of cleavable water-solubilizing groups. Simple acyl groups and phosphate groups at C-7 are not readily cleaved under physiological conditions

(*154*) and thus do not make good prodrugs, but various phosphatase-activated prodrugs have been prepared and found to generate taxol under physiological conditions. These include the simple derivative **4.1.1.17** (*155*) and the masked lactone **4.1.1.18** (*156*).

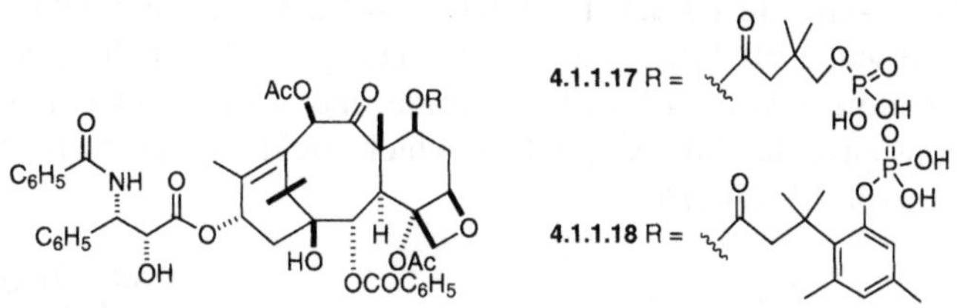

Various ether derivatives of taxol have also been prepared. Simple C-7 silyl ethers have poor *in vitro* cytotoxicity (*148*), but other ether derivatives sometimes have improved activity. Thus treatment of 2′-protected taxol derivatives with chloromethyl methyl ether followed by deprotection yields the acetal derivative **4.1.1.19** (*157*), while treatment of baccatin III with dimethyl sulfide in the presence of benzoyl peroxide followed by coupling with side chain gives thiomethyl derivatives such as **4.1.1.20** (*158*). This latter compound has been selected for development by Bristol-Myers Squibb, and is currently in clinical trials (*159*). Other ethers such as the amino acid derivative **4.1.1.21** have also been prepared by treatment of **4.1.1.20** with chloroacetic acid, N-iodosuccinimide, and silver triflate followed by N-methylpiperazine (*158*).

4.1.1.19 X = O
4.1.1.20 X = S

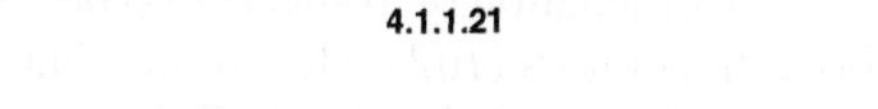

4.1.1.21

4.1.2. Epimerization, Oxidation, and Reduction

Epimerization of taxol and baccatin III was reported in the previous review in this series (*10*), and 7-*epi*taxol (**4.1.2.1**) is readily formed by treatment of taxol with sodium hydride or (less efficiently) with various other bases. The epimerization of taxol was reinvestigated by Fang and his colleagues who suggested that stabilization of the 7-epi-hydroxyl group by the 13α-hydroxyl group may play an important role in the equilibrium process (*160*). Deoxygenation of the C-7 hydroxyl group has been achieved by application of Barton's radical deoxygenation reaction, and this chemistry can also be used to convert 7-*epi*taxol to taxol. Thus treatment of either 2′-triethylsilyl taxol or 2′-triethylsilyl-7-*epi*taxol (**4.1.2.2**) with carbon disulfide, sodium hydride, and methyl iodide, gave the same 7-xanthate **4.1.2.3**. Treatment of **4.1.2.3** with tributyltin hydride in the presence of AIBN and wet toluene gave 2′-(triethylsilyl)taxol, while treatment with tributyltin hydride and AIBN under anhydrous conditions gave the 7-deoxy product, which could be deprotected to give 7-deoxytaxol **4.1.2.4** (*161*).

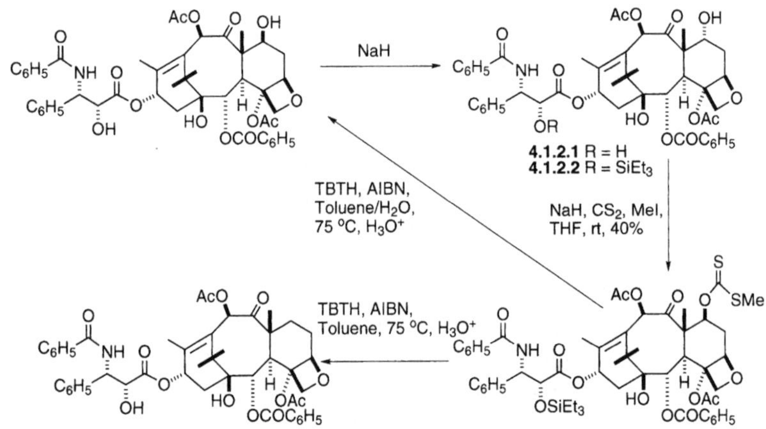

7-Deoxytaxol and 7,10-dideoxytaxol have also been prepared by a similar route starting from baccatin III (*162*); this group also prepared 7-epi-10-deoxytaxol (*163*). 7-Deoxytaxol has a cytotoxicity comparable to taxol, but 7,10-dideoxytaxol is significantly less cytotoxic to HCT116 cells (*162*).

Deoxygenation of 10-deacetoxybaccatin III was studied by Chen and his collaborators (*162*). They found that deoxygenation of the 7-xanthate gave a mixture of the simple 7-deoxy product **4.1.2.5**, the methyl ether **4.1.2.6**, and the three rearranged products **4.1.2.7**, **4.1.2.8**, and **4.1.2.9**. The product distribution varied with the reductant used, with Bu_3SnH giving **4.1.2.5** and **4.1.2.7**, Ph_3SnH giving **4.1.2.5**, **4.1.2.6**, and **4.1.2.7**, and $(TMS)_3SiH$ giving **4.1.2.7**, **4.1.2.8**, and **4.1.2.9**. Reasonable mechanisms for the formation of all the products, including the surprising methyl ether **4.1.2.6**, were suggested.

Deoxygenation at C-7 was also achieved by electrolytic reduction of the 7α-iodide of docetaxel, prepared from the corresponding 7β-triflate by treatment with sodium iodide in acetonitrile. Electrolytic reduction

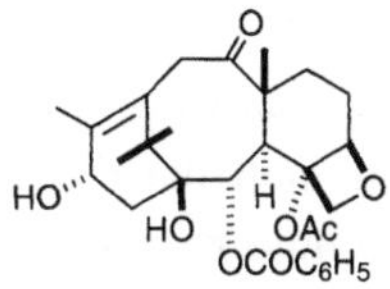

4.1.2.5

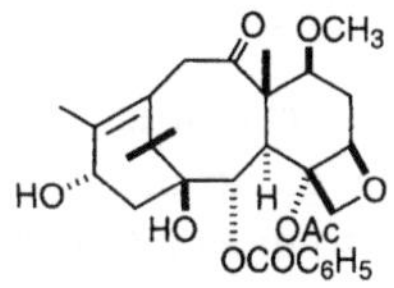

4.1.2.6

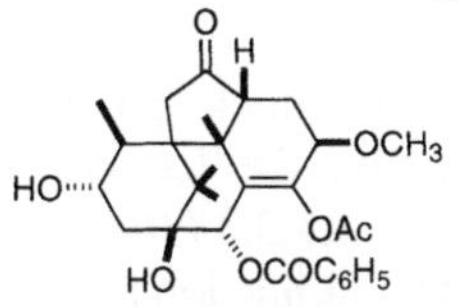

4.1.2.7

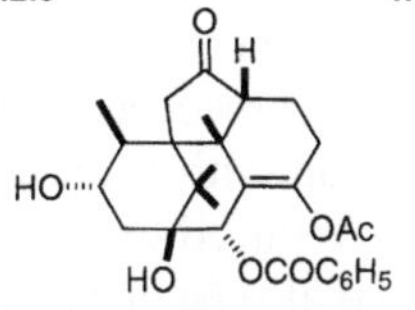

4.1.2.8

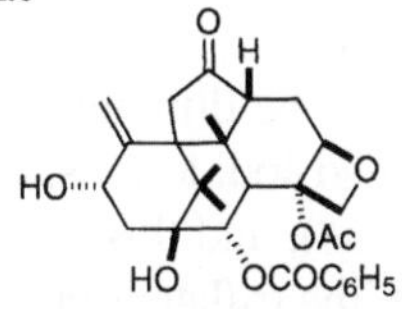

4.1.2.9

with a reduction potential of −1.7V gave the 7-deoxy compound **4.1.2.10** and the cyclopropane derivative **4.1.2.11**; this latter compound was converted to **4.1.2.10** on treatment with dilute hydrochloric acid. Further reduction of **4.1.2.10** with a reduction potential of −1.9V and different catholyte and anolyte solutions gave the 7,10-dideoxy compound **4.1.2.12**. Compound **4.1.2.10** was slightly more cytotoxic to P388 cells than docetaxel, while both **4.1.2.11** and **4.1.2.12** were approximately eight times less cytotoxic (*164*).

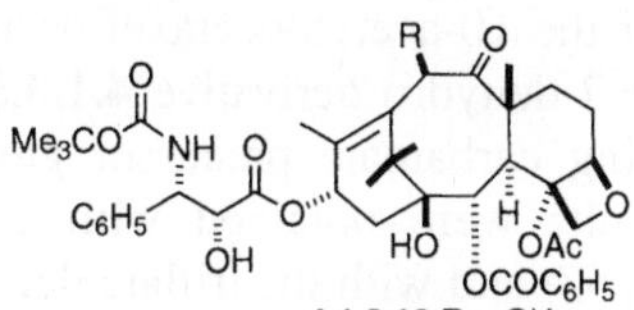

4.1.2.10 R = OH
4.1.2.12 R = H

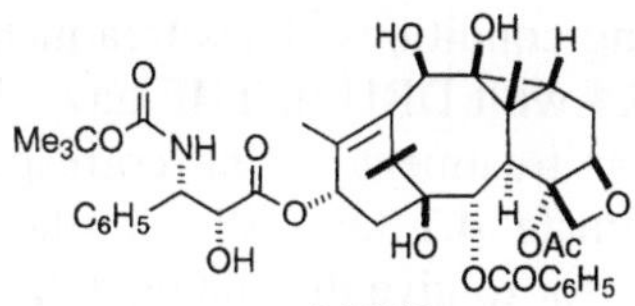

4.1.2.11

Oxidation at the 7-position was reported in the previous review (*10*) to yield the unstable 7-oxotaxol, which underwent facile oxetane ring-opening on treatment with mild base (*165*). Oxidation of 10-deacetyl-baccatin III under Swern conditions gave the unstable 7-oxo derivative, which underwent ring opening under the reaction conditions to give the enone **4.1.2.13** (*167*, *168*). The Dess-Martin reagent was less selective, giving a 7,13-dioxo product (*42*).

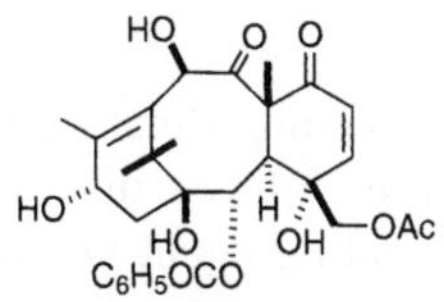

4.1.2.13

4.1.3. Displacement of a Leaving Group at C-7

Some interesting chemistry has emerged from studies of taxol analogs carrying a leaving group at C-7. Although the C-7 mesylate derivative proved unreactive to moderate basic conditions (*166*) the 7-triflate proved to be reactive enough for nucleophilic displacement and elimination reactions. Nucleophilic displacements on the 7β-triflate of taxol (**4.1.3.1**) proved to be difficult, and only proceeded with good nucleophiles such as azide and iodide ions. Reaction with azide ion gave the 7α-azide **4.1.3.2** (*167, 168*), and this could be reduced to the amine **4.1.3.3** (*168*). As noted earlier, the triflate group could also be displaced by iodide ion to give a 7α-iodide derivative (*164*).

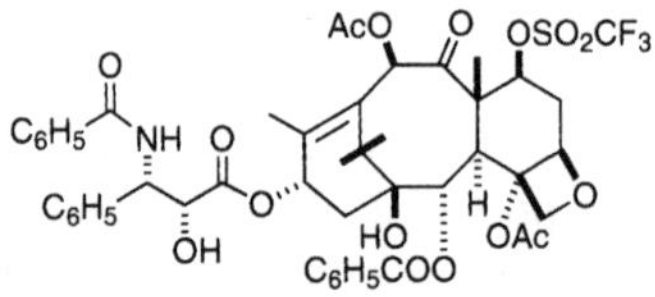

4.1.3.1

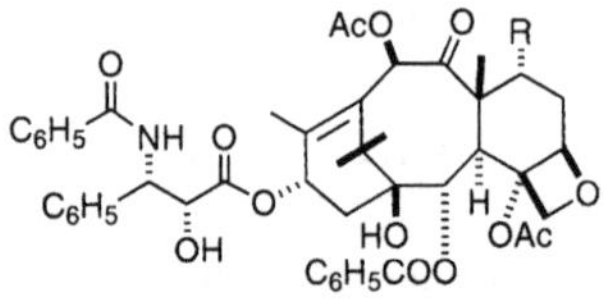

4.1.3.2 R = N_3
4.1.3.3 R = NH_2

Elimination of a C-7β triflate derivative to give 6,7-dehydrotaxol analogs can be achieved by reaction with non-nucleophilic bases under forcing conditions. Thus treatment of the 10-acetyldocetaxel derivative **4.1.3.4** with DBU in THF gave the 6,7-dehydro derivative **4.1.3.5**, and similar treatment of the corresponding carbamate precursor gave the carbamate **4.1.3.6** (*59*). Similar results were obtained with a taxol precursor to give the analog **4.1.3.7** (*167*) and with the triflate derivative of 2-acetyl-7-*epi*-taxol (*168*), which also gave **4.1.3.7** after deprotection. Derivative **4.1.3.7** was less cytotoxic than taxol but was slightly more effective at stabilizing polymerized tubulin (*167*).

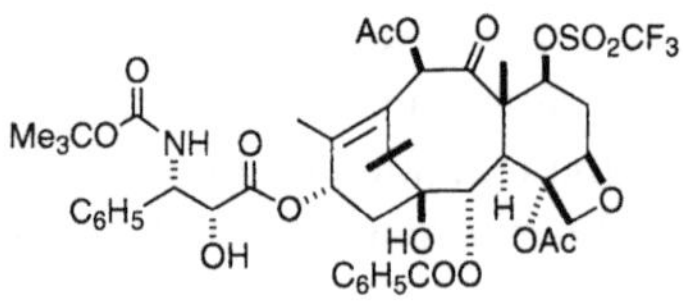

4.1.3.4

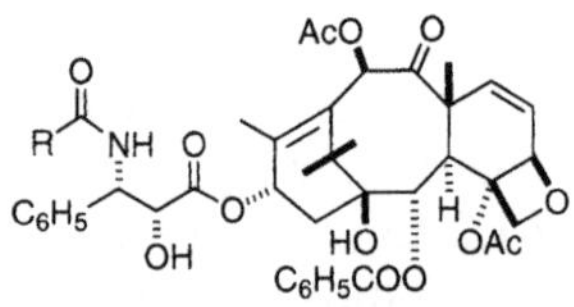

4.1.3.5 R = Me_3CO
4.1.3.6 R = Me_3CNH
4.1.3.7 R = C_6H_5

Perhaps the most interesting displacement reactions at the C-7 position came from the serendipitous discovery that treatment of taxol with the well-known fluorinating agent DAST gave 7,8-cyclopropataxol products. Thus treatment of 2′-(benzyloxycarbonyl)-7-*epi*-taxol with

DAST gave the A-nor-product **4.1.3.8** as the major product, but the 19-nor-7,8-cyclopropataxol **4.1.3.10** could also be obtained in modest yield (after deprotection at 2′) (*169*). Further investigation revealed that the 7α-fluoro derivative **4.1.3.11** is also produced (*170*), and another study reported the additional isolation of the sulfinamide **4.1.3.9** (*171*). These results were confirmed independently by Johnson and his colleagues using methyl DAST, but these investigators found that 6,7-dehydrotaxols such as **4.1.3.7** were also produced in this reaction; it was this finding that led to their discovery of the triflate chemistry described earlier (*59*).

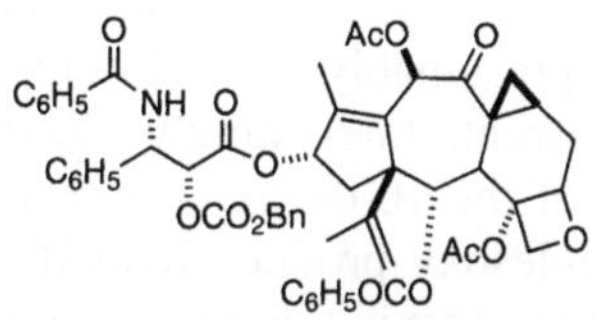

4.1.3.8

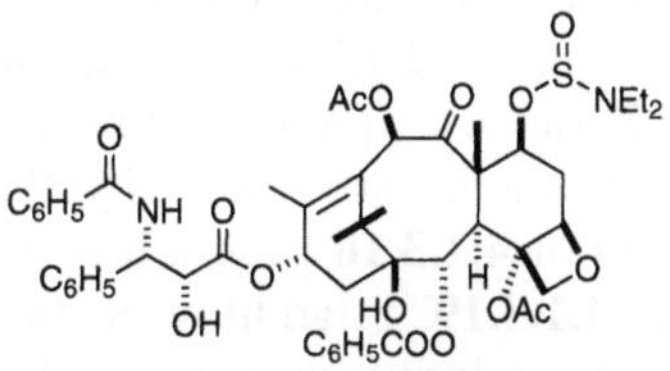

4.1.3.9

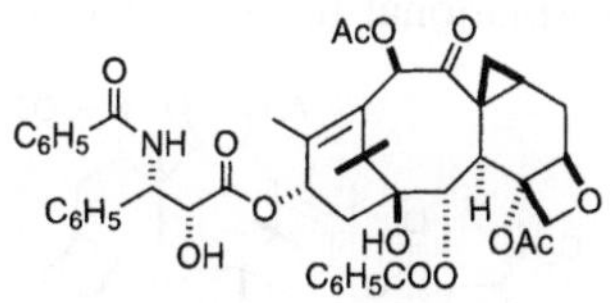

4.1.3.10

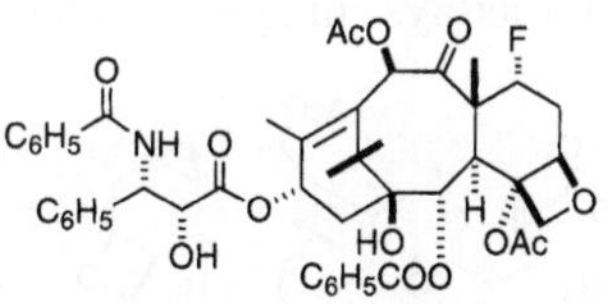

4.1.3.11

The bioactivity of derivatives **4.1.3.10** and **4.1.3.11**, as well as the dehydro compound **4.1.3.7** were reported by Chen and his colleagues (*148*); the latter compound was also reported by Johnson and his colleagues (*59*). The dehydrotaxol **4.1.3.7** and the fluorotaxol **4.1.3.11** were both more active than taxol in a tubulin-assembly assay, but the cyclopropane derivative **4.1.3.10** was less active. None of the three compounds was quite as cytotoxic to HCT 116 cells as taxol, although the differences were not great.

Johnson and his colleagues have prepared 6,7-dehydro derivatives of their 12,13-isotaxols by appropriate modifications of the dehydration procedure. The isotaxoids **4.1.3.12** and **4.1.3.13** were both prepared, and proved to be more active than taxol in several assays with analog **4.1.3.13** showing the best activity against the B16 melanoma in an *in vivo* mouse model (*172*). Cyclopropane derivatives of 12,13-isotaxanes were also prepared in good yield by treatment of the 7β-triflate of 12,13-isotaxol with silica gel in 1,2-dichloroethane at 60°C. The resulting analogs **4.1.3.14** and **4.1.3.15** were slightly more active than taxol and docetaxel in the B16 *in vivo* mouse model (172).

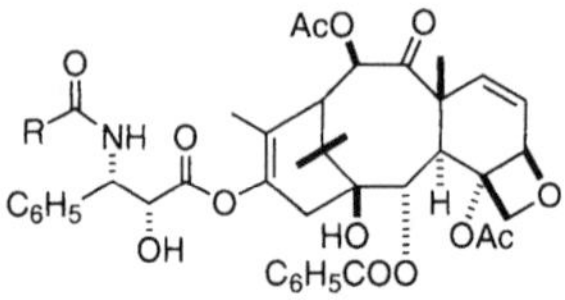

4.1.3.12 R = Me_3CO
4.1.3.13 R = Me_3CNH

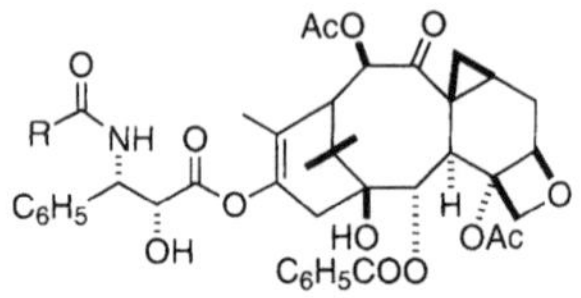

4.1.3.14 R = Me_3CO
4.1.3.15 R = Me_3CNH

An alternate route for the formation of 7,8-cyclopropataxol analogs was discovered by Commerçon and his colleagues, who treated the 7β-triflate of a masked docetaxel precursor with excess sodium azide in acetonitrile. The major product, formed in 95% yield, gave the cyclopropane analog **4.1.3.16** after deprotection. The acetyl amino product **4.1.3.17** was also formed in low yield, presumably through trapping of an intermediate carbocation by the solvent. Interestingly electrolytic reduction of **4.1.3.16** gave a low yield of the 10-desoxy ring-expanded product **4.1.3.19**, in addition to the 10-desoxy product **4.1.3.18** (*173*). This C-homo derivative was completely inactive in tubulin-assembly and cytotoxicity assays, in spite of its conformational resemblance to taxol.

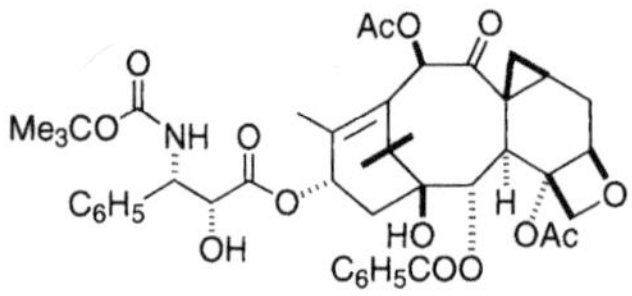

4.1.3.16

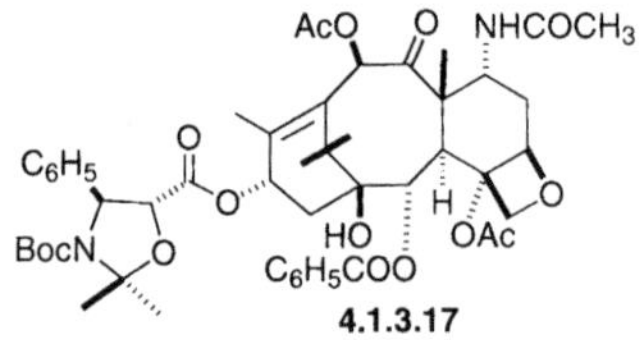

4.1.3.17

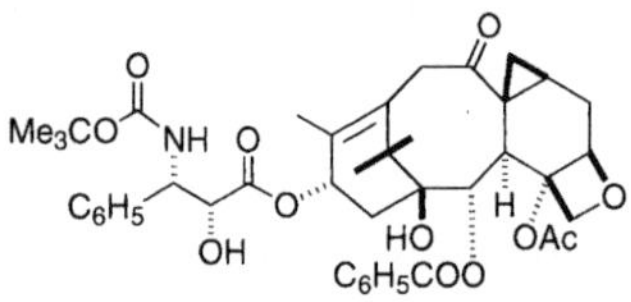

4.1.3.18

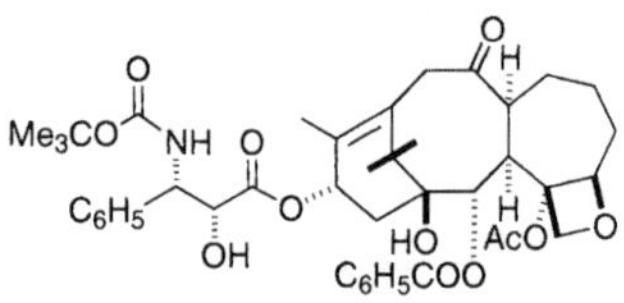

4.1.3.19

Cyclopropataxol analogs have also been prepared from 9-dihydrobaccatin III. Treatment of 13-acetyl-9-dihydrobaccatin III with triflic anhydride in pyridine gave a B-nor-analog which will be discussed later, but treatment of the protected dihydrotaxol analog **4.1.3.20** under the same conditions also gave a low yield of the cyclopropataxol analog

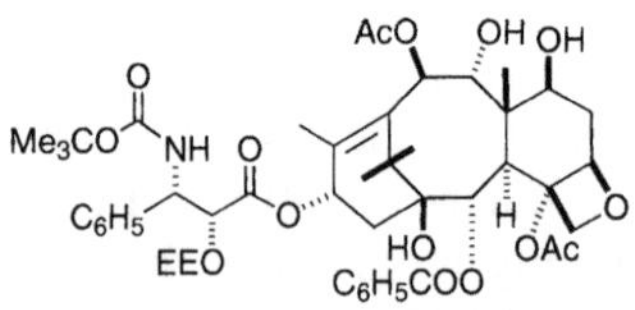

4.1.3.20

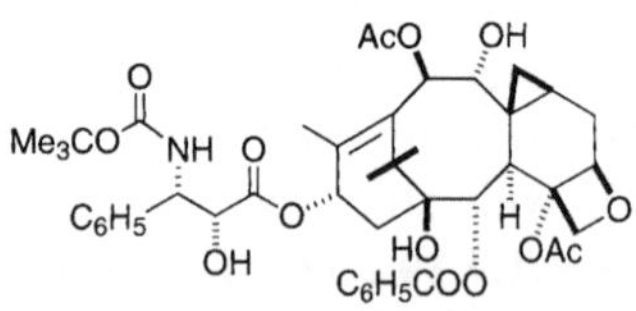

4.1.3.21

4.1.3.21 after deprotection (*174*). Compound **4.1.3.21** was twice as cytotoxic as taxol against the HT29 cell line and eight times as cytotoxic against the B16F10 cell line.

4.2. Products Derived from 6,7-Dehydrotaxol

4.2.1. Addition Products

The availability of the 6,7-dehydro derivative **4.1.3.7** opened up routes to analogs derivatized at C-6 as well as to analogs with other modifications. Hydroxylation of **4.1.3.7** gave the α-diol **4.2.1.1** (*167*) while treatment with dimethyldioxirane gave the α-epoxide **4.2.1.2** (*167, 168*); a conflict between the β-stereochemistry (*167*) and the α-stereochemistry was resolved in favor of the latter by an x-ray crystal structure determination (*175*).

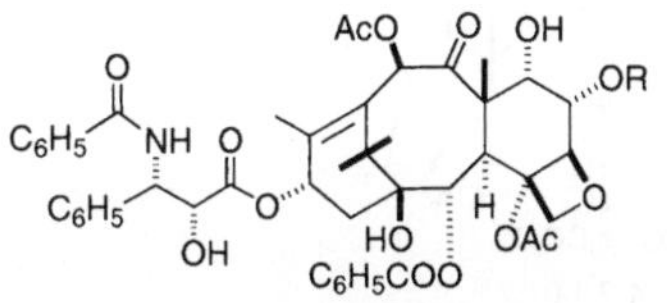

4.2.1.1 R = H
4.2.1.4 R = COR[1]

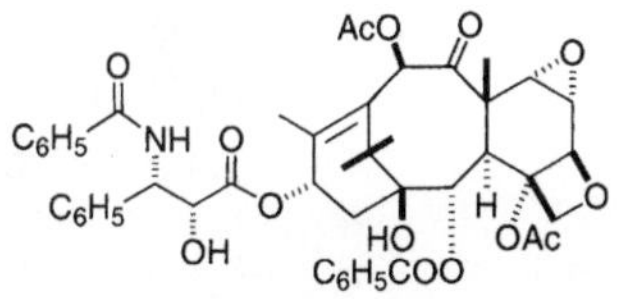

4.2.1.2

Treatment of 6,7-dehydro derivative **4.1.3.7** with bromine in chloroform did not give any addition product, but led instead to the D-seco product **4.2.1.3** (*168*).

The epoxide **4.2.1.2** proved resistant to nucleophilic attack, but the diol **4.2.1.1** could be converted into a number of new analogs. These included simple analogs such as various C-6 acyl derivatives **4.2.1.4** (*167*) but also various derivatives produced by deoxygenation at C-7. Thus diol **4.2.1.1** could be protected as its 2′-(triethylsilyl)ether, converted to its cyclic 6,7-α-thiocarbonate, and deoxygenated to the 7-deoxyanalog **4.2.1.5**. Analog **4.2.1.5** can be oxidized to the ketone **4.2.1.6**, which can then be reduced to the β-alcohol **4.2.1.7** (*176*). Both compounds **4.2.1.5** and **4.2.1.7** were as active as taxol in a tubulin-assembly assay, but were 4 and 10 times less potent than taxol in a cytotoxicity assay with the HCT-116 human colon tumor cell line (*176*).

The diol **4.2.1.1** has also been converted into 6-α-hydroxytaxol **4.2.1.8**, the major human metabolite of taxol. In the first approach, direct treatment of diol **4.2.1.1** (protected as its 2′-*tert*-butyldimethylsilyl derivative) with DBU in xylene at 80°C gave **4.2.1.8** in 15% yield after

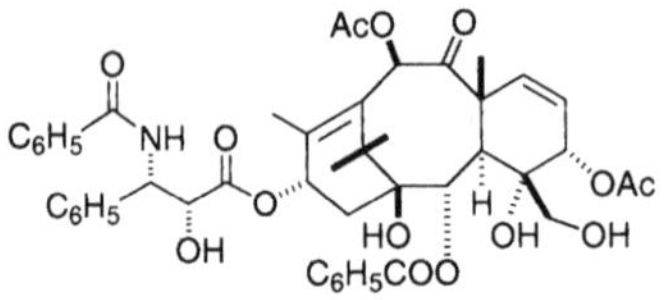

4.2.1.3

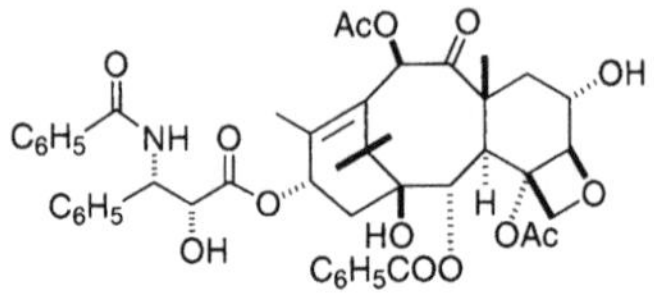

4.2.1.5

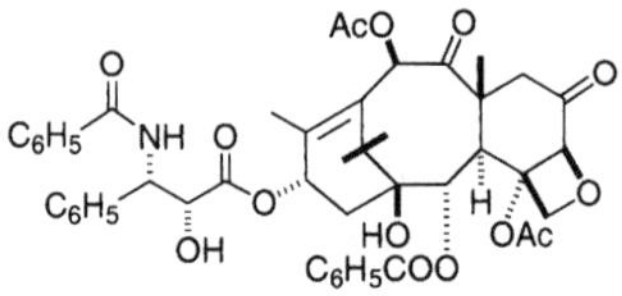

4.2.1.6

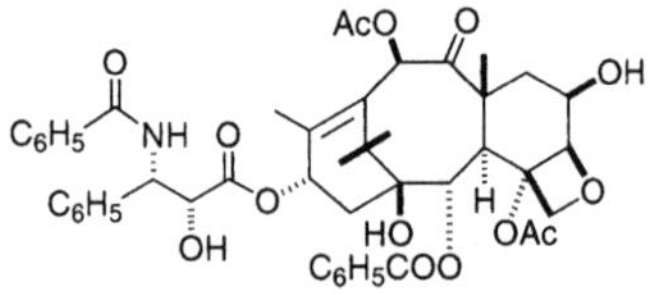

4.2.1.7

deprotection; other conditions and other bases were less effective. Although the conversion was low, the starting material could be recovered and reused (*177*).

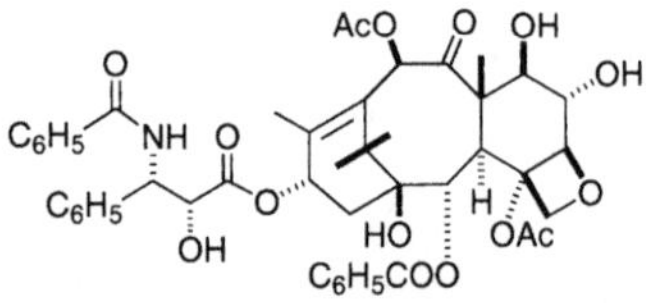

4.2.1.8

A recent publication gives an improved method by which the diol **4.2.1.1** (as its 2′-triethylsilylether) is oxidized selectively to its 6-oxo derivative, which was then epimerized to the 7β-hydroxy derivative **4.2.1.9**. Reduction of **4.2.1.9** with sodium triacetoxyborohydride gave the 6β,7β-diol, and this was converted to the α-nitrate **4.2.1.10** by treatment with thionyl chloride and then sodium periodate to give a cyclic sulfate, followed by tetrabutyl ammonium nitrate. Hydrogenolysis of **4.2.1.10** and deprotection gave the metabolite **4.2.1.8** in 37% overall yield from **4.2.1.1** (*178*).

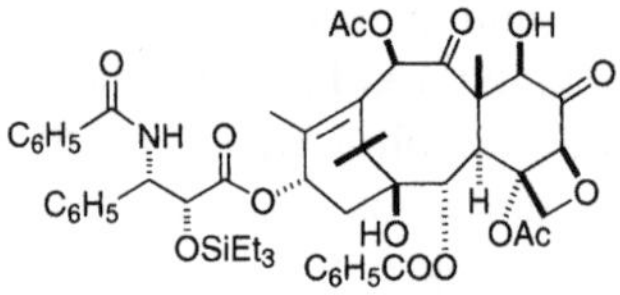

4.2.1.9

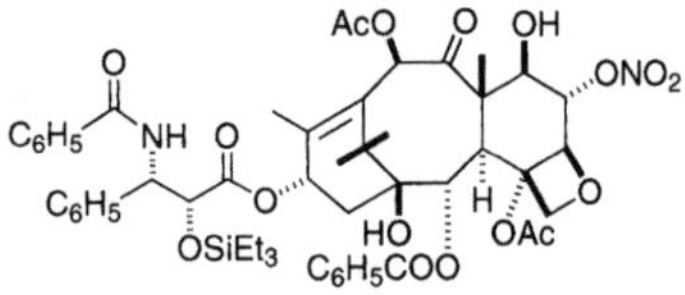

4.2.1.10

4.2.2. Ring Contracted and Related Analogs

Several ring contraction reactions of the diol **4.1.4.1** and its derivatives have been observed. Thus treatment of the C-6 triflate **4.2.2.1** or the corresponding tosylate with even weak bases led to the C-ring contracted product **4.2.2.2**. The deprotected form of this product **4.2.2.3** was completely inactive in tubulin-assembly and cytotoxicity assays (*179*, *180*), consistent with previous observations that the oxetane ring is an important factor in taxol's activity.

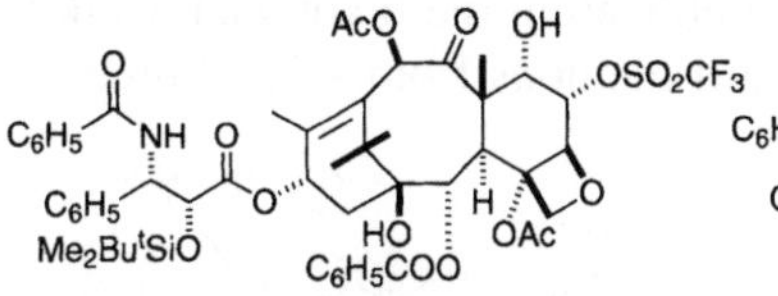

4.2.2.1

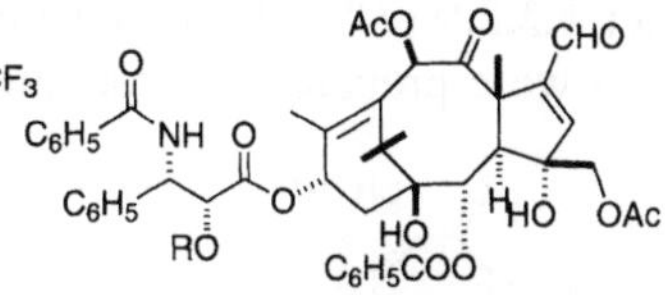

4.2.2.2 R = Me_2Bu^tSi
4.2.2.3 R = H

A C-nor derivative that retained the oxetane ring was formed by oxidation of 2′-protected diol with tetrapropyl-ammonium perruthenate (TPAP), which oxidized the C-6 hydroxyl group selectively to give an unstable 7-hydroxy-6-keto derivative. This derivative then underwent a retro-aldol reaction followed by an aldol reaction to give the C-nortaxol **4.2.2.4**; unfortunately the deprotected final product **4.2.2.5** was unstable and could not be completely characterized (*179*).

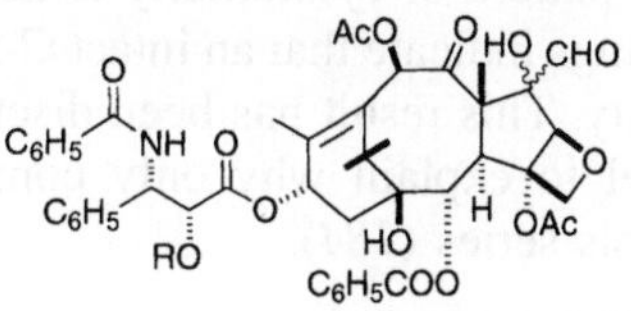

4.2.2.4 R = Me_2Bu^tSi
4.2.2.5 R = H

The most interesting analog in this series was prepared by oxidation of diol **4.2.1.1** with lead tetracetate. This reaction gave the C-seco product **4.2.2.6** and the C-nor product **4.2.2.7**, which was deprotected to the C-nortaxol **4.2.2.8**. Somewhat surprisingly, since molecular models showed a strong similarity to the shape of taxol, the C-nor derivative **4.2.2.8** was significantly less active than taxol in a tubulin polymerization assay, although it was only slightly less cytotoxic to HCT116 cells (*179*). The analog **4.2.2.8** was also prepared in essentially the same way by Wender (*181*).

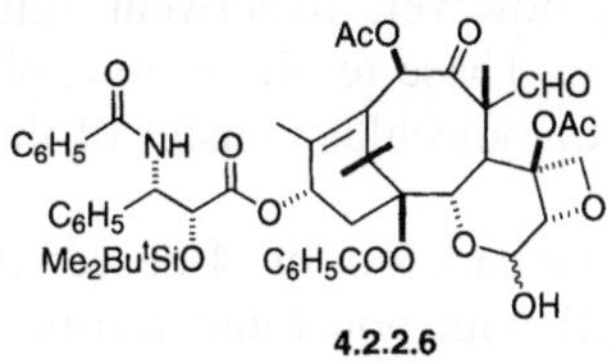

4.2.2.6

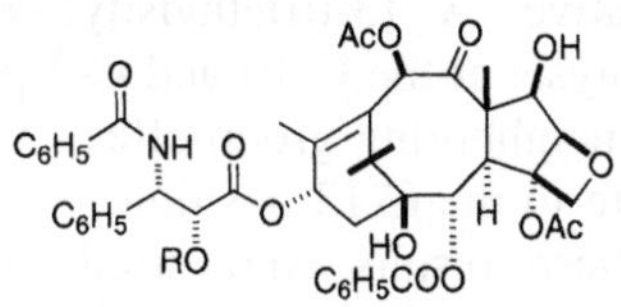

4.2.2.7 R = Me_2Bu^tSi
4.2.2.8 R = H

Various C-seco analogs have been prepared in addition to **4.2.2.6** described above. Thus treatment of 10-deacetylbaccatin III with cupric acetate gave the 10-oxo derivative as described previously (*42*), and treatment of this compound with sodium borohydride and ceric chloride gave a mixture of the 10-*epi*-10-deacetylbaccatin III and the D-seco product **4.2.2.9** (*182*). In a later publication the taxol analogs **4.2.2.10–4.2.2.13** were prepared by standard methods from **4.2.2.9** (*183*).

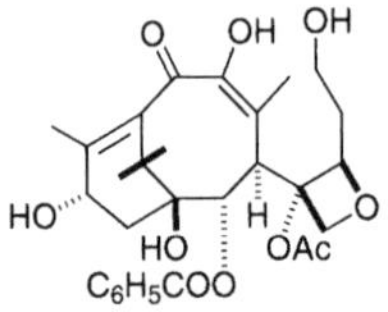

4.2.2.9

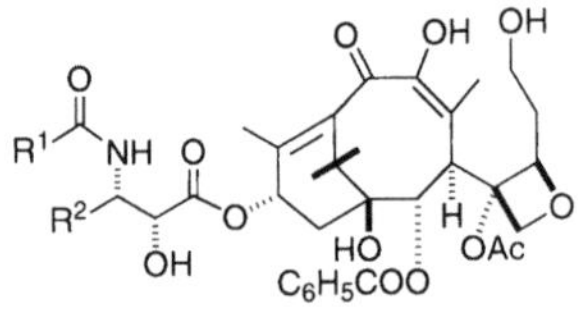

4.2.2.10 $R^1 = C_6H_5$, $R^2 = C_6H_5$
4.2.2.11 $R^1 = C_6H_5$, $R^2 = Me_3CO$
4.2.2.12 R^1 = *i*-Butyl, $R^2 = Me_3CO$
4.2.2.13 R^1 = *i*-Butyl, R^2 = *n*-Hexyl

Surprisingly, the analog **4.2.2.12** was almost half as cytotoxic as taxol in two different cell lines and in a tubulin-assembly assay, and it also showed a similar pattern of cytotoxicity to taxol in the NCI 60-cell line panel. These findings indicate that an intact C-ring is not an absolute requirement for activity. This result has been discussed by Snyder, who has provided a model to explain why only compounds with 3′-alkyl chains are active in this series (*184*).

4.3. Modifications at C-4

4.3.1. Deacetylation of Taxol and Baccatin III

Selective deacetylation of the C-4 acetate of taxol or baccatin III proved to be a challenging task because this acetate group is in a very hindered location. The first clue to a method for selective deacetylation came in studies of the hydrolysis of hexahydrobaccatin III (*58*) in which it was found that hexahydrobaccatin III itself underwent reasonably selective deacylation at C-10 and C-4 to give the 4,10-bis-deacetyl derivative. A 13-(triethylsilyl)-analog, however, underwent selective hydrolysis at the C-10 and C-2 positions. These results were explained by a neighboring group effect, with intramolecular transfer of the C-4 acetate to the C-13 position.

These results were used in the preparation of 4-deacetyltaxol. Treatment of 7-(triethylsilyl)baccatin III with potassium *tert*-butoxide gave the di-deacetyl product **4.3.1.1**, which was converted to 4-deacetyltaxol **4.3.1.2** by acylation with protected side chain, acetylation, and deprotection (*185*). Similar results were reported by Georg and her collaborators, who reported that treatment of 7-(triethylsilyl)baccatin III with potassium *tert*-butoxide gave the 4-deacetyl derivative **4.3.1.3** in 58% yield, while treatment of 7,13-bis(triethylsilyl)baccatin III under the same conditions gave the 2-debenzoyl derivative **4.3.1.4** in 69% yield (*186*). The 4-deacetyl analog **4.3.1.3** was subsequently converted to 4-deacetyltaxol **4.3.1.2** and 4-deacetyldocetaxel **4.3.1.5** by analogous methods to those used by Kingston (*187*). The resulting 4-deacetyl taxol was significantly less active than taxol (*185, 187*).

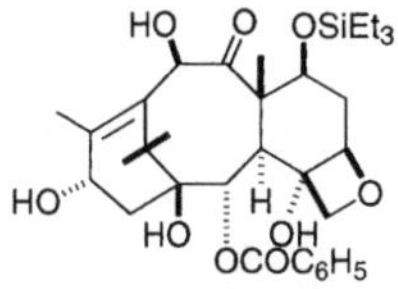

4.3.1.1

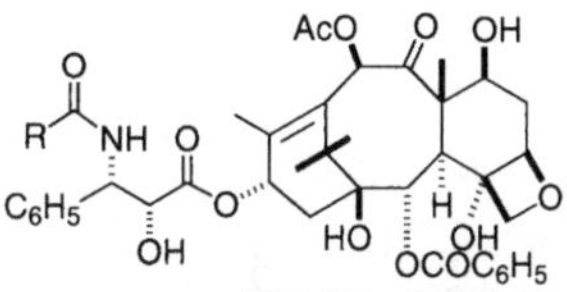

4.3.1.2 R = C_6H_5
4.3.1.5 R = Me_3CO

A slightly different approach was adopted by Chen and his colleagues, who protected baccatin III as its 1-dimethylsilyl derivative **4.3.1.6**. Treatment of this compound with Red-Al gave a 66% yield of 4-deacetyl product **4.3.1.7**, together with a small amount of dideacetyl product (*188*).

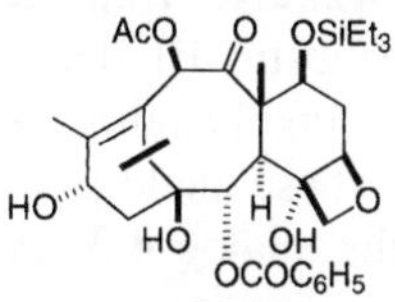

4.3.1.3

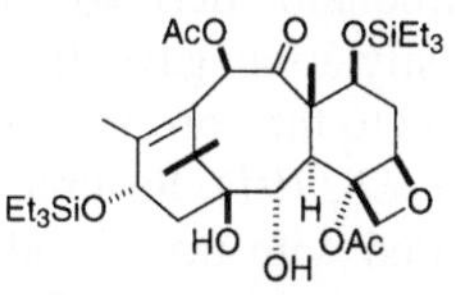

4.3.1.4

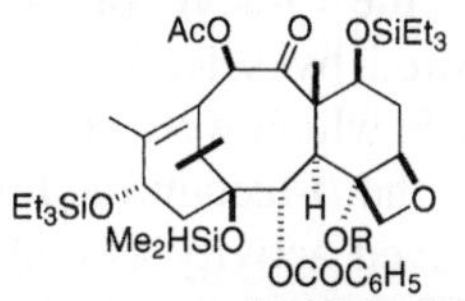

4.3.1.6 R = Ac
4.3.1.7 R = H

A final approach was reported by Kim (*189*) following an earlier patent disclosure by Holton (*107*). In this approach 7,10-bis(triethylsilyl)-13-(trimethylsilyl)-10-deacetylbaccatin III (**4.3.1.8**) was treated with lithium aluminum hydride in ether at a low temperature to give its 2-debenzoyl-4-deacetyl derivative **4.3.1.9**. This was then protected as its cyclic 1,2-carbonate **4.3.1.10**, which could be converted to 4-acyl analogs as described below.

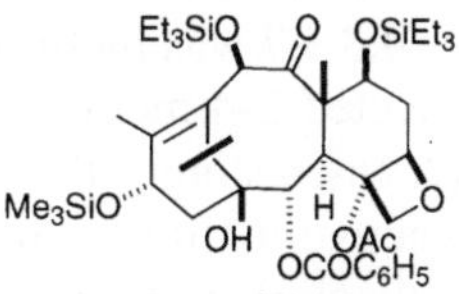

4.3.1.8

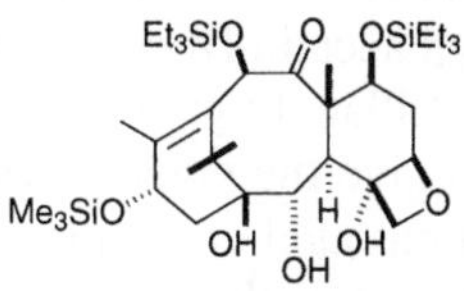

4.3.1.9

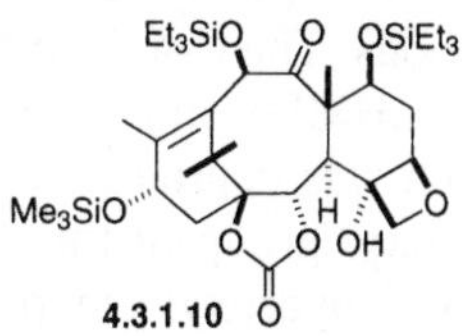

4.3.1.10

4.3.2. Acylation and Alkylation at C-4

Reacylation at the C-4 position is also difficult because of the hindered nature of this position, but various methods have been developed to overcome this problem. The first C-4 analog of taxol was reported by Georg, who converted a 2′,7-protected taxol to its 2,4-dideacyl derivative and thence to the cyclic carbonate **4.3.2.1**. Esterification at C-4 was achieved with excess isobutyric anhydride and DMAP to give the 4-isobutyrate derivative **4.3.2.2**, which was converted to the taxol analog **4.3.2.3** with phenyllithium followed by deprotection. The corresponding docetaxel analog was also prepared (*190*).

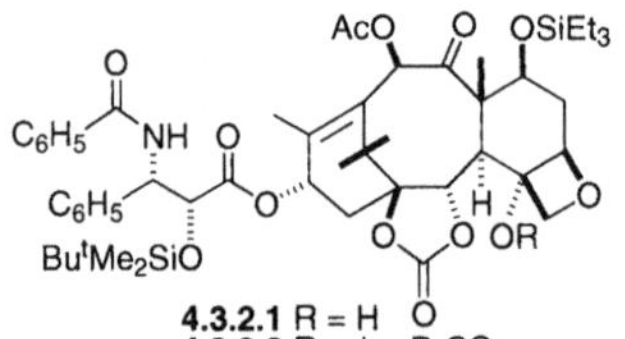

4.3.2.1 R = H
4.3.2.2 R = *iso*-PrCO

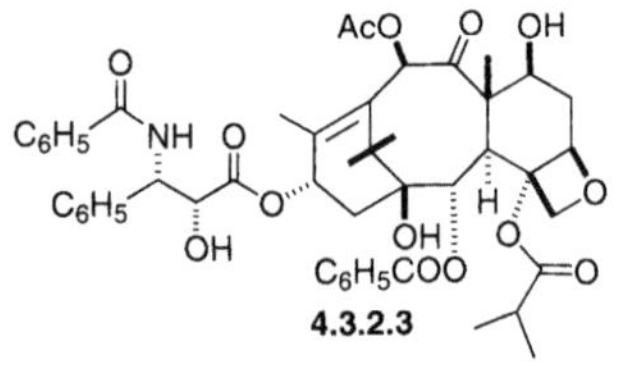

4.3.2.3

Chen and his colleagues have made a number of 4-acyl analogs of taxol. Their general route is based on the selective C-4 deacetylation described above. Treatment of the resulting 4-deacetylbaccatin III derivative **4.3.1.7** with LHMDS and an acyl chloride or a chloroformate gave the 4-acyl or 4-carbonato derivatives **4.3.2.4**. Deprotection followed by selective C-7 silylation gave the baccatin III derivatives **4.3.2.5**, which were converted to the corresponding taxol analogs **4.3.2.6** using the β-lactam method (*188*, *191*). Several of these analogs showed improved activity as compared with taxol, and the analog **4.3.2.6** with R=CH_3OCO is currently in clinical trials (*159*).

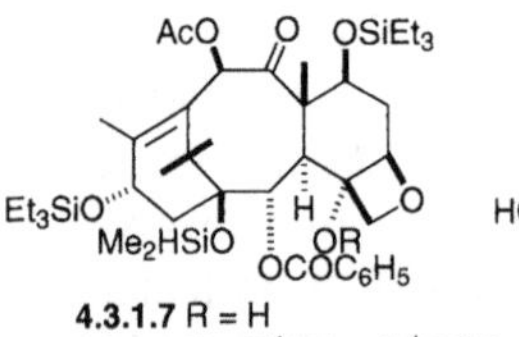

4.3.1.7 R = H
4.3.2.4 R = R[1]CO or R[1]OCO

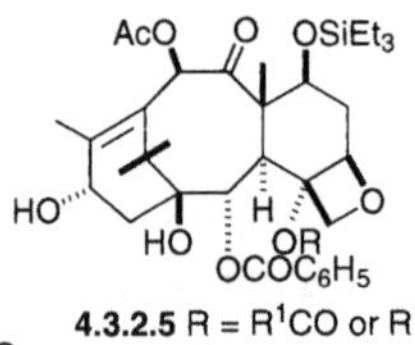

4.3.2.5 R = R[1]CO or R[1]OCO

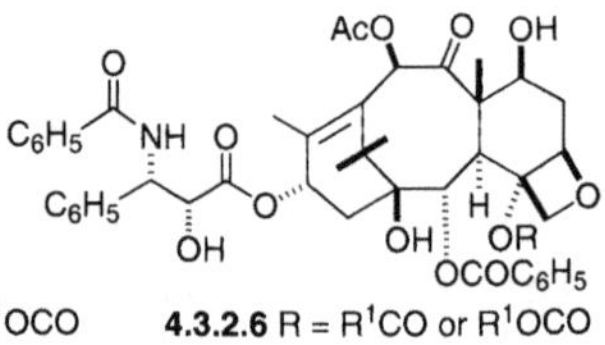

4.3.2.6 R = R[1]CO or R[1]OCO

A C-4 methyl ether derivative of taxol has been made by similar methods. The C-4 deacetylbaccatin III derivative **4.3.1.7** was methylated with LHMDS and methyl triflate to give the protected C-4 methyl ether, which was deprotected to give the ether **4.3.2.7**. Interestingly the C-13 hydroxyl group of this compound was more reactive to silylation than the C-7 hydroxyl group, which is the reverse of the normal order due to the absence of hydrogen bonding with the C-4 acetate. The baccatin **4.3.2.7** was thus converted to its C-7 troc derivative by the process of silylation at C-13, acylation at C-7, and desilylation at C-13. Final coupling with side chain and deprotection gave the C-4 ether **4.3.2.8**, which was about 17 times less cytotoxic than taxol (*192*).

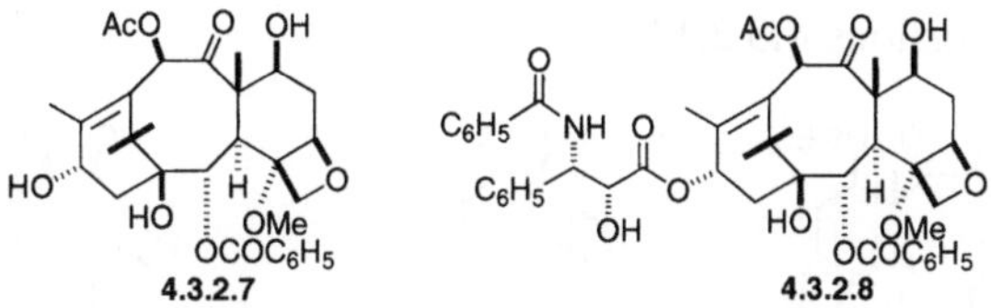

4.3.2.7 **4.3.2.8**

Acylation of the triol **4.3.1.9** was reported by Kim to give acylation at different positions depending on the acylating reagents used (*108*). Acylation with LDA followed by acetyl chloride gave the 4-acetate, but a similar reaction with benzoyl chloride gave the 2-benzoate. The triol **4.3.1.9** did not react with ketene.

Two studies to date have reported on modifications at both the C-2 and the C-4 positions. In the first of these the taxol isomer with a benzoate at C-4 and an acetate at C-2 was prepared. The 4-deacyl compound **4.3.1.7** was benzoylated at C-4 and then converted to its 2-debenzoyl derivative by reduction with Red-Al. Acetylation at C-2 and side chain attachment completed the synthesis of the analog **4.3.2.9**. The product was found to be completely inactive in both tubulin-assembly and cytotoxicity assays (*193*).

A second paper describing the synthesis of some 2,4-diacyl analogs of taxol has appeared recently (*194*). The analogs were prepared by modifications of the routes already described, and some of them showed improved activity as compared with taxol. The most interesting one was the 2-(2,4-difluorobenzoyl)-4-(cyclopropylcarbonyl) analog **4.3.2.10** which was significantly more active than taxol against the drug-resistant HCT116/VM46 cell line.

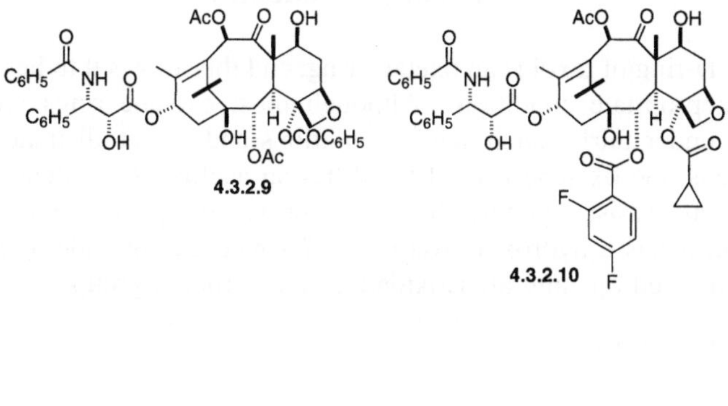

4.3.2.9 **4.3.2.10**

An interesting new approach to the preparation of C-4 derivatives of baccatin III has been developed by Soga and his colleagues (*195*). In this approach the 7-(triethylsilyl)-13-oxo-10-deacetylbaccatin III (**4.3.2.11**) is treated with a strong base and an alkylating agent. The base deprotonates the acetate methyl group selectively, and alkylation yields the C-4 acyl derivatives **4.3.2.12**. Reduction of **4.3.2.12** yields the 10-deacetylbaccatin III analogs **4.3.2.13**.

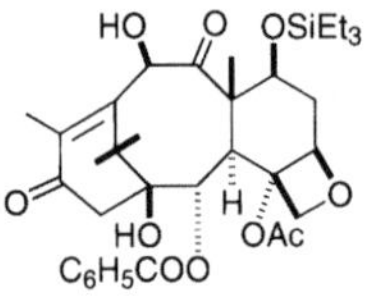

4.3.2.11

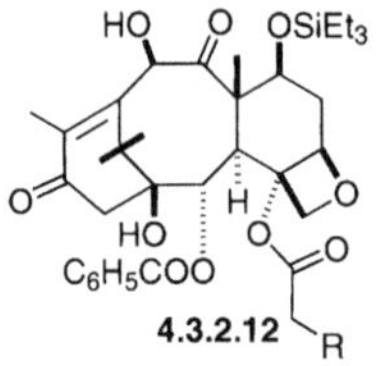

4.3.2.12

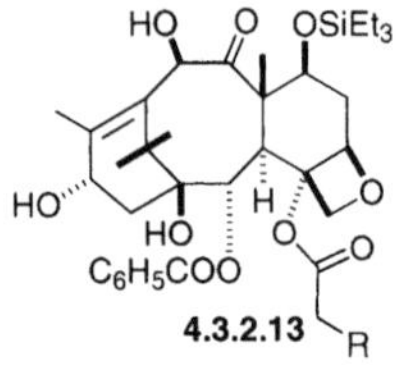

4.3.2.13

4.3.3. Deoxygenation at C-4

Deoxygenation at C-4 has been achieved by application of the methods described. The carbonate **4.3.2.1** was converted to its 4-xanthate derivative, **4.3.3.1**, and this underwent smooth deoxygenation to its 4-deoxy derivative. Treatment of this derivative with phenyllithium followed by deprotection gave the 4-deacetoxy taxol **4.3.3.2**. The product **4.3.3.2** was only marginally active in a cytotoxicity assay against Burkitt CA46 cells, with about 20-fold less activity than taxol (*196*).

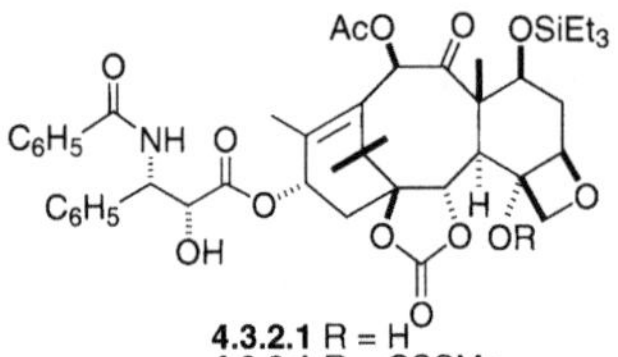

4.3.2.1 R = H
4.3.3.1 R = CSSMe

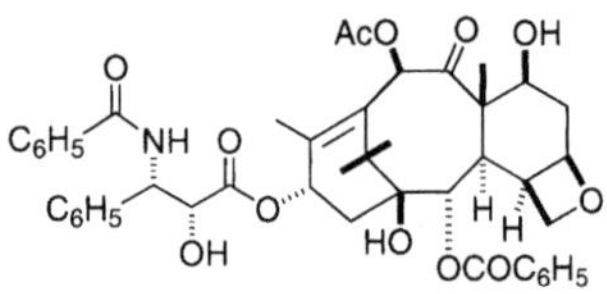

4.3.3.2

5. D-Ring Chemistry

The D-ring of taxol is an oxetane ring, and this allows it to be opened under appropriate conditions. Although most oxetane rings are only opened under fairly harsh acidic conditions and not at all under basic conditions, the oxetane ring of taxol has an acetoxy substituent at C-4 which can assist opening by neighboring group participation. In addition, it has a hydroxyl group at C-7 which can provide a route to base-promoted opening after oxidation to a carbonyl group.

5.1. Ring Opening by Electrophiles

The most common ring-opening reactions of the oxetane ring occur under electrophilic conditions. The first example was discovered using Meerwein's reagent (*57*) which led to the formation of a product **5.1.1**; the reaction was proposed to occur through neighboring-group participation by the C-4 acetoxy group. The mechanism of this reaction was studied in more detail by Chen *et al.* (*197*), who proposed that the two cationic intermediates **5.1.2** and **5.1.3** were involved when $SnCl_4$ was used as the electrophile. This suggestion was supported by the isolation of the ring-opened baccatins **5.1.4** and **5.1.5** when baccatin III was treated with trifluoroacetic acid in the presence of phenyldimethylsilane. A final piece of evidence in favor of intermediate **5.1.3** was the isolation of the bridged product **5.1.6** as a minor product when 2′-protected taxol was treated with $SnCl_4$ for an extended period of time.

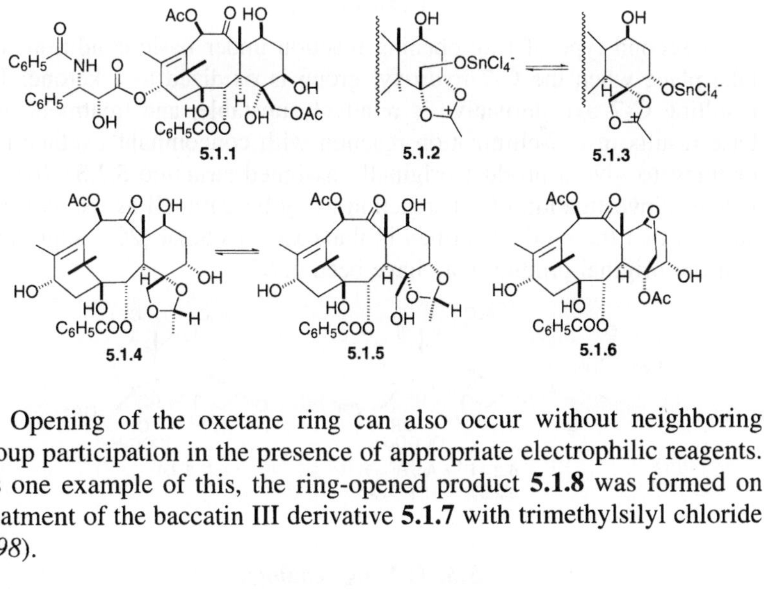

Opening of the oxetane ring can also occur without neighboring group participation in the presence of appropriate electrophilic reagents. As one example of this, the ring-opened product **5.1.8** was formed on treatment of the baccatin III derivative **5.1.7** with trimethylsilyl chloride (*198*).

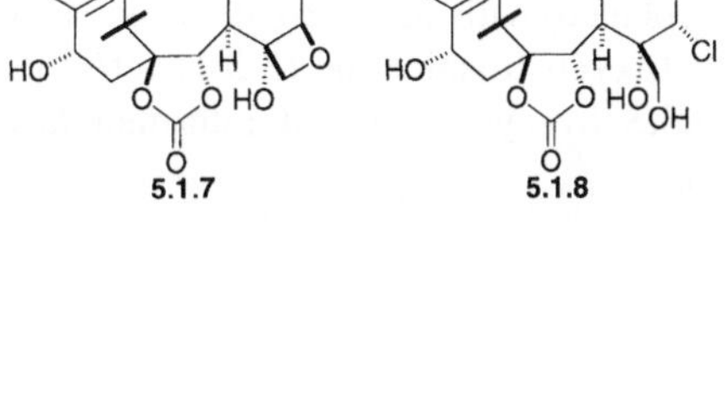

5.2. Ring-Opening by Nucleophiles

Although oxetane rings are normally reasonably stable to nucleophilic reagents, the oxetane ring of taxol can be opened by the neighboring 2-hydroxyl group when this is present. The first example of this was discovered independently by Farina (*199*) and Wahl (*200*). Thus treatment of 7-(triethylsilyl)-10-deacetylbaccatin III with base produced a mixture of 2- and 4-deacyl products as well as the tetrahydrofuran derivative **5.2.1** (*200*). Compound **5.2.1** is formed by nucleophilic attack of the C-2 hydroxyl group (as an alkoxide ion) on the oxetane. A similar reaction with tributyltin methoxide on 7,13-diacetylbaccatin III gave the product **5.2.2** (*201*).

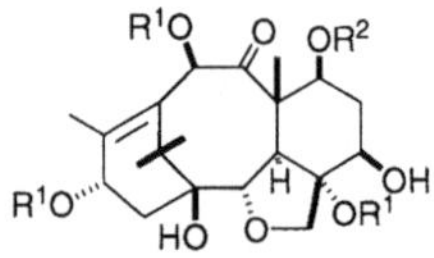

5.2.1 R^1 = H, R^2 = $SiEt_3$
5.2.2 R^1 = R^2 = Ac

A second type of ring-opening reaction under basic conditions can take place when the C-7 hydroxyl group is oxidized to a ketone. The resulting C-7 oxo derivative is relatively unstable, and treatment with base results in a β-elimination reaction with concomitant oxetane ring opening to give a product originally assigned structure **5.2.3** (*165*). A recent reinvestigation of this reaction using baccatin III as the substrate showed that the product formed in this case was **5.2.4** (*201*) suggesting that the original product may have been **5.2.5**.

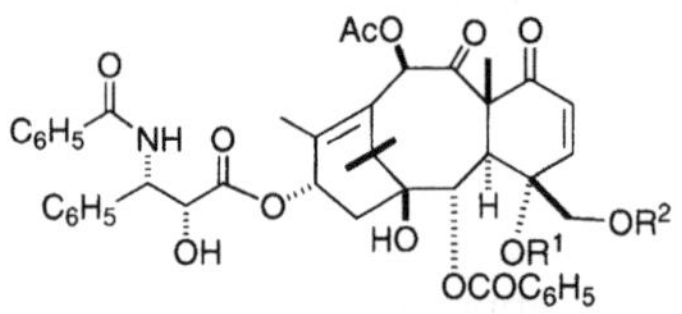

5.2.3 R^1 = Ac, R^2 = H
5.2.5 R^1 = H, R^2 = Ac

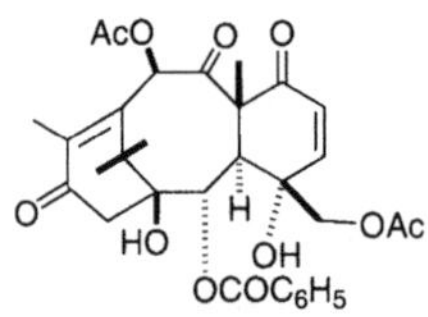

5.2.4

5.3. D-Ring Analogs

Several investigators have prepared taxol analogs in which the oxetane ring is replaced with an alternate ring. Thus Fenoglio *et al.* converted 2′-deacetoxyaustrospicatine (**5.3.1**) into the azetidine derivatives **5.3.4** and **5.3.5** by cyclization of intermediates **5.3.2** and **5.3.3**, respectively (*202*).

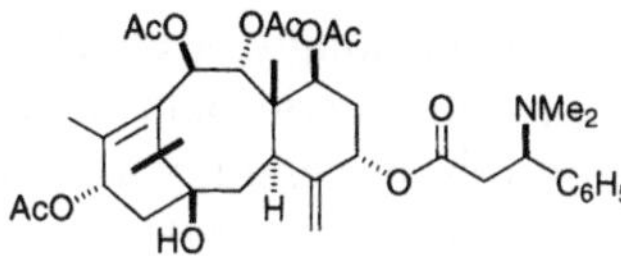

5.3.1

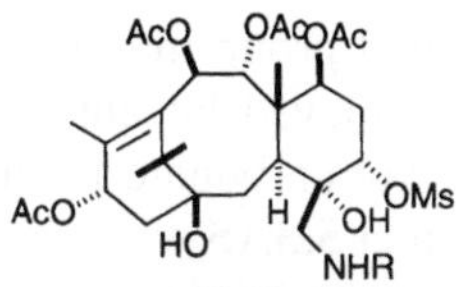

5.3.2 R = H
5.3.3 R = Ac

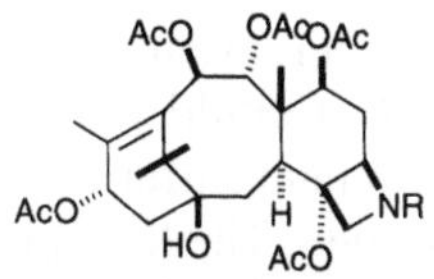

5.3.4 R = H
5.3.5 R = Ac

A closer analog to taxol was prepared by Marder-Karsenti *et al.* (*203*), who converted 10-deacetylbaccatin III into the key intermediate **5.3.6** by ring-opening using boron trifluoride etherate and tetrabutyl-ammonium bromide as the electrophilic reagent. Compound **5.3.6** was then converted into the docetaxel analog **5.3.7** by condensation with benzylamine to give the corresponding benzylimine, reduction to the amine, cyclization, functional group manipulation, and side chain attachment (*203*). The resulting analog **5.3.7** was 16 times less active than docetaxel in a tubulin polymerization assay and was not cytotoxic to KB cells.

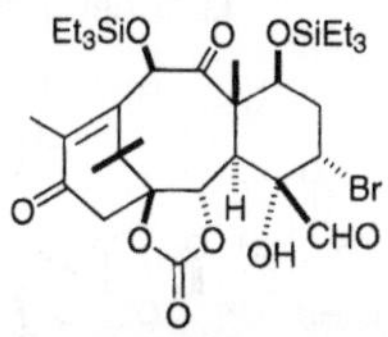

5.3.6

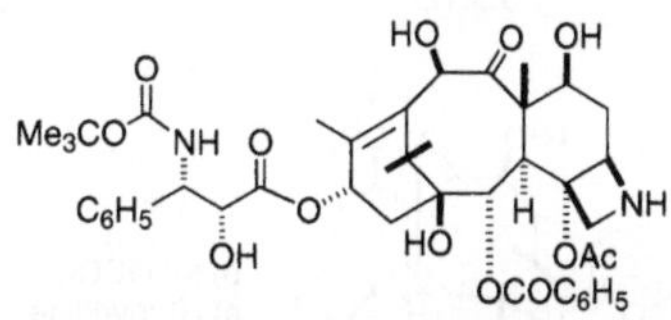

5.3.7

A thietane analog of 7-deoxobaccatin III was prepared from taxine B by Poupat and her colleagues (*204*). Taxine B was converted to **5.3.8** by standard chemistry, and the thietane ring was constructed through intermediates **5.3.9** and **5.3.10** to give the analog **5.3.11**.

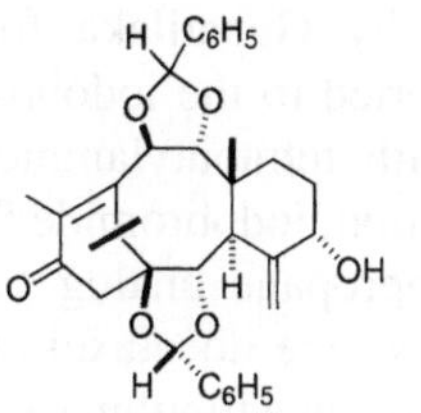

5.3.8

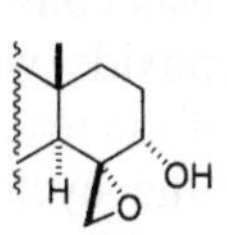

5.3.9

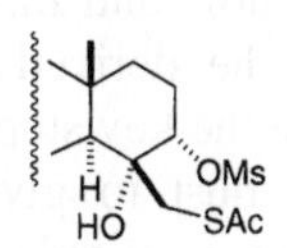

5.3.10

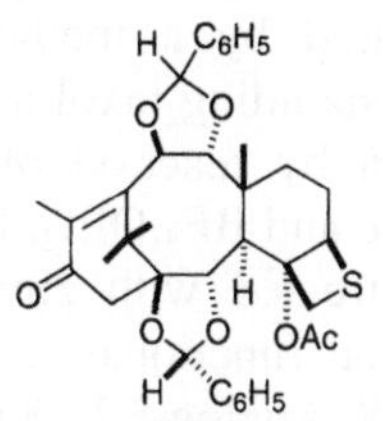

5.3.11

A thietane analog of taxol itself was prepared by Gunatilaka and Kingston *et al.* by a similar route, although historically this work pre-dated the Poupat studies described above (*205*). Taxol was converted to the key epoxy iodide **5.3.14** by a ring-opening reaction on the 2-debenzoyl-4-deacetyl-1,2-carbonate **5.3.12** to give the iododiol **5.3.13**, followed by treatment with mesyl chloride and DBU. Treatment of **5.3.14** with dilithium sulfide under carefully controlled conditions gave a sulfetane derivative, and this was converted to the taxol analog **5.3.15** by appropriate functional group manipulation. Interestingly, acetylation at C-4 could not be achieved because of the extra steric bulk of the sulfetane sulfur atom, but carbonate formation was successfully accomplished. The final product **5.3.15** was about 8 times less active than taxol in a tubulin-assembly assay, but it was over 200 times less active in two different cytotoxicity assays. Very recently an alternate synthesis of a sulfetane analog of taxol has been reported in which the C-4 acetate group is incorporated at an earlier stage, and the analog **5.3.16** was thus obtained (*206*).

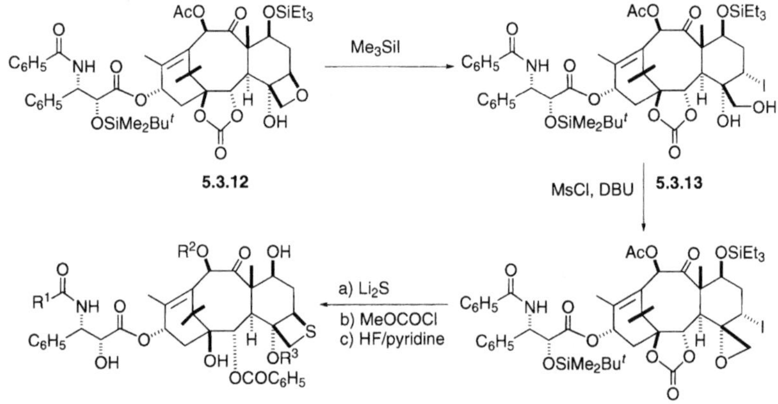

5.3.15 $R^1 = C_6H_5$, $R^2 = Ac$, $R^3 = COOCH_3$
5.3.16 $R^1 = OCMe_3$, $R^2 = H$, $R^3 = COCH_3$

A final and very interesting modification is the replacement of the oxetane ring with a cyclopropane ring (*207*). The bromodiol **5.3.15** was prepared by a method similar to that used by Gunatilaka for the corresponding taxol analog, and this was converted to the iodobromide **5.3.16** by reaction of the derived epoxide with tetrabutylammonium iodide and $BF_3.OEt_2$. In the key step of the reaction, iodobromide **5.3.16** was treated with zinc dust to give the cyclopropane analog **5.3.17**. Normal functional group manipulation then gave the docetaxel analog **5.3.18**. Analog **5.3.18** was almost as active as taxol in a tubulin assembly assay; this is the first analog lacking an oxetane ring that has shown this level of activity.

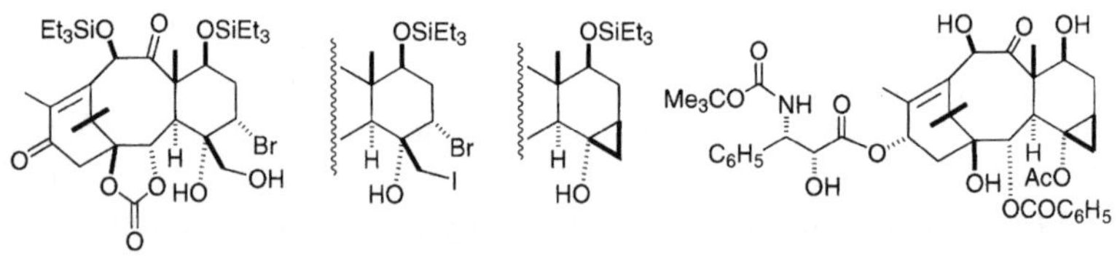

5.3.15 **5.3.16** **5.3.17** **5.3.18**

5.4. Spectroscopic and Theoretical Studies

The fact that taxol analogs with opened oxetane rings are much less active than taxol was noted early in studies on the chemistry of taxol (*57, 165*) and there has been a lot of discussions as to why this is so. It was noted that the oxetane ring serves as a conformational "lock" for the taxol ring system (*57*) and this observation has been followed up with a detailed NMR investigation (*208*). This showed that the oxetane ring influences the relative orientation of the C-2 benzoyl group and the side chain and that the A-ring changes from a distorted boat conformation in taxol to a pseudo envelope conformation in D-secotaxol **5.1.1**.

Although the conformational lock provided by the oxetane ring is clearly a factor in taxol's three-dimensional structure, it may not be a crucial component of taxol's bioactivity. The evidence that a rigid ring C-D combination is not essential for activity is provided by the fact that certain C-seco analogs such as **4.2.2.12** show significant activity (*183*). Other factors that could play an important role include the hydrogen bond acceptor capacity of the oxetane oxygen and the overall size of the oxetane ring.

These questions have recently been addressed from a theoretical and modeling perspective by Snyder and his colleagues (*184*); the approach uses a minireceptor model of the binding site of taxol on tubulin. This study showed that hydrogen bond acceptor properties and the rigidification of taxol by the oxetane ring both play a role in stabilizing the taxol-tubulin complex. However, these functions are not exclusive to the oxetane ring, and other rings such as the cyclopropane ring and a putative epoxide ring were calculated to serve almost as well as the oxetane ring. The authors conclude "we believe the current protocol to be sufficiently sound to suggest that the oxetane ring should be unseated as an essential ingredient of Taxol SAR" (*184*).

6. Rearrangements and Related Reactions

The complex structure of taxol and related taxoids have made them fertile sources of rearrangement reactions, and a number of interesting and even spectacular rearrangements have been reported. Classification of these reactions is somewhat arbitrary, but is based on the ring involved in the major structural change. This section also includes coverage of photochemical reactions where the major change is the formation of a new bond and is thus not strictly a rearrangement reaction.

6.1. Rearrangements Leading to 11(15→1)-abeo Taxols (A-nortaxols)

The first major rearrangement in this area was found when taxol was treated with acetyl chloride, leading to the formation of the A-nortaxol 11(15→1)abeo-5-epi-20,0-secotaxol **6.1.1**. (*57*). Treatment of 2′,7-bis(triethylsilyl)taxol with mesyl chloride and triethylamine gave the A-nortaxol **6.1.2** without concomitant ring D opening. Interestingly the deprotected analog **6.1.3** had good tubulin-assembly activity, although it was not significantly cytotoxic.

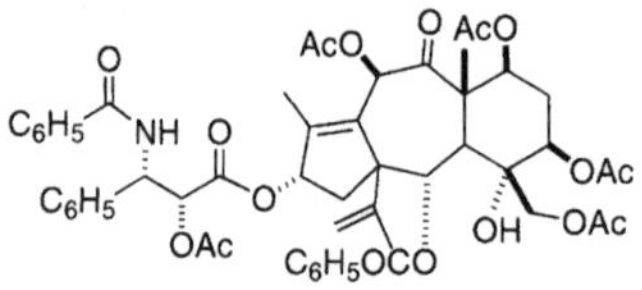

6.1.1

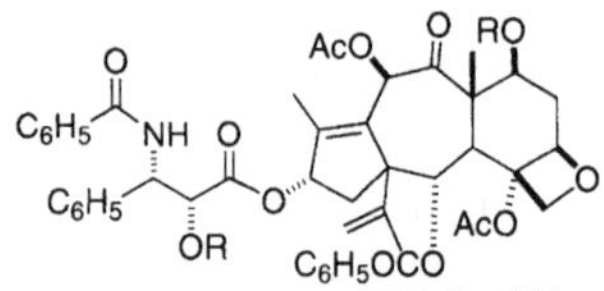

6.1.2 R = SiEt3
6.1.3 R = H

Many other workers have observed similar reactions leading to A-nortaxols; in some cases these reactions are coupled with reactions involving electrophilic opening of the oxetane ring. Thus treatment of 7,10-di(trichloroethoxycarbonyl)-10-deacetylbaccatin III with $ZnCl_2$ gave the A-nor product **6.1.4** and the A-nor-D-secoproduct **6.1.5**; product **6.1.4** gave the cyclic carbonate **6.1.6** on treatment with alumina (*200*). The docetaxel analog **6.1.7** of **6.1.6** was prepared by treating 7,10-di-Troc docetaxel with trifluoroacetic acid, followed by deprotection; it was 40 times less active than docetaxel in a microtubule disassembly assay (*209*). The 14-hydroxybaccatin III derivative **6.1.8** gave the A-nor product **6.1.9** on heating in the presence of pyridinium p-toluene sulfonate; in this case the oxetane ring opened with concomitant benzoate migration to the C-14 hydroxyl (*210*).

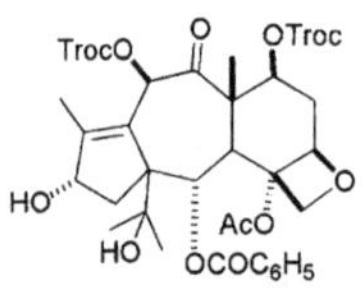

6.1.4

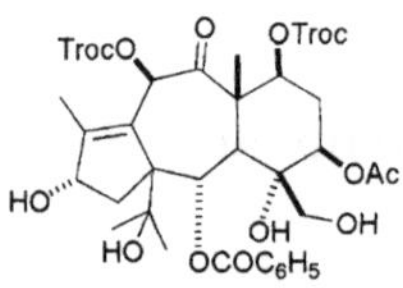

6.1.5

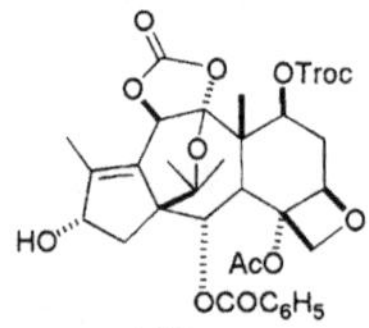

6.1.6

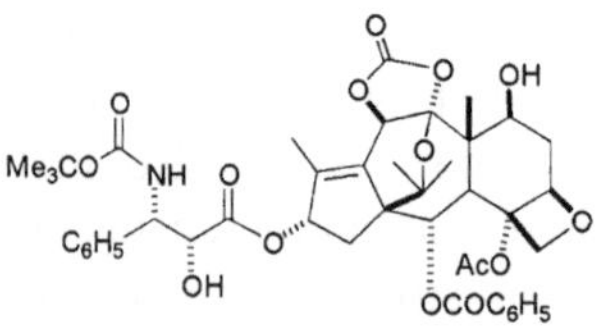

6.1.7

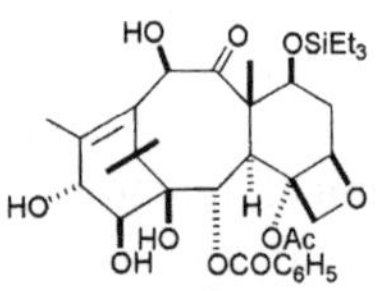

6.1.8

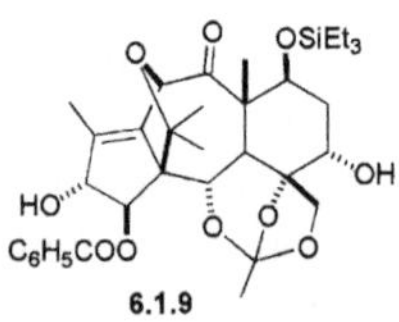

6.1.9

Rearrangements of the reduced baccatin III derivative 9R-dihydrobaccatin III were investigated by Zamir and her colleagues. Thus treatment of (9R)-dihydro-13-acetylbaccatin III with p-toluenesulfonic acid in methanol gave the A-nor-D-seco product **6.1.10**. In acetone, however, the ketal **6.1.11** and two isomers were formed (*211*, *212*).

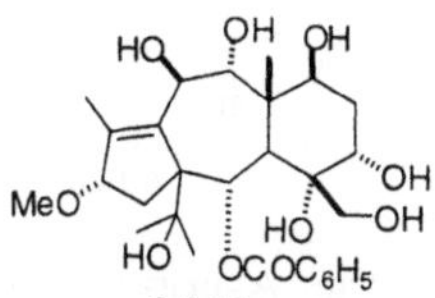

6.1.10

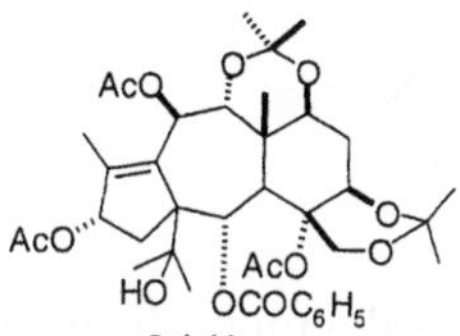

6.1.11

Treatment of 10-deacetoxybaccatin III with SeO_2 gave the A-nor analog **6.1.12** in which selenium has been incorporated into a selenophene ring. Attachment of the taxol side chain gave the analog **6.1.13**, which was about 40 times less cytotoxic to J774.2 cells than taxol (*181*).

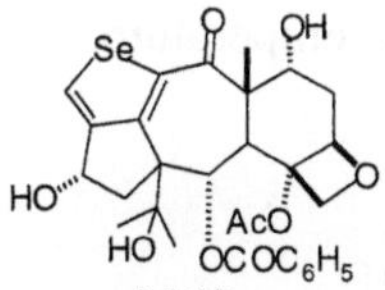

6.1.12

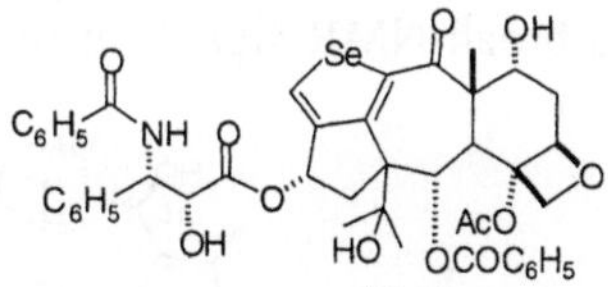

6.1.13

Several investigators have prepared analogs of A-nortaxols. A series of compounds of general structure **6.1.14** was prepared by Chordia and Kingston (*213*); the analogs **6.1.15** and **6.1.16** showed promising *in vitro* activity, but were not significantly active against the M109 tumor in mice. The brevifoliol analog **6.1.17** was prepared by Georg and her collaborators by standard methods; unsurprisingly it was essentially completely inactive in cytotoxicity assays (*214*).

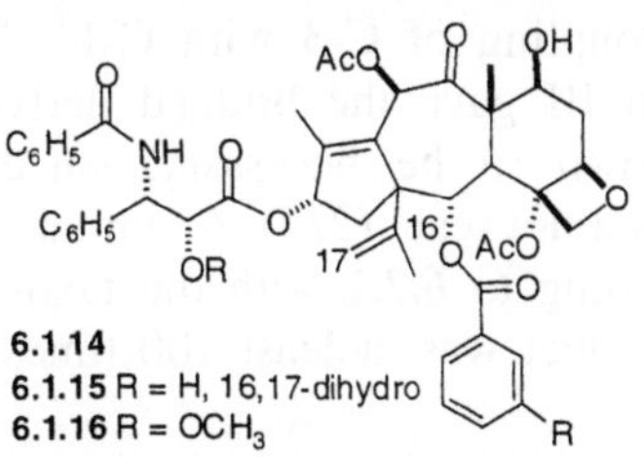

6.1.14
6.1.15 R = H, 16,17-dihydro
6.1.16 R = OCH_3

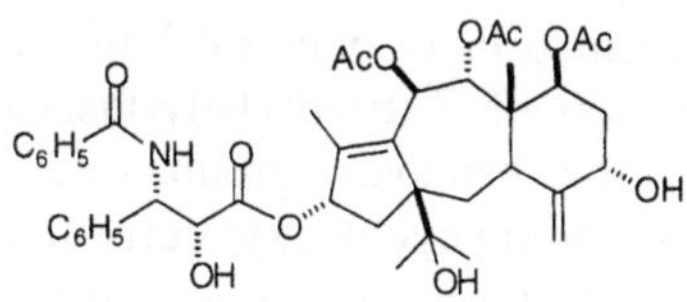

6.1.17

The aza-A-nortaxol analogs **6.1.20** where Ar is phenyl, *m*-methoxyphenyl, or *m*-chlorophenyl, were prepared from baccatin III by treatment with $SOCl_2$ to give the A-nor-chloro compound **6.1.18**, which was converted to the azide **6.1.19** and thence to **6.1.20**. None of the analogs **6.1.20** showed any significant cytotoxicity or tubulin-assembly activity (*215*).

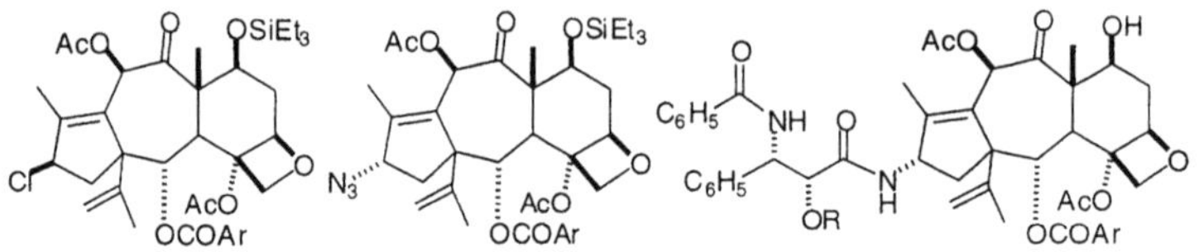

6.1.18 **6.1.19** **6.1.20**

The functional group chemistry of the A-nortaxol class of compounds has been investigated, and was found to differ in some respects from that of baccatin III. Thus the ease of silylation of the hydroxyl groups of the A-nortaxane **6.1.21** was 13 > 5 ≫ 10, which is very different from that of baccatin III where the C-13 hydroxyl group is by far the least reactive (*216*). These investigators also noted that A-nortaxanes are flexible and can adopt two conformations in solution, giving rise to broad NMR signals at room temperature.

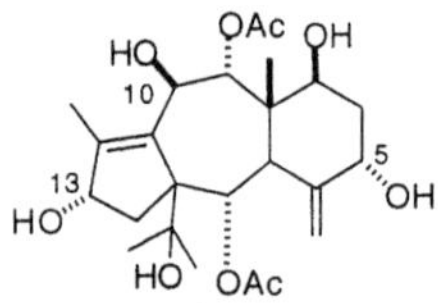

6.1.21

6.2. Rearrangements Involving the B-Ring

Several investigators have investigated the photochemistry of taxoids, and have found that the most important photochemical transformation is one involving a coupling of C-3 with C-11. Thus irradiation of 7-(triethylsilyl)baccatin III gave the bridged derivative **6.2.1**; the benzoate group was shown to be necessary, since the corresponding cyclohexyl derivative did not react (*217*). A similar study with taxol gave an analog corresponding to **6.2.1** with the taxol side chain attached at C-13 (*218*); the product was at least 100 times less active than taxol.

Photochemistry of taxine B gave the bridged product **6.2.2** in which the dimethylamino substituent on the C-5 side chain was also lost reductively (*219*). The isomer of **6.2.2** with the acetate at C-10 was also obtained.

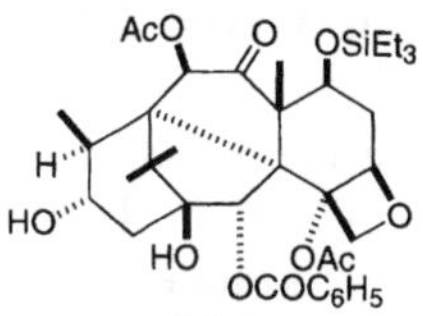

6.2.1

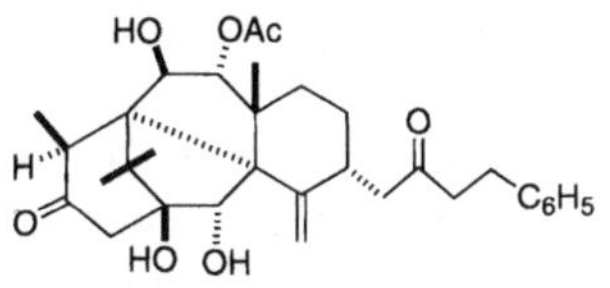

6.2.2

Treatment of the 7-xanthate of 10-deacetoxybaccatin III with Bu_3SnH and AIBN at 80°C gave the expected 7-deoxyproduct, but also gave a new product **4.1.2.7** formed by a radical coupling reaction. When $(TMS)_3SiH$ was used as a hydrogen source, the three internally cross-linked products **4.1.2.7**, **4.1.2.8**, and **4.1.2.9** were all obtained, as previously discussed (*220*).

Various reactions in which ring B is converted to a seven-membered ring have been discovered. Thus treatment of 13-acetyl-9-dihydrobaccatin III with trifluoromethane sulfonic anhydride in pyridine caused it to rearrange to the B-nor-analog **6.2.3** (*221*). Conversion of **6.2.3** to the corresponding taxol and docetaxel analogs **6.2.4** and **6.2.5** was achieved by standard methods; the analog **6.2.5** had comparable cytotoxicity to taxol in three different cell lines (*221*).

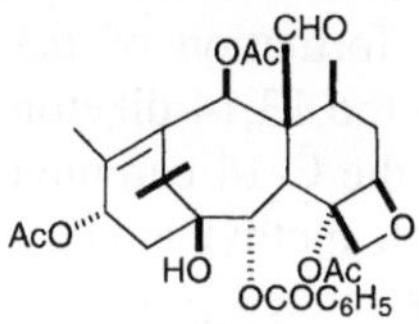

6.2.3

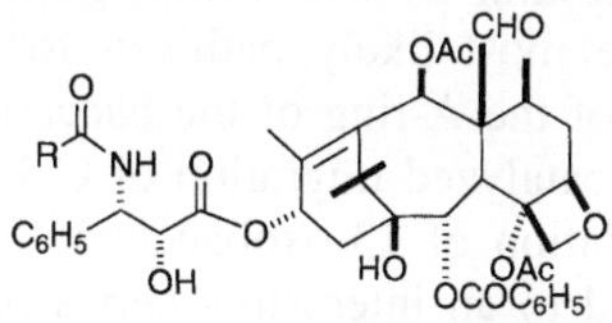

6.2.4 R = C_6H_5
6.2.5 R = OBu^t

An alternate B-nortaxol analog was prepared from 7-(triethylsilyl)-13-oxobaccatin III (**6.2.7**). Treatment of this substrate with trifluoroacetic anhydride in pyridine gave a product **6.2.6** related to those discussed in the previous section. Treatment with DBU, however, yielded the B-nor product **6.2.8**. This product was postulated to arise by an initial C-2 to C-1 benzoyl migration, followed by a pinacol-type rearrangement (*222*).

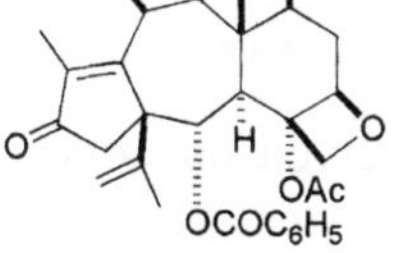

6.2.6

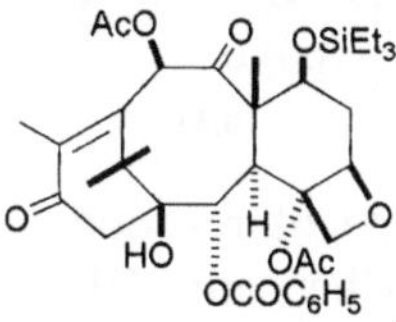

6.2.7

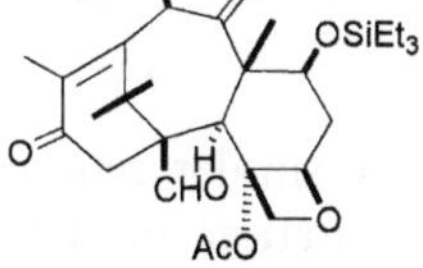

6.2.8

6.3. Rearrangements Involving Both A- and B-Rings

Several rearrangement reactions are more deep-seated, and involve both the A and B rings. Thus treatment of taxicin I (**6.3.1**) or its precursor (a mixture of cinnamoyltaxicin I acetates) with sodium methoxide gave the bisacetal **6.3.2** as the major product. Compound **6.3.2** is postulated to be formed by a homoretroaldol reaction on the A-ring of **6.3.1**, followed by formation of two hemiacetal units (*223*). An analogous product **6.3.3** was formed by treatment of *O*-cinnamoyltaxicin I with potassium carbonate. This product rearranged to the ketal **6.3.4** on treatment with p-toluenesulfonic acid (*224*).

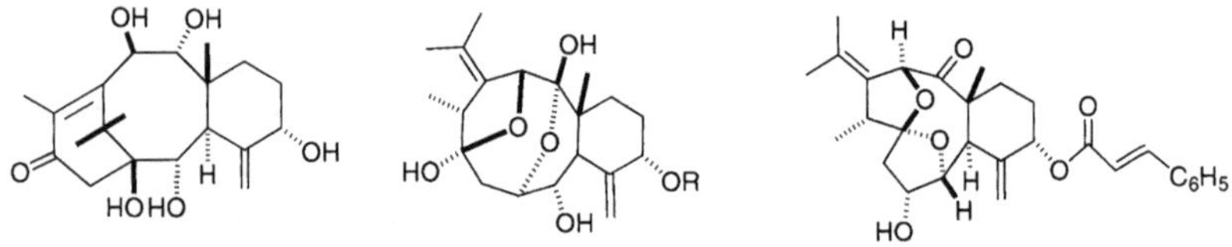

6.3.1

6.3.2 R = H
6.3.3 R = COCH=CHC$_6$H$_5$

6.3.4

Treatment of protected 10-deacetylbaccatin III or 14β-hydroxy-10-deacetylbaccatin III with MnO_2 gave the A-nor-B-homo product **6.3.5** (*225*). The most likely pathway for the formation of **6.3.5** involves oxidation of the A-ring of the baccatin to the 13,14-diketone, followed by a base-catalyzed migration of C-15 to the C-14 carbonyl carbon.

Methylation of 13-oxobaccatin III with methyl iodide and sodium hydride led to an interesting series of rearranged products. The initial product **6.3.6** is presumably formed by a homoretroaldol type of reaction initiated by methylation of the C-13 carbonyl group. Continued treatment with methyl iodide and base gave rise to an intramolecular aldol reaction, leading to the tetracyclic product **6.3.7** and two less methylated intermediates (*226*).

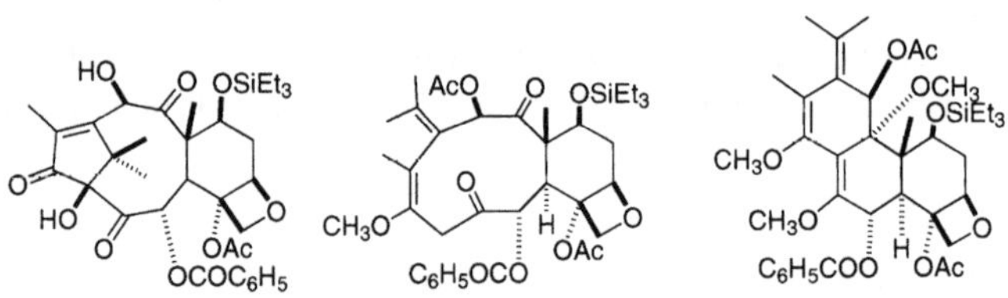

6.3.5 **6.3.6** **6.3.7**

A similar retroaldol type of reaction was observed on enzymatic dehydration of 10-deacetylbaccatin III with *Pseudomonas cepacia* lipase, which led to the seco analog **6.3.8** (*227*). A related reaction was observed on treatment of 2-debenzoyl-4,13-dideacetyl baccatin VI (**6.3.9**) with camphor sulfuric acid; in this case the oxetane ring was converted to a tetrahydrofuran ring by intramolecular attack of the C-2 hydroxyl group, as discussed previously in section 3.4, and the product was **6.3.10**.

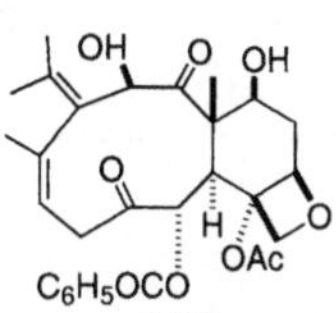
6.3.8

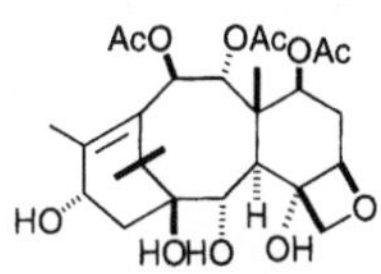
6.3.9

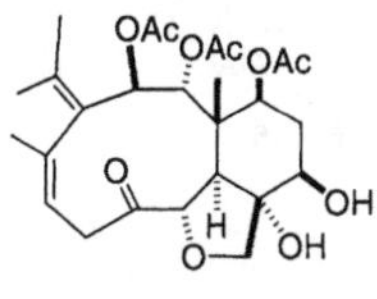
6.3.10

Protection of 13-acetyl-9-dihydrobaccatin III as its 7,9-ketal, followed by hydrolysis of the C-13 acetate, oxidation at C-13, and treatment with sodium methoxide gave the bridged ketal **6.3.11**. Continued treatment with base then led to the A/B secobaccatin **6.3.12**, in which a homo-retroaldol reaction is followed by cleavage of a diketone and lactone formation (*228*).

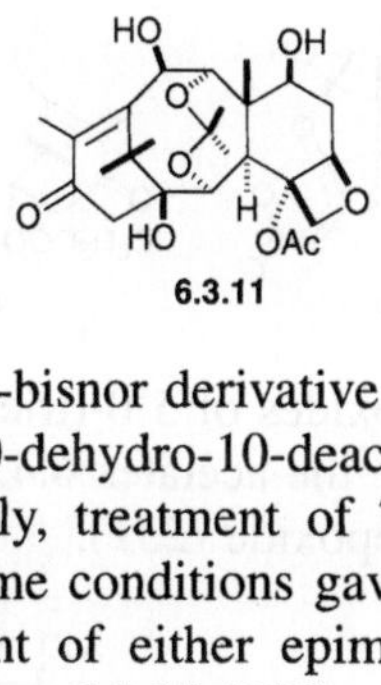
6.3.11

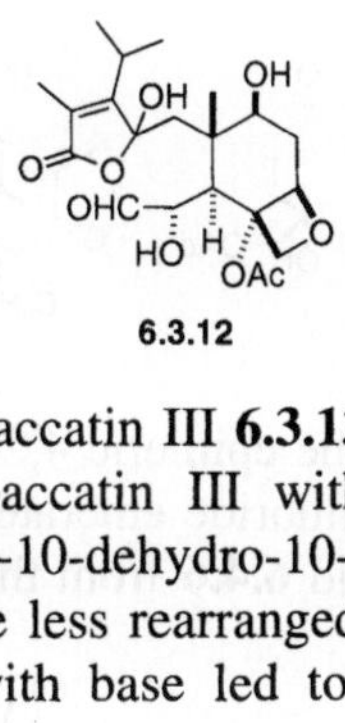
6.3.12

The A-nor-B-bisnor derivative of baccatin III **6.3.13** was formed by treatment of 10-dehydro-10-deacetylbaccatin III with trichloroacetic acid. Interestingly, treatment of 7-epi-10-dehydro-10-deacetylbaccatin III under the same conditions gave the less rearranged product **6.3.14**. Finally, treatment of either epimer with base led to the extensively rearranged product **6.3.15** (*229*).

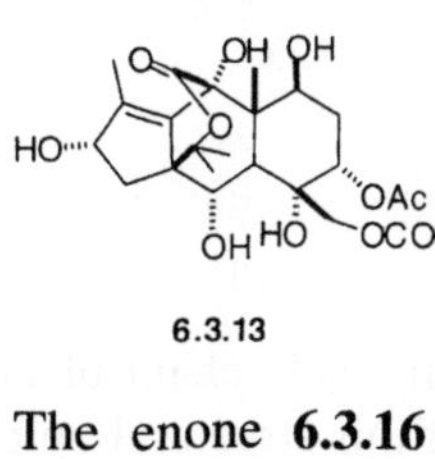
6.3.13

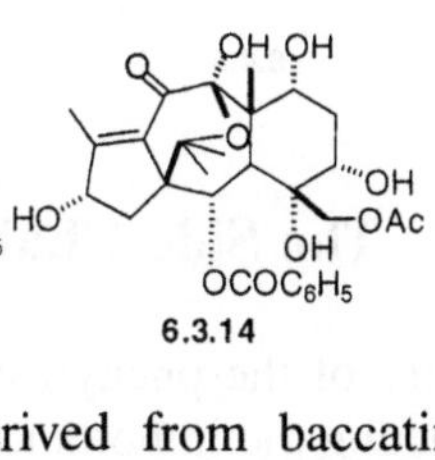
6.3.14

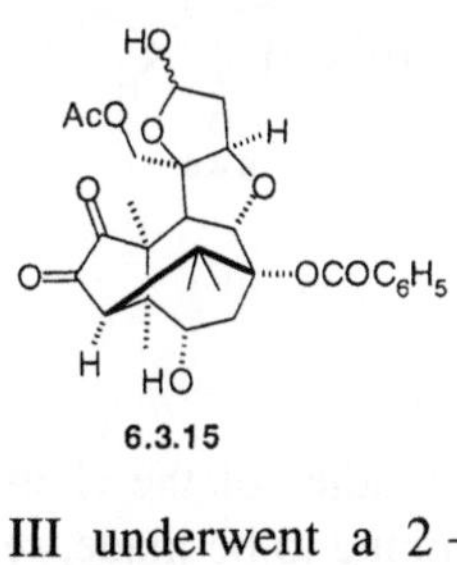
6.3.15

The enone **6.3.16** derived from baccatin III underwent a 2 + 2 cycloaddition reaction to give the hexacyclic product **6.3.17**. This product was converted to the two taxol analogs **6.3.18** (epimeric at C-

13), but neither analog showed any activity in a microtubule binding experiment (*230*).

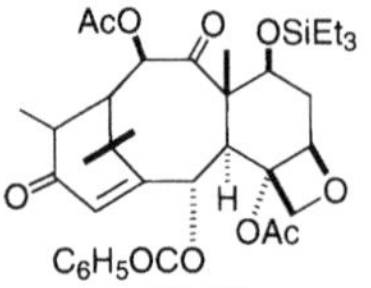

6.3.16

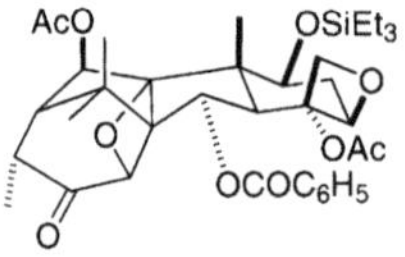

6.3.17

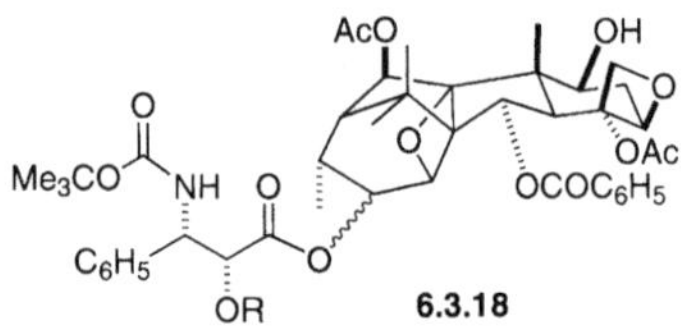

6.3.18

6.4. Rearrangements of Ring C

Some rearrangement reactions of ring C were described in section 5 where the chemistry of ring C was discussed; these rearrangements involved various ring contractions and one ring expansion reaction. A few additional rearrangements involving ring C are noted here. Thus the D-secobaccatin **6.4.1** underwent rearrangement in base to the lactones **6.4.2** and **6.4.3** (*201*).

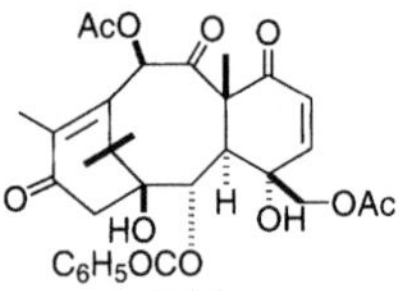

6.4.1

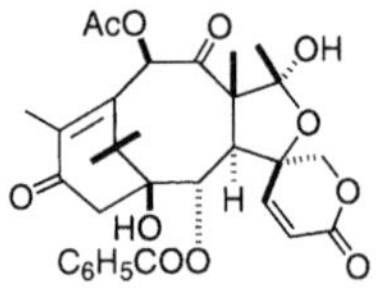

6.4.2

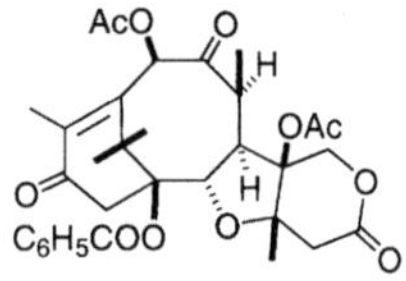

6.4.3

Reaction of the epimeric 4,20-epoxides of 5-0-(triethylsilyl)taxinine A with boron trifluoride etherate gave the acetates **6.4.4** and **6.4.5** from the β-epoxide and **6.4.6** from the α-epoxide (*231*).

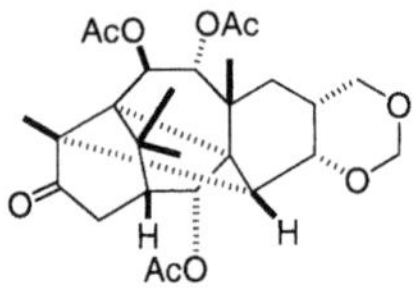

6.4.4

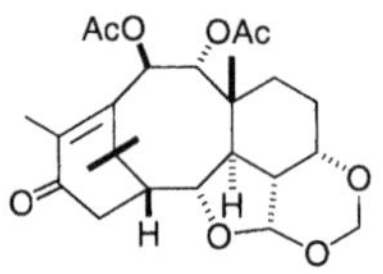

6.4.5

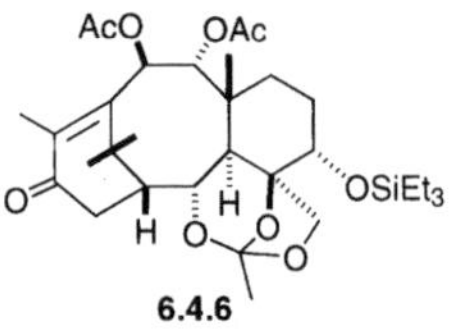

6.4.6

7. The Side Chain

Studies of the chemistry of the phenylisoserine side chain of taxol continue to be made, reflecting the importance of this structural unit for the synthesis of taxol from baccatin III and the potential of preparing improved analogs of taxol by modification of the side chain. Synthetic approaches to the taxol side chain and related aminoacids have been reviewed (*232*, *233*, *234*, *235*), and the preparation of taxoids with modified side chains has also been reviewed recently (*236*). This section will focus initially on synthetic approaches to the side chain, followed by methods of coupling the side chain to baccatin III. A short section on unusual side chains will be followed by a final section on analogs with side chains modified on the 3′- and N-acyl positions.

7.1. Synthesis of the Side Chain by the β-lactam Approach

Since the last review in this series in 1993 (*10*) and the publication of Ojima's paper summarizing his approach to the β-lactam (*237*), a number of new approaches for the asymmetric synthesis of the β-lactam taxol side chain have emerged. For example, chiral sulfonamide esters have been used as chiral auxiliaries to achieve diastereoselective cycloaddition with N-TMS-imine, to give lactam **7.1.1** in 94% yield. Benzoylation of **7.1.1** yielded the ready-to-couple lactam **7.1.2** in 96% yield (*238*).

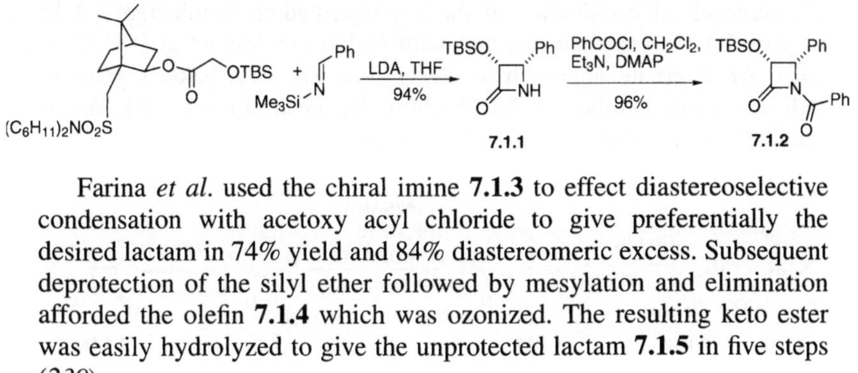

Farina *et al.* used the chiral imine **7.1.3** to effect diastereoselective condensation with acetoxy acyl chloride to give preferentially the desired lactam in 74% yield and 84% diastereomeric excess. Subsequent deprotection of the silyl ether followed by mesylation and elimination afforded the olefin **7.1.4** which was ozonized. The resulting keto ester was easily hydrolyzed to give the unprotected lactam **7.1.5** in five steps (*239*).

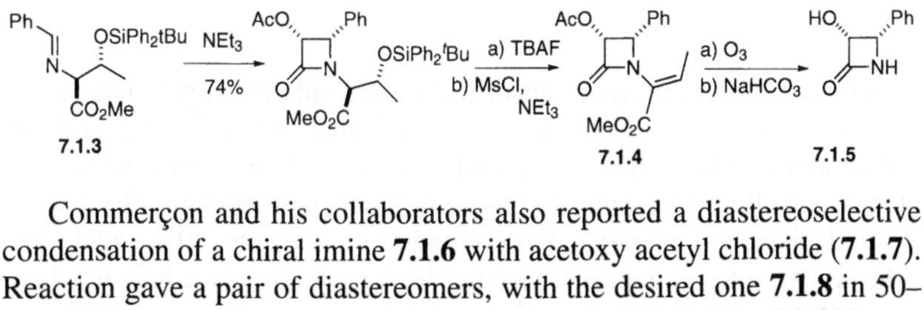

Commerçon and his collaborators also reported a diastereoselective condensation of a chiral imine **7.1.6** with acetoxy acetyl chloride (**7.1.7**). Reaction gave a pair of diastereomers, with the desired one **7.1.8** in 50–60% diastereomeric excess. After cleavage of the acetate, cis-β-lactam

7.1.9 was isolated in optically pure form by simple crystallization. In this chemistry, the product β-lactam was hydrolyzed to give the open side chain, and the chiral auxiliary was then removed regioselectively by hydrogenation. After conversion to its N-Boc derivative, **7.1.10**, this side chain was protected as the cyclic acetonide **7.1.11** and was used for the preparation of docetaxel analogs (*240*).

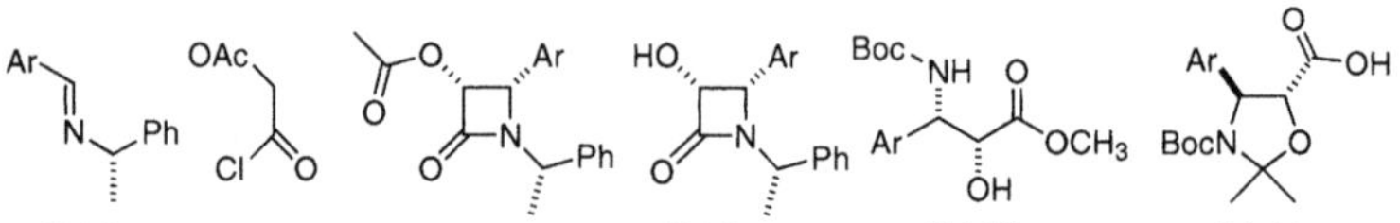

7.1.6 7.1.7 7.1.8 7.1.9 7.1.10 7.1.11

In another report, the traditional oxazolidinone auxiliary was attached to the acyl chloride for the [2+2] cycloaddition with an imine. Thus, both Holton's and Palomo's groups reported the reaction of **7.1.12** and **7.1.13** to yield lactam **7.1.14** in a virtually complete stereoselective fashion. The asymmetry at the C-3 position of the lactam was destroyed by hydroxylation or chlorination at this position. Hydrolysis of the auxiliary was achieved with silica gel (SiO_2) or silver nitrate ($AgNO_3$). Diastereoselective reduction of the key intermediate ketolactam **7.1.15** to give **7.1.16** was then effected with $NaBH_4$ in MeOH at 0°C (*241, 242*). An alternate approach to chiral β-lactam uses baker's yeast to reduce racemic ketolactam **7.1.15** (*243*). The desired isomer **7.1.16** was obtained in up to 80% ee at 50% conversion.

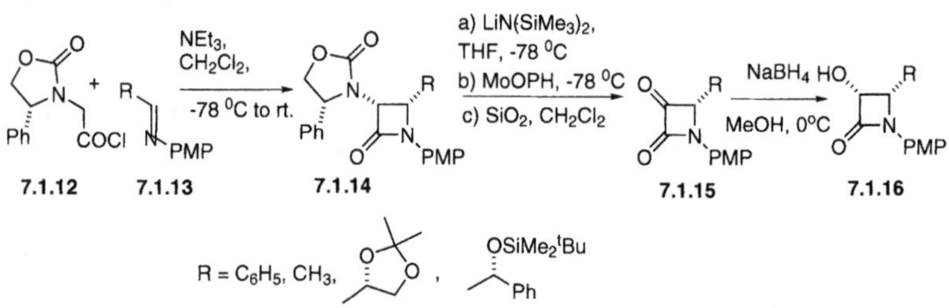

7.1.12 7.1.13 7.1.14 7.1.15 7.1.16

Diastereoselective synthesis of a β-lactam side chain precursor using a chiral imine has also been accomplished by Fujisawa and his collaborators who condensed a chiral imine with an ester enolate in a stereocontrolled way. Thus, reaction of **7.1.17** with enolate afforded the desired cislactam in 84% diastereomeric excess and 90% yield. The other enantiomer of the cis β-lactam (diastereomeric when R is attached) was not detected (*244*).

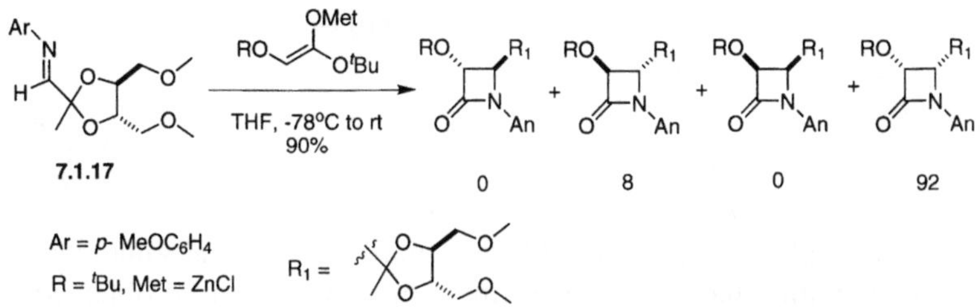

Enzymatic hydrolytic resolution of racemic cis β-lactams provided an alternative to the use of chiral auxiliaries for the synthesis of enantiomerically pure β-lactam side chain. Sih and his collaborators found that P-30 specifically hydrolyzed the undesired 3-acetate group of the β-lactam, affording the acetate of the desired enantiomer in 99.5% enantiomeric excess and 47% yield. Manipulations using known chemistry proceeded smoothly to give optically active taxol side chain (*245*).

Functional group transformations of the β-lactam form of the taxol side chain precursor have also been investigated. For example, both the Ojima and the Georg groups showed that N-(*t*-Boc) β-lactams can be converted to other N-acyl β-lactams in very good yield without losing optical purity. These transformations allowed preparation of various side chain modified taxol or docetaxel analogs (*246*, *247*, *248*).

Recently Lee and his collaborators reported a new route to the β-lactam taxol side chain precursor through a β-bromo amide (*249*). Thus, Sharpless asymmetric dihydroxylation of a cinnamide gave the diol **7.1.18** in 94–96% ee. After converting the diol to β-bromo amide **7.1.19**, treatment with TBAF or NaH gave the β-lactam **7.1.20** in 93–94% yield without significant loss of optical purity.

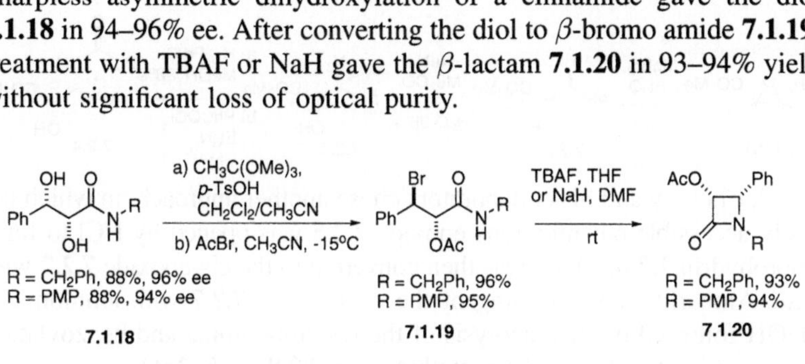

Optical resolution of racemic β-lactam has also been achieved by Bose and his collaborators via a stereospecific glycosylation. Thus, reaction of racemic **7.1.22** with the anhydrosugar **7.1.21** gave the desired β-lactam **7.1.23** in 24% yield along with 36% of the other diastereomer (*250*).

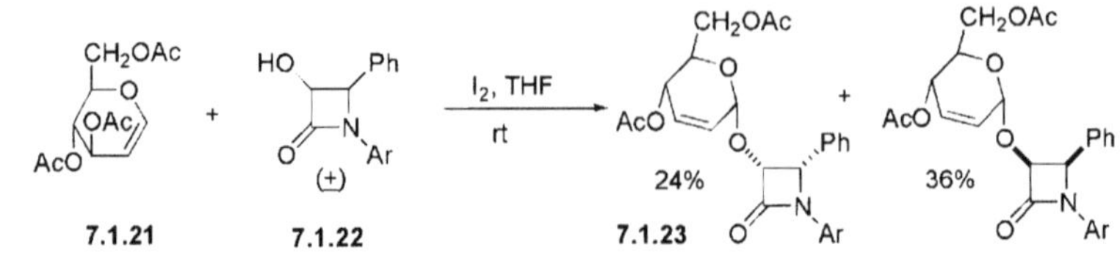

7.1.21 7.1.22 7.1.23

Interestingly, resolution of β-lactam may not be an absolute necessity, since Ojima has shown that kinetic resolution can be obtained in coupling racemic β-lactams with baccatins (*251*).

7.2. Synthesis of the Side Chain via an Epoxide Intermediate

Since Greene's first enantioselective synthesis of the taxol side chain through an epoxide intermediate prepared by Sharpless epoxidation of *cis*-cinnamyl alcohol (*252*), much work has been published concerning alternative pathways to the *cis* epoxide and its regioselective opening. In one of these, asymmetric Sharpless epoxidation of *trans*-cinnamyl alcohol gave a *trans* epoxide, which was oxidized and methylated to the corresponding glycidic ester **7.2.1**. This was reacted with $MgBr_2$ to give the bromoester **7.2.2**, which was then substituted by azide ion to give azide **7.2.3** with the correct stereochemistry of the taxol side chain. Reduction and benzoylation gave the final product **7.2.4**. This procedure has the advantage of using readily available *trans*-cinnamyl alcohol instead of the scarcer *cis* isomer (*253*, *254*).

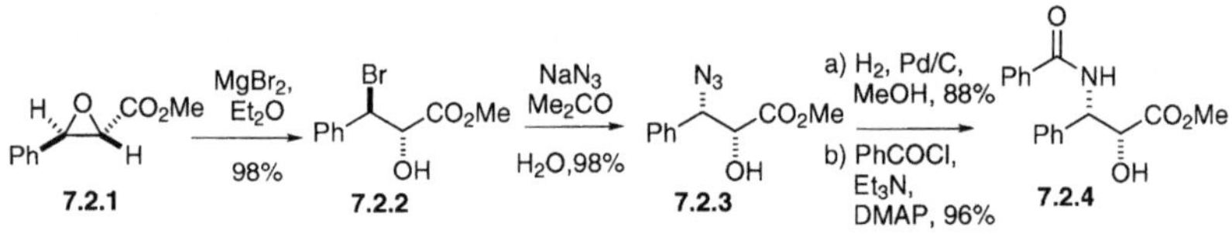

7.2.1 7.2.2 7.2.3 7.2.4

McChesney and his collaborators chose another approach, in which the easily accessible racemic trans epoxide **7.2.5** was opened by HCl to form chlorohydrin **7.2.6**, which was then converted to the cis-epoxide **7.2.7** with basic Amberlite 400. Opening of the cis-epoxide **7.2.7** with ammonia in MeOH followed by methanolysis of the resulting amide and benzoylation gave racemic taxol side chain methyl ester **7.2.8** (*255*, *256*).

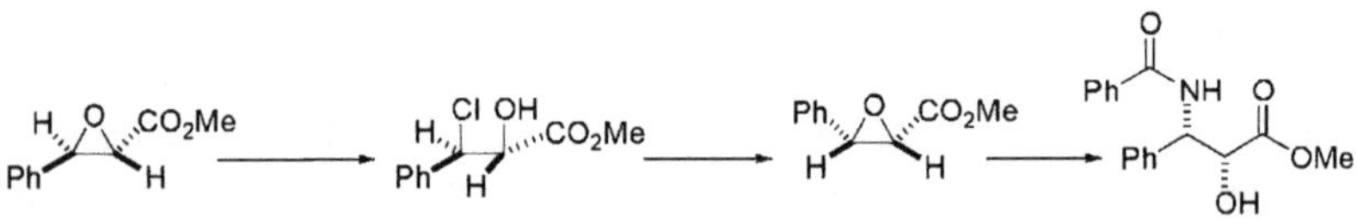

7.2.5 7.2.6 7.2.7 7.2.8 (±)

Catalytic asymmetric synthesis of the taxol side chain through an epoxide has also been achieved by Jacobsen and his collaborators. Thus ethyl *cis*-cinnamate, obtained from catalytic partial hydrogenation of the corresponding alkyne, was oxidized with sodium hypochlorite and 4-PPNO in the presence of 6 mol% of the readily available (R,R)-salen-Mn(III) complex **7.2.9**, yielding the cis-epoxide **7.2.10** in 95–97% enantiomeric excess. Subsequent transformations, which included regio-specific opening of the epoxide (*257*), gave the taxol side chain **7.2.11** in good yield (*258*).

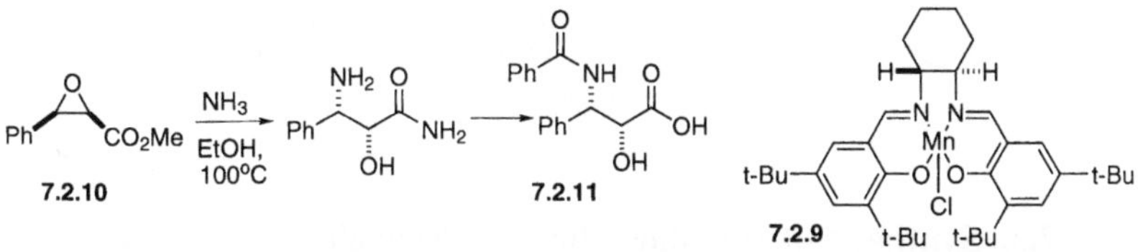

As an alternative functional group to an epoxide, cyclic sulfates and sulfites have also been used as intermediates to prepare the taxol side chain. Thus, asymmetric dihydroxylation of cis methyl cinnamate using AD-mix β gave the diol **7.2.12**, which was converted to the cyclic sulfate **7.2.13** with thionyl chloride followed by oxidation. The opening of **7.2.13** with azide proceeded in a similar fashion to that of an epoxide, giving the azide **7.2.14**, which was converted to the taxol side chain **7.2.15** (*259*).

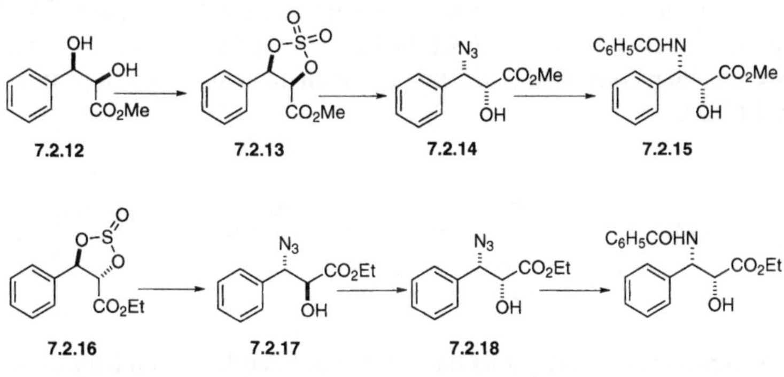

In another approach, commercially available ethyl *trans*-cinnamate was used as starting material, and similar chemistry to that described above was performed, except that the cyclic sulfite **7.2.16** was opened directly with azide instead of being oxidized to sulfate; the resulting azidoalcohol **7.2.17** had the opposite stereochemistry at C-2. The stereochemistry at C-2 was inverted by the Mitsonobu procedure to give azidoalcohol **7.2.18**, which was converted by standard methods to the taxol and docetaxel side chain esters (*260*).

Another synthesis of the taxol side chain started from ethyl cinnamate and used AD-mix α to generate the diol **7.2.19**. Conversion of **7.2.19** to the cyclic sulfite **7.2.20** was followed by ring opening with NH_4Br (*259*) or $MgBr_2$ (*261*) to give bromide **7.2.21**, which was converted by standard chemistry to the taxol side chain.

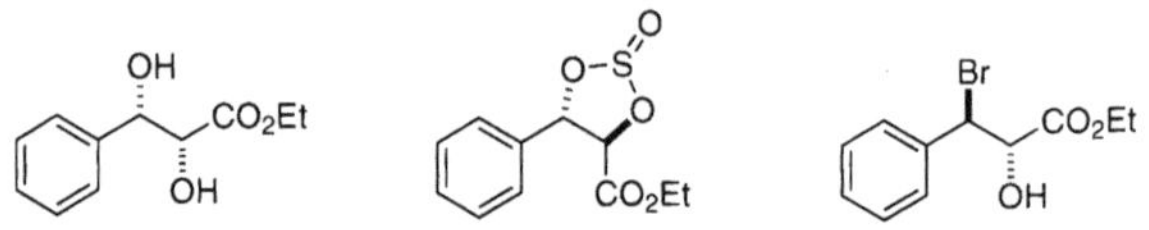

7.2.19 **7.2.20** **7.2.21**

7.3. Synthesis of the Side Chain by Sharpless Asymmetric Aminohydroxylation

Sharpless catalytic asymmetric aminohydroxylation (AA) provides a direct and efficient route to the taxol side chain as well as a number of other natural products and synthetic ligands. The required asymmetry of the taxol side chain was controlled by $(DHQ)_2$-PHAL, and hydroxylation was achieved with potassium osmate ($K_2OsO_2(OH)_4$). Various N-haloamide salts were investigated, and yields and percentage ee were good to excellent under different conditions. For example, reaction of methyl cinnamate under Sharpless catalytic AA conditions gave (2*R*,3*S*)-N-tosyl-3-phenyl isoserine **7.3.1** in 82% ee in 69% yield. Compound **7.3.1** was then converted to the taxol side chain acid in two steps in good yield (*262*).

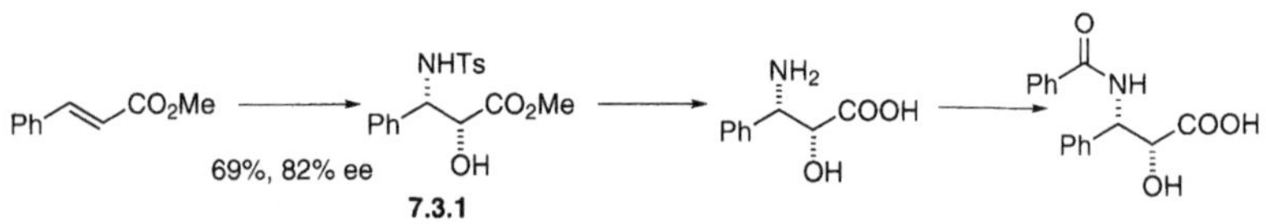

Several other investigators have used the Sharpless chemistry to make the taxol side chain or analogs of it. Thus an N-benzoyl side chain was prepared in one step from *trans*-isopropyl cinnamate using benzamide-based AA in 60% yield with 3:1 regioselectivity and 97% ee (*263*). Side chains with heterocyclic rings at C-3 have been prepared by AA; thus for example furan (*264*) and pyridine (*265*) analogs of the side chain have both been prepared by this method.

The Sharpless tosylamide **7.3.2** can be converted directly to the oxazolidine **7.3.3**, which can be coupled to baccatin III as described below (*266*).

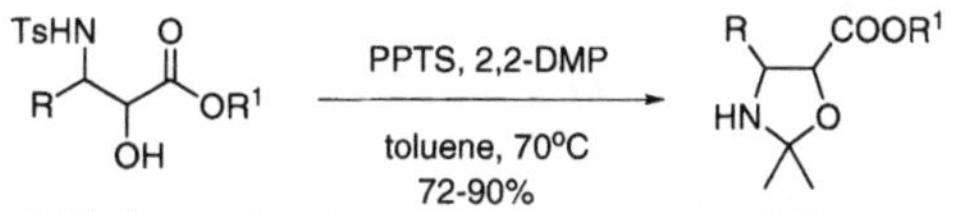

7.3.2 (R = alkyl, aryl) **7.3.3**

7.4. Synthesis of the Side Chain by Coupling Reactions

The approaches to the taxol side chain described so far have used either a cycloaddition reaction to form a β-lactam, or have installed functional groups on a preexisting carbon skeleton. Another possible approach is to couple two "halves" of the side chain in some sort of stereocontrolled coupling reaction, and several examples of this chemistry have been published.

In the absence of strong base and a good leaving group, a silyl enol ether reacts with an imine to give an open chain adduct instead of lactam. The outcome of the stereochemistry of the product can be controlled by the use of either chiral substrate or chiral catalyst. Yamamoto and Leon reported an enantioselective synthesis of the taxol side chain and other β-amino esters using double stereodifferentiation with both chiral imine and chiral catalyst. Thus, reaction of **7.4.1** and *Z*-**7.4.2** in the presence of a chiral boron reagent gave the desired (2*R*,3*S*) adduct **7.4.3** in 98% diastereomeric excess. Hydrogenation and benzoylation easily converted **7.4.3** to the taxol side chain methyl ester in good yield (*267, 268*).

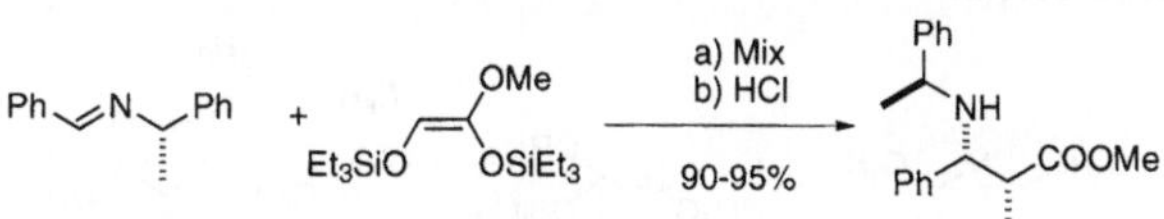

7.4.1 **7.4.2** **7.4.3**

Recently, Ueno *et al.* used the chiral zirconium catalyst **7.4.4** for asymmetric induction. Thus, reaction of imine **7.4.5** and silyl enol ether **7.4.6** in the presence of 10 mol% *S*-**7.4.4** gave the desired phenyl isoserine derivative **7.4.7** in 94% ee and 95% yield. The α-hydroxy-phenylamine which was required for the catalytic asymmetric induction was then cleaved oxidatively using CAN after being converted to an α-methoxy-phenylamine derivative. Aqueous hydrolysis of the resulting aminoester **7.4.8** yielded the amino acid quantitatively, which was benzoylated to give the taxol side chain (*269*).

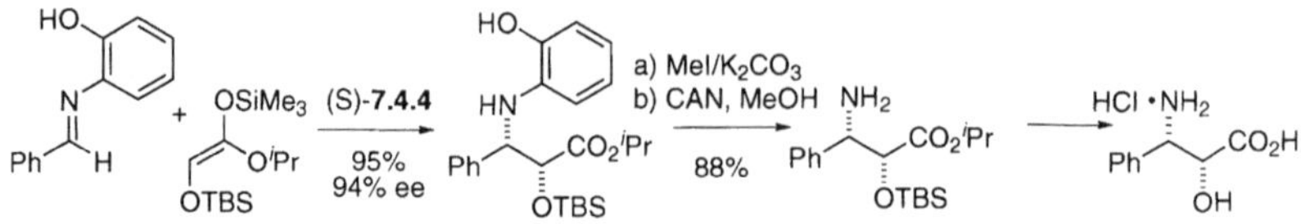

7.4.5 7.4.6 7.4.7 7.4.8

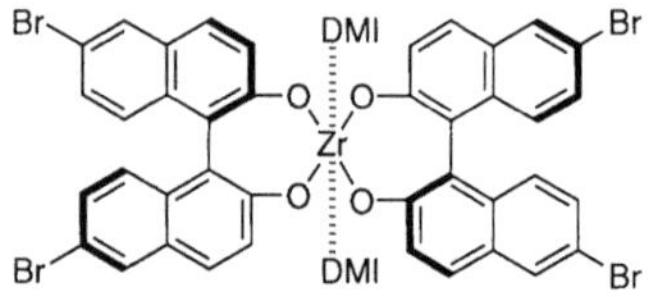

DMI = 1,2-dimethylimidazole

7.4.4

Greene *et al.* used a chiral camphor sulfonamide to achieve diastereoselective condensation of imine **7.4.9** and enolate of **7.4.10** to give **7.4.11**. This gave the C-2 benzoyl protected docetaxel side chain **7.4.12** in greater than 99% enantiomeric excess and 45% chemical yield (*270*). A *p*-methylbenzylidene acetal protected taxol side chain was also synthesized in a similar way in 99% enantiomeric excess and 45% yield (*271*).

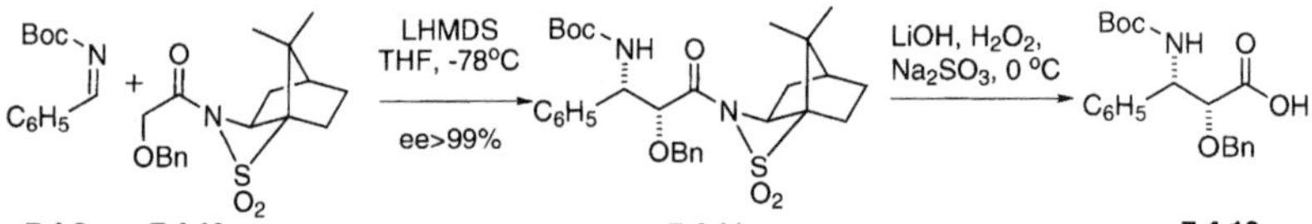

7.4.9 7.4.10 7.4.11 7.4.12

An enolate-imine coupling reaction has been used to give the side chain in 59% yield in the key step (*272*). In this approach the imine **7.4.13** was coupled with the enol **7.4.14** in the presence of $MgBr_2$ to give the adduct **7.4.15**; this compound was converted to side chain methyl ester in three steps.

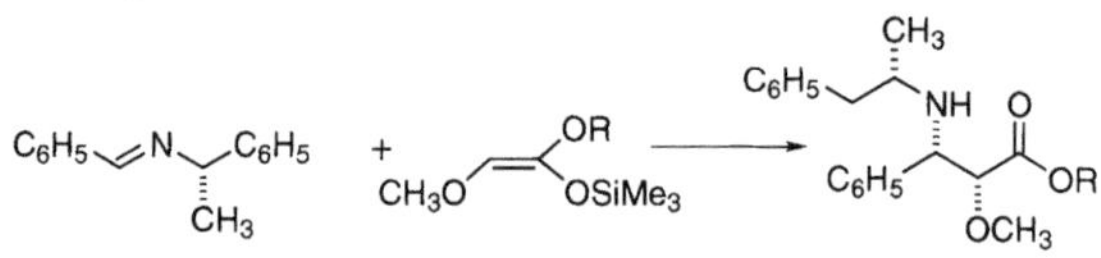

7.4.13 7.4.14 7.4.15

In another approach, lithiated allene **7.4.16** was coupled with aldehyde **7.4.17** to give adduct **7.4.18**, which could be ozonized to give side chain methyl ester in 28% overall yield (*273*).

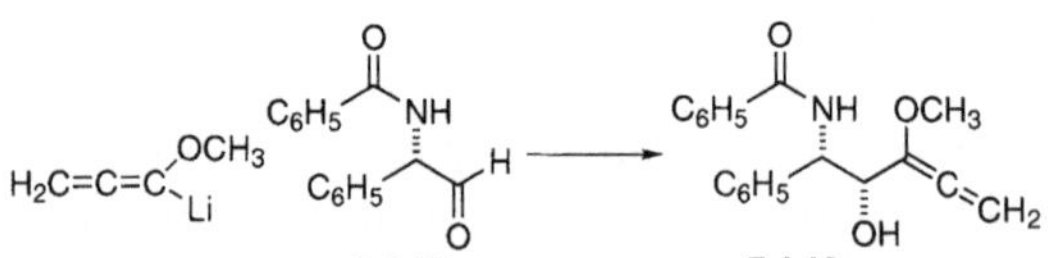

7.4.16 7.4.17 7.4.18

Finally in this area, addition of phenyl magnesium bromide to the N-benzylnitrone **7.4.19** gave the hydroxylamine **7.4.20**, which was converted to the enantiomer of the docetaxel side chain (*274*).

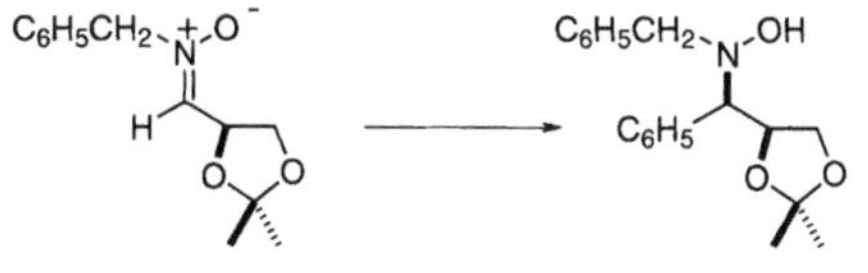

7.4.19 **7.4.20**

Coupling reactions have also been carried out in a reversal of the imine strategy, by adding a carbanion to an appropriate aldehyde. Thus addition of anion **7.4.21** to acrolein gave racemic adduct **7.4.22** which could be protected, oxidized, and resolved with (+)-ephedrine to give the protected acids **7.4.23** and **7.4.24** (*275*).

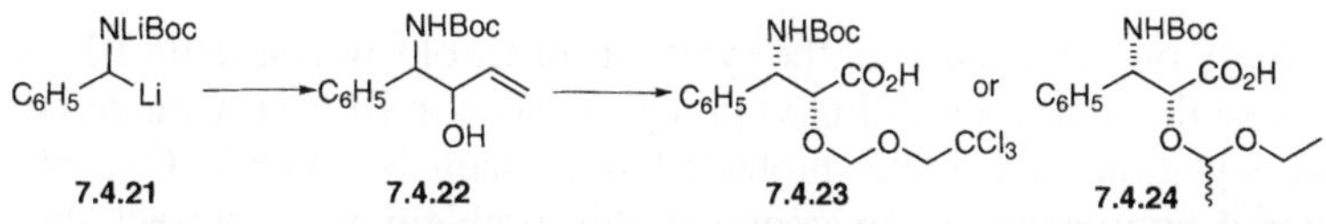

7.4.21 **7.4.22** **7.4.23** **7.4.24**

Addition of an enolate to a chiral aldehyde has also been used. In this interesting approach, addition of the enolate of **7.4.26** (formed in the presence of $TiCl_4$ and Et_3N) to the chiral benzaldehyde **7.4.25** gave the adduct **7.4.27**. The chromium was removed from **7.4.27** by treatment with tetrabutylammonium fluoride followed by photolysis, and the azide group was introduced by a Mitsunobu-type reaction to give azide **7.4.28**. This compound was then converted to the taxol side chain methyl ester by standard chemistry (*276*, *277*).

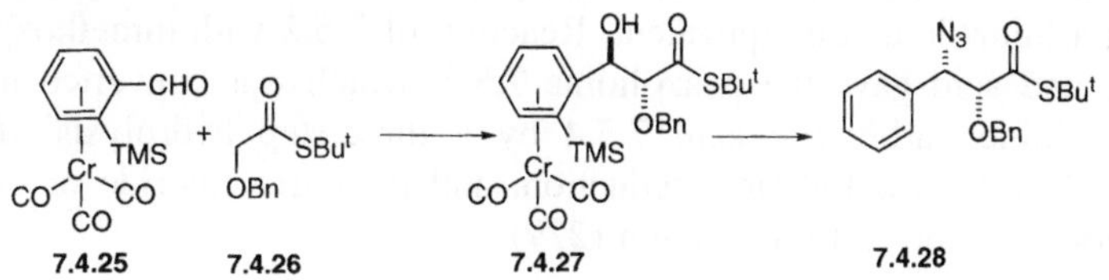

7.4.25 **7.4.26** **7.4.27** **7.4.28**

Mukaiyama and coworkers developed a new synthesis of the taxol side chain as a part of their synthesis of taxol (*278*). Reaction of the enol silyl ether **7.4.29** with benzaldehyde in the presence of the chiral diamine **7.4.30** and $Sn(OTf)_2$ and $n Bu_2Sn(OAc)_2$ gave the aldol adduct **7.4.31** in good yield and stereoselectivity. Conversion of **7.4.31** to the amine **7.4.32** was accomplished by Mitsunobu inversion and reduction of the resulting azide to the amine. Benzoylation gave **7.4.33**, and debenzylation with $SnCl_4$ followed by transesterification gave the methyl ester **7.4.34**.

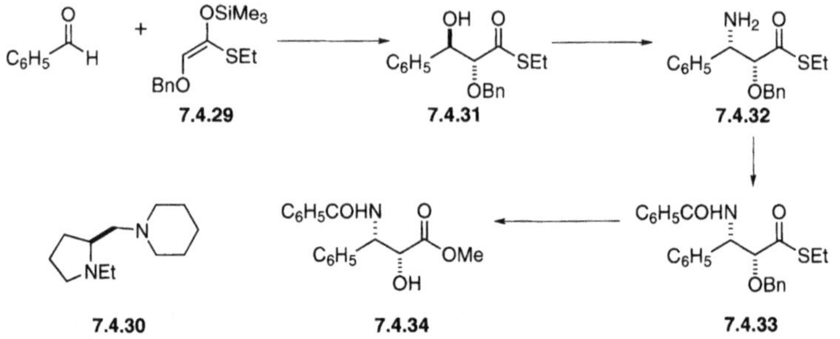

7.5. Synthesis of Cyclically Protected Side Chains

Early on in the study of the synthesis of taxol from baccatin III it was realized that the C-13 hydroxyl group of baccatin III is very hindered, so that acylation by a bulky protected side chain is difficult. One of the favored approaches to overcoming this problem is to protect the 2′-hydroxyl group through a cyclic structure, usually as an oxazolidine or an oxazoline. The synthesis of these protected side chains thus becomes an important goal.

The first cyclic protecting group introduced in this area was the oxazolidine group, and side chains with this protecting group built in from an early stage have been prepared by Dondoni and his colleagues using thiazole chemistry. Thus the N-acylphenylglycine ester **7.5.1** was reduced with DIBAL to the corresponding aldehyde, which was converted with good stereoselectivity to the adduct **7.5.2** by reaction with a thiazole anion equivalent. Reaction of **7.5.2** with dimethoxypropane and acid gave the oxazolidine **7.5.3**, which was converted to the oxazolidine carboxylic acid **7.5.4** by a three-step hydrolysis of the thiazole ring (methylation, reduction, and mercuric chloride or cupric chloride) followed by oxidation (*279*).

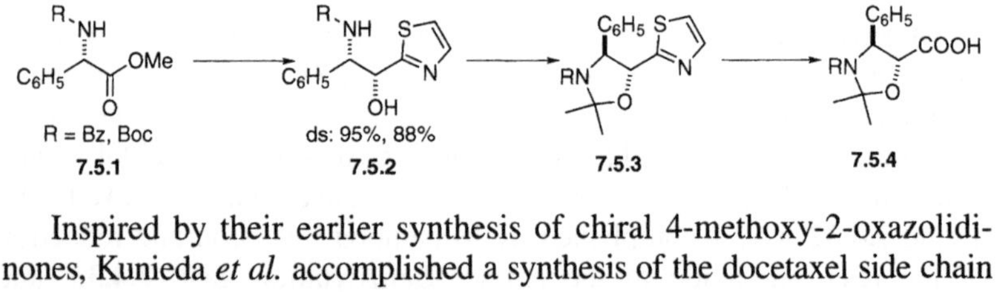

Inspired by their earlier synthesis of chiral 4-methoxy-2-oxazolidinones, Kunieda *et al.* accomplished a synthesis of the docetaxel side chain using this synthon. Thus bromomethoxylation of the chiral oxazolidone **7.5.5** gave 4-methoxy-5-bromo-oxazolidinone derivative **7.5.6** in 96% diastereomeric excess. Subsequent substitution of both the bromo and the methoxy group using organo-metallic reagents afforded **7.5.7** with retention of stereochemistry at C-4 and C-5. Further manipulation and protection of the amino alcohol as an acetonide gave **7.5.8** as a ready-for-coupling docetaxel side chain. The disadvantage of this route is the somewhat cumbersome introduction of the carboxylic acid functional group (*280*).

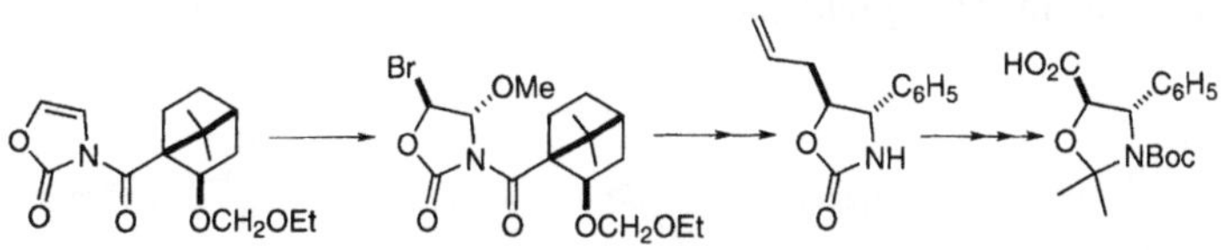

7.5.5 **7.5.6** **7.5.7** **7.5.8**

Following the original report that the oxazoline carboxylic acid **7.5.12** can serve as an excellent side chain precursor (*281*), several syntheses of this compound have been reported. Thus racemic oxazoline was prepared from protected cinnamyl alcohol by addition of phenylselenyl triflate and azide ion to give the adduct **7.5.9**, which was then converted to oxazoline **7.5.10** and ester **7.5.11** (*282*). A similar but more efficient route used diphenyldiselenide and ammonium persulfate to give the racemic oxazoline **7.5.10** in 95% yield from cinnamyl acetate (*283*).

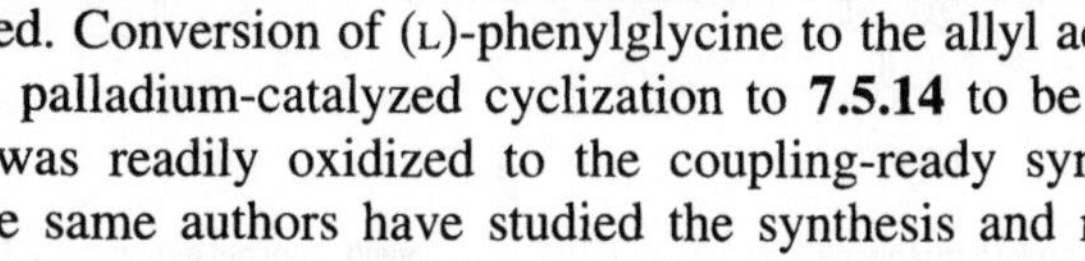

7.5.9 **7.5.10** **7.5.11** R = CH_3
7.5.12 R = H

The synthesis of chiral oxazolines from (L)-phenylglycine has been investigated. Conversion of (L)-phenylglycine to the allyl acetate **7.5.13** allowed a palladium-catalyzed cyclization to **7.5.14** to be carried out, and this was readily oxidized to the coupling-ready synthon **7.5.12** (*284*). The same authors have studied the synthesis and ring-opening reactions of oxazolines, and have used them as precursors to sulfur and nitrogen analogs of the side chain (*285*).

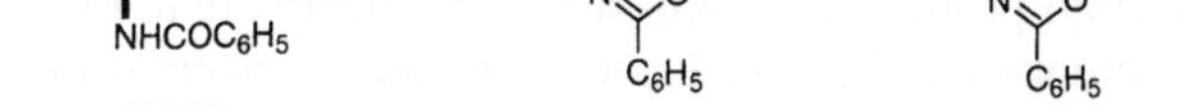

7.5.13 **7.5.14** **7.5.12**

7.6. Miscellaneous Syntheses

It was originally suggested by Swindell that acylation of baccatin III with protected N-benzoyl-3-phenylisoserine proceeds through an oxazinone intermediate (*286*). He then developed a synthesis of the side chain methyl ester by hydrolysis of a protected oxazinone intermediate **7.6.3**, which was itself prepared in 70% yield by a Diels-Alder type reaction between the imine **7.6.1** and the chiral ketene acetal **7.6.2**. Hydrolysis of **7.6.3** and removal of protecting groups gave the side chain methyl ester in 93% yield (*287*).

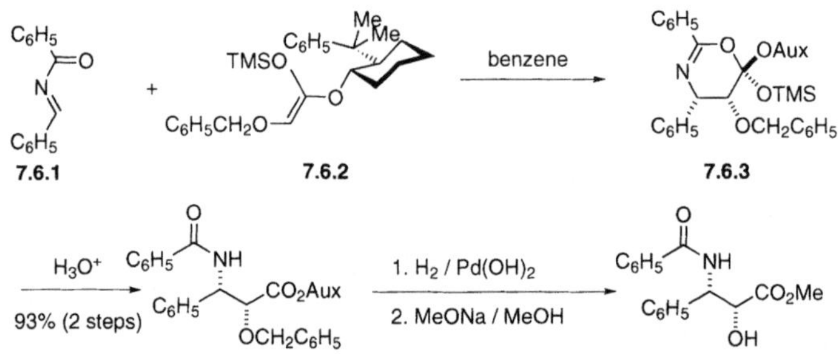

Several microbial processes have been applied to the synthesis of taxol side chain. Thus Patel *et al.* demonstrated that racemic α-keto-β-amino esters such as **7.6.4** can be reduced by enzymes isolated from *Hansenula* species to give the side chain ester **7.6.5** in 80–85% yield and 90–98% enantiomeric excess (*288*).

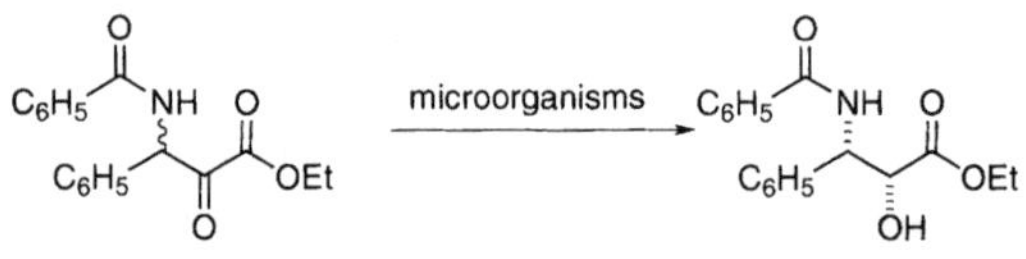

7.6.4 **7.6.5**

microorganisms	yield	optical purity
Hansenula polymorpha SC 13865	80%	99%
Hansenula fabianii SC 13894	85%	95%

Similarly, baker's yeast has also been shown to reduce α-keto-β-amino-3-phenyl propionate (**7.6.6**) diastereoselectively to give **7.6.7** as a single diastereomer in 72% yield; this was then converted to side chain methyl ester **7.6.8** (*289*).

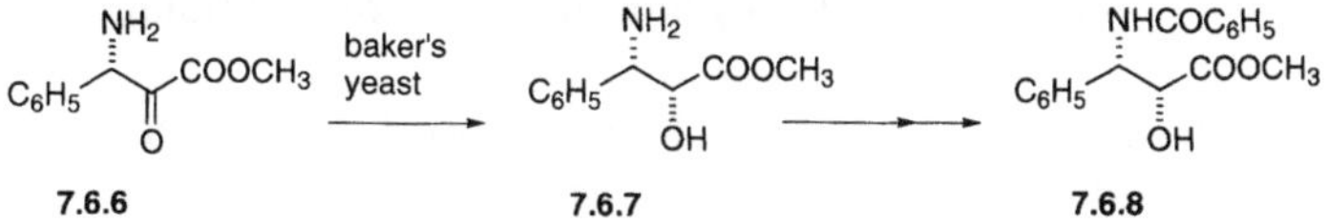

Reduction of the keto azide **7.6.9** with baker's yeast gave the diastereomeric alcohols **7.6.10** and **7.6.11** in about a 1:1 ratio with high (98%) ee (243).

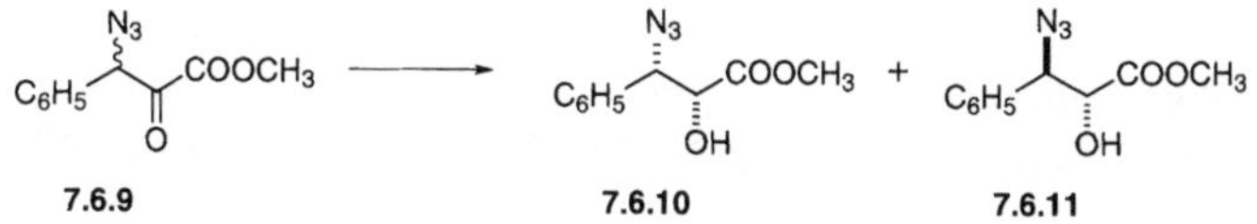

Lee and Kim used a lipase catalyzed transesterification to resolve racemic syn-1,2-diols of trans cinnamate **7.6.12**. Both enantiomers of the resolved compounds **7.6.13** and **7.6.14** were then converted to side chain methyl ester **7.6.19** in good yield (*290*). One route proceeded from diol monoacetate **7.6.13** through the tosylate **7.6.15** to the epoxide **7.6.16** and thence to the azide **7.6.17**. The second route converted the diol **7.6.14** to the bromoacetate **7.6.18** and thence to the azide **7.6.17**. Conversion of **7.6.17** to the final product **7.6.19** by standard methods completed the synthesis.

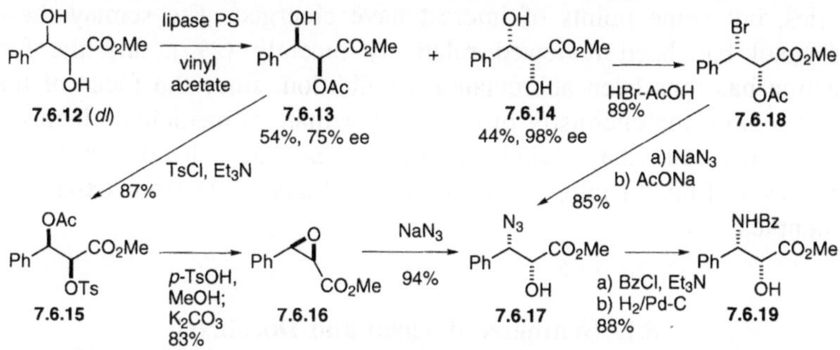

An enzymatic strategy was also used by Wuts and his collaborators in a strategy starting with the chloro-oxopropionate **7.6.20**. Reduction with MAP-10 lipase (out of six different enzymes tested) gave the best combination of yield and stereoselectivity for formation of chloroalcohol **7.6.21**, and subsequent steps followed established pathways to the epoxide **7.6.22** and thence to the amino ester **7.6.23** (*291*).

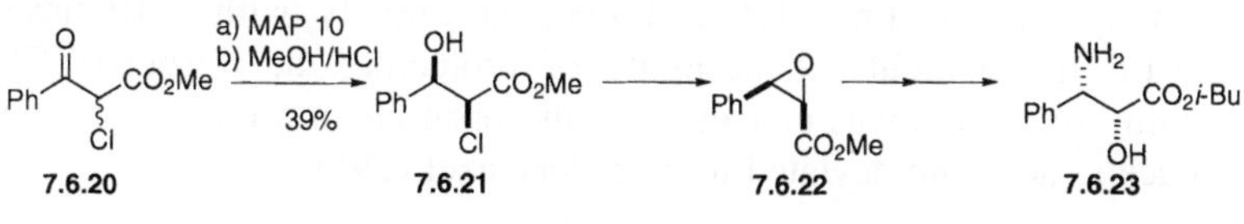

7.7. Syntheses of Phosphonate Side Chain Analogs

The phosphonate analog **7.7.3** of the taxol side chain has been prepared by addition of dimethyl phosphite to racemic *N*-Boc-phenylglycinal **7.7.1**, followed by crystallization to give the *syn*-diastereomer **7.7.2** in 50% yield. Resolution to give **7.7.3** was effected via the (*S*)-O-methylmandelate esters (*292*, *293*). Similar syntheses have been applied to the docetaxel side chain (*294*).

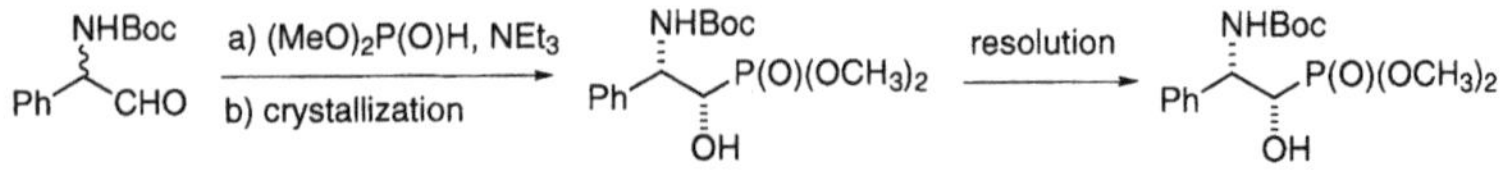

7.7.1 **7.7.2** (racemic) **7.7.3**

8. Synthesis of Taxol and Taxol Side Chain Analogs from Baccatin III

The synthesis of taxol or docetaxel requires the coupling of a protected side chain with a protected baccatin III. The procedures for this have not changed significantly since the previous review in this series, but some points of interest have emerged. The semisynthesis of taxol has been reviewed relatively recently (*295*), and the first section has thus been abbreviated. In addition, since the focus of this review is on the chemistry of taxol rather than its medicinal chemistry and structure-activity relationships, the sections dealing with the syntheses of taxol analogs will present illustrative rather than exhaustive examples.

8.1. Synthesis of Taxol and Docetaxel

The β-lactam synthon approach pioneered by Holton, Ojima, and Georg remains the most widely used method, and is the method currently used by Bristol-Myers Squibb for the commercial synthesis of taxol.

The oxazolidine approach has undergone some refinement, particularly in the nature of the protecting groups used. Thus trichloromethyl-substituted oxazolidines **8.1.2** and **8.1.3** could be prepared from the debenzoyl side chain **8.1.1** and coupled directly with 7,10-ditroc-baccatin III (without the necessity of protecting the nitrogen). The resulting oxazoline was converted to the open-chain product with zinc and acetic acid and acylated to give docetaxel (*296*).

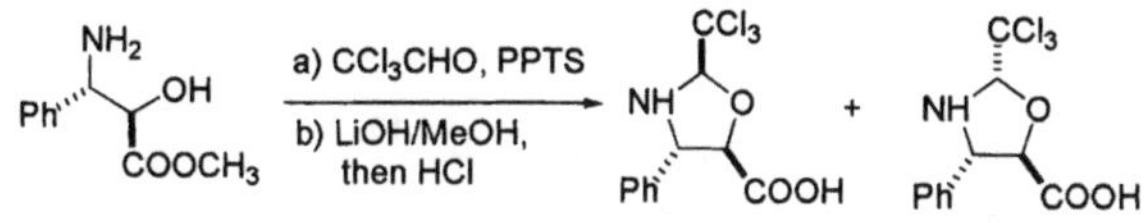

8.1.1 8.1.2 8.1.3

In another approach, Greene *et al.* found that compounds such as **8.1.4**, which had the "wrong" stereochemistry at the C-2′ position, gave a product **8.1.5** with the taxol configuration at C-2′ when coupled with 7,10-bis-troc-10-deacetylbaccatin III under DCC/DMAP conditions (*297*).

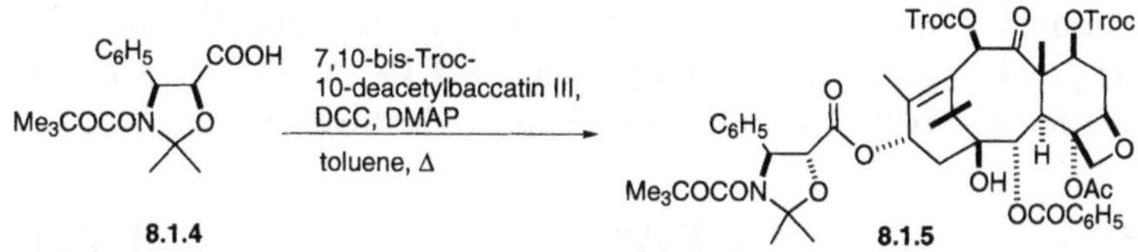

8.1.4 8.1.5

As noted earlier, the oxazoline **7.5.12** can serve as an excellent side chain precursor. It can be coupled with protected baccatin III in good yield, and the resulting cyclic product **8.1.6** can be converted to taxol under mild conditions. An attempt to modify this approach by the use of dioxo-oxathiazolidine **8.1.7** came to a surprising conclusion, in that coupling of **8.1.7** with protected baccatin III gave the same product **8.1.6** as had previously been prepared. Subsequent studies showed that the oxathiazolidine **8.1.7** was unstable and rearranged to the oxazoline **7.5.12** under the conditions of the reaction (*298*).

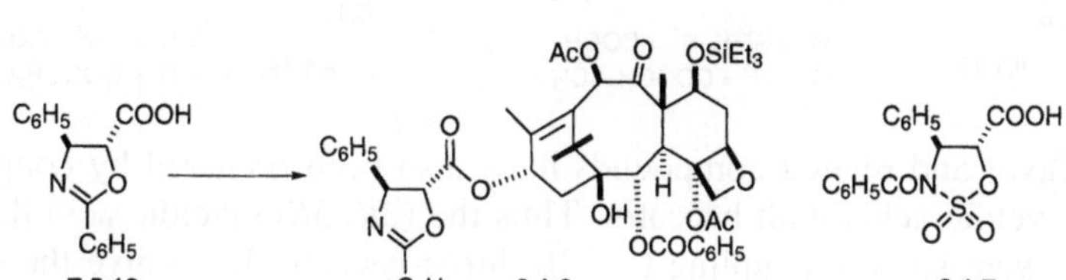

7.5.12 8.1.6 8.1.7

In a series of reports Gennari *et al.* demonstrated the synthesis of taxol side chain thioesters protected as oxazolines and oxazolidines. Thus reaction of the enolate of thioester **8.1.8** and TMS-imine **8.1.9** in the presence of chiral borate **8.1.10** gave the side chain thioester **8.1.11** in 60% yield. Compound **8.1.11** was then protected as its acetonide **8.1.12**. The oxazoline analog **8.1.13** was synthesized by a similar strategy with an inversion to achieve the desired stereochemistry at C-2 during the formation of the oxazoline ring (*299*).

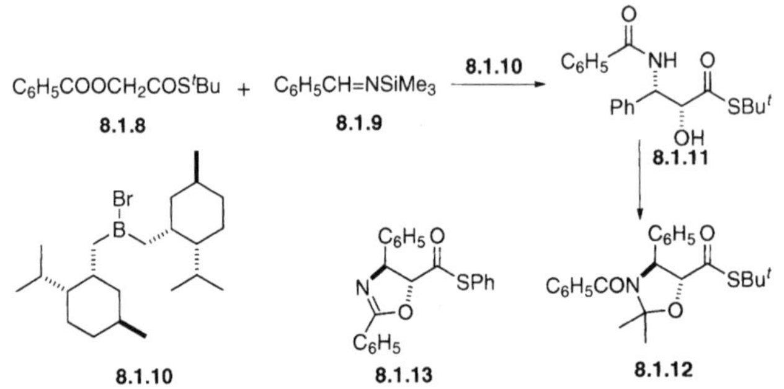

Coupling of the thioesters with protected baccatin III proceeded in good yield to give the coupled products **8.1.14** and **8.1.15**, which were converted to taxol or docetaxel by known chemistry (*299*, *300*).

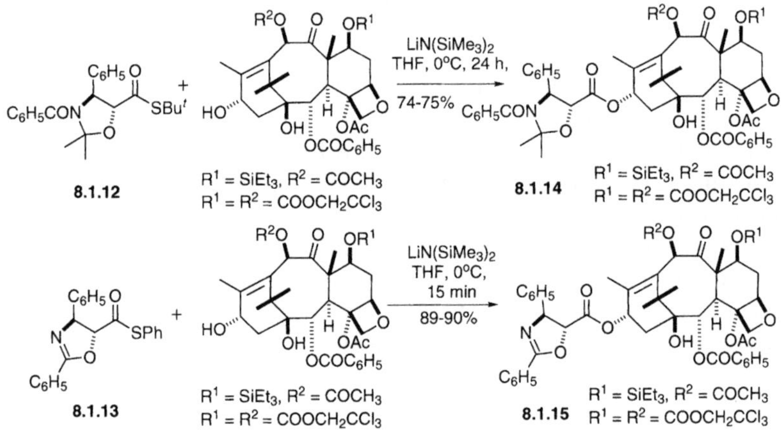

Taxol and related compounds have also been prepared by coupling cis-glycidic acids with baccatin. Thus the (2*R*, 3*S*) glycidic acid **8.1.16** underwent smooth coupling to 7,10-ditroc-baccatin III to give the derivative **8.1.17**, which could be converted to docetaxel in good yield by the successive steps of reaction with sodium azide, carbonate formation, and deprotection (*301*).

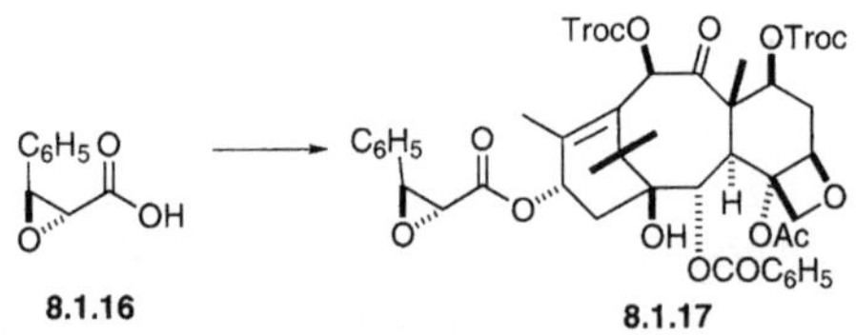

Another side chain attachment pathway takes advantage of the fact that *β*-ketoesters can be coupled readily with baccatin III (*302*). The *β*-ketoester **8.1.18** was converted to its O-benzyloxime, and then in a key step to the diazo derivative **8.1.19**. Reductive hydrolysis of the diazo group gave epimers **8.1.20** and **8.1.21** in a 1:2 ratio, and the major epimer **8.1.21** was successfully converted to taxol using oxazine ring formation to invert the stereochemistry at C-2 (*303*).

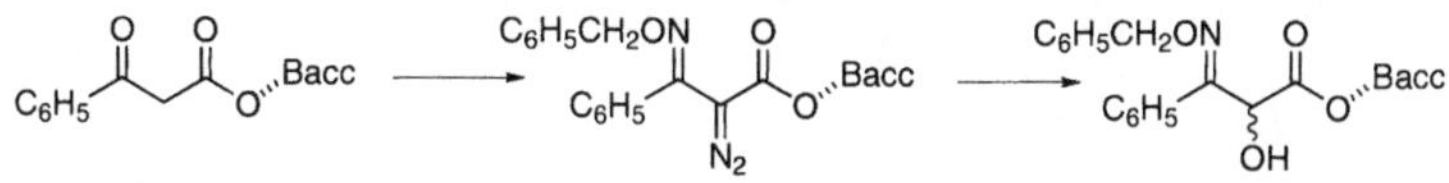

8.1.18 **8.1.19** **8.1.20** (α–OH) **8.1.21** (β–OH)

A procedure that has so far only appeared in the patent literature was developed by Wuts and Kelly (*304*). Amino acid **8.1.22** was protected as its sulfonamide derivative and then converted to the oxazolidine **8.1.23**, where Ar is 9-anthracenyl, 2-benzothiazolyl, or 4-nitrophenyl. Coupling of **8.1.23** with a 7-silyl-baccatin III gave the coupled product **8.1.24**, which could be converted to taxol by treatment with PhSH / DIPEA / DMF followed by benzoylation and desilylation.

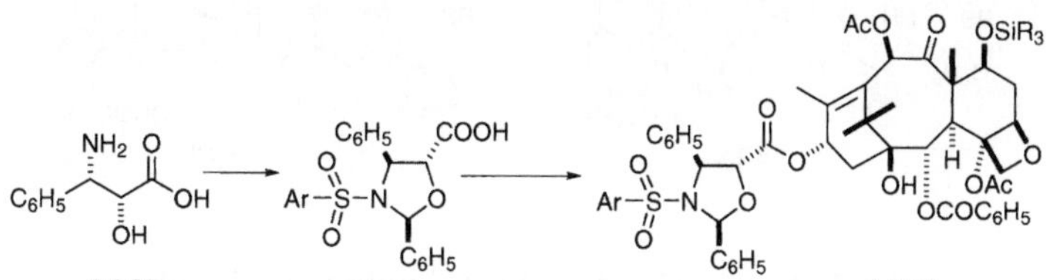

8.1.22 **8.1.23** **8.1.24**

8.2. Synthesis of N-Acyl Analogs of Taxol

The best-known N-acyl analog of taxol is of course docetaxel (Taxotere™, **1.1.2**), which differs from taxol in having a *t*-butoxycarbonyl group in place of the N-benzoyl group of taxol and in lacking the 10-acetate group of taxol. Docetaxel was discovered as a by-product of a synthetic approach to taxol (*41*), and it has taken its place beside taxol as a clinically used anticancer drug. Its discovery and chemistry have been reviewed (*305*).

A number of other simple N-acyl derivatives of taxol have been made. Thus Georg prepared various analogs with substituted N-benzoyl groups and analyzed their activity by the Topliss approach (*306*). She used two different synthetic approaches for this work, employing either an appropriately substituted β-lactam side chain or deprotecting the N-*t*-Boc group of 10-acetyldocetaxel with TFA or formic acid and reacylating on nitrogen. More recently Song and his collaborators have reported the synthesis of the analogs **8.2.1**, with the cyclopentenyl analog being significantly more cytotoxic than taxol (*307*).

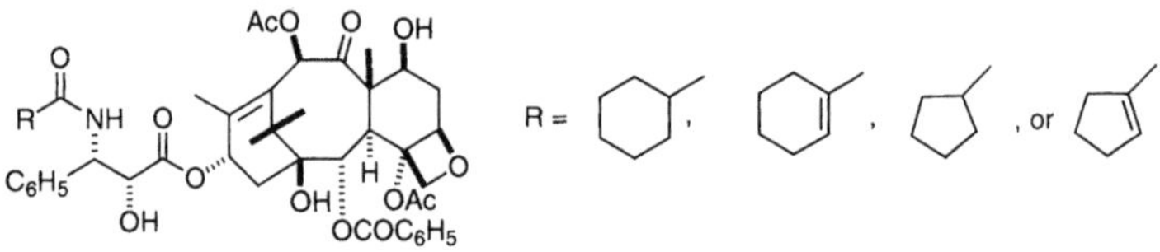

8.2.1

The Bristol-Myers Squibb group has reported on the synthesis and bioactivity of various 3′-*N*-thiourea and 3′-*N*-thiocarbamate-substituted taxol analogs (*308*). The thiocarbamate analogs of general type **8.2.2** and **8.2.3** proved to be the most active, with greater activities in tubulin-assembly and cytotoxicity assays than taxol. The corresponding thiourea analogs were all less active than taxol.

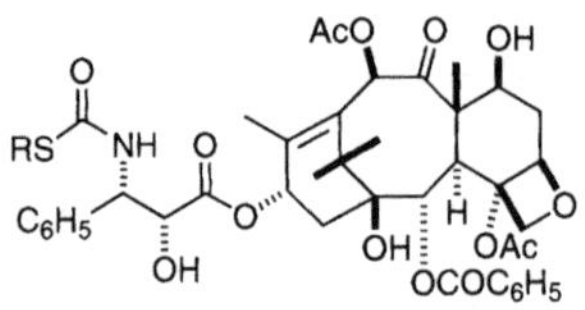

8.2.2

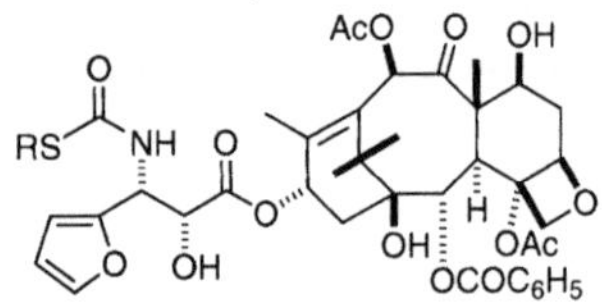

8.2.3

8.3. Synthesis of 3′-Aryl Analogs of Taxol

A significant number of 3′-aryl analogs of taxol have been made; many of these are reported only in the patent literature, which is not fully covered in this review. As one example, Holton used the β-lactam methodology to prepare a number of analogs such as the naphthyl analog **8.3.1** (*309*) and the furan and thiophene analogs **8.3.2** and **8.3.3** (*310*). The synthesis of 3′-furyl as well as 3′-pyridyl analogs of taxol was described by Georg (*311*), while the synthesis of 3′-pyridyl analogs was also reported by Holton (*312*). Ojima has reported the synthesis of a large number of 3′-analogs of taxol, including several with improved activity as compared with the parent drug; the β-lactam methodology was used in almost all of these cases. Thus 3′-alkyl and 3′-alkynyl analogs have been prepared (*313*), 3′-cyclopropa-alkyl and 3′-epoxy-alkyl analogs have been made (*314*), and various fluorine-containing analogs of paclitaxel and docetaxel were also made (*315*, *316*). In the

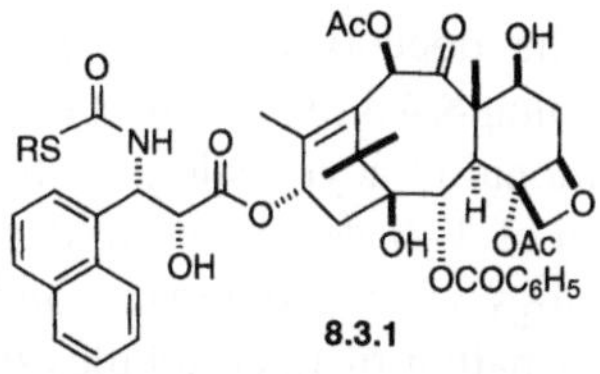

8.3.1

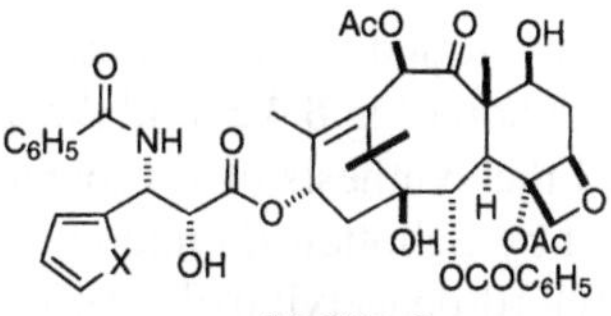

8.3.2 X = O
8.3.3 X = S

case of the 3′-trifluoromethyl analogs the chiral products could be obtained from racemic β-lactam by kinetic resolution (*317*). Bioassay results on the fluorine-containing taxoids showed that 3′-(4-fluorophenyl)-10-acetyldocetaxel (**8.3.4**) and 3′-(4-fluorophenyl)docetaxel (**8.3.5**) possessed improved cytotoxicities in several cell lines as compared with taxol, and the 3′-trifluoromethyl analogs also had improved activity, especially against the drug-resistant MCF7-R cell line (*316*). Molecular modeling and conformational studies based on NMR suggested that the fluorine of the 3′-(4-fluorophenyl) group of **8.3.4** and **8.3.5** significantly contributed to the formation of a "hydrophobic cluster" in protic solvents. This and other work from the Ojima group on fluorine-containing taxoids has been reviewed (*318*). Docetaxel analogs with modified 3′-phenyl groups such as **8.3.6**–**8.3.9** have also been prepared by Commerçon and his colleagues using the oxazolidine approach described earlier (*319*). The synthesis of some radiosensitizing 3′-analogs of taxol has also been described in the patent literature (*320*).

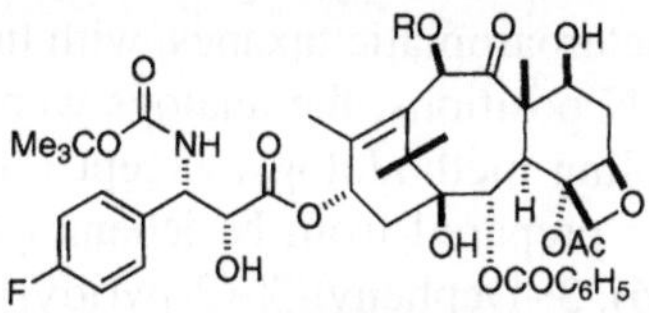

8.3.4 R = Ac
8.3.5 R = H

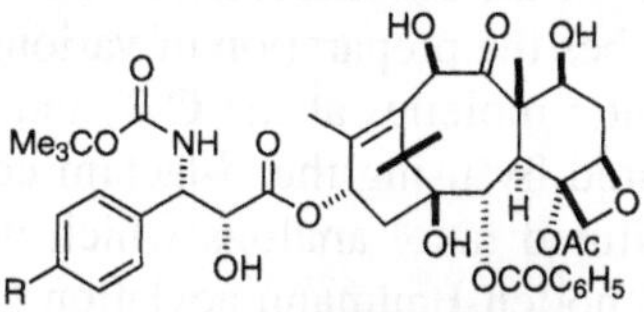

8.3.6 R = C_6H_5
8.3.7 R = $CH{=}CH_2$
8.3.8 R = CHO
8.3.9 R = CH_2OH

8.4. Taxol Analogs with Modified N-Acyl and 3′-Aryl Groups

Several workers have looked at the effects of modifying both the N-acyl and the 3′-aryl groups of taxol. Thus Swindell prepared analogs with either hydrogen or phenyl at the 3′-position and various N-acyl substituents (*321*), and Holton has also prepared many doubly substituted analogs, although most of these are reported only in the patent literature (*322*). More accessibly, Georg reported the synthesis of a range of 3′-(*t*-butyl)-N-acyl analogs of taxol using the oxazolidine route. Analogs were tested for cytotoxicity against B16 melanoma cells; the most active derivatives found in this study were the docetaxel analog **8.4.1**, the docetaxel homolog **8.4.2**, and the urea analog **8.4.3** (*323*). The same group reported the synthesis of seventeen additional N-acyl analogs of butitaxel prepared by Schotten-Baumann acylation of 3′-*N*-debenzoyl-3′-dephenyl-3′-*t*-butyl-10-deacetyltaxol. Various analogs were synthesized and evaluated for their ability to stimulate the formation of microtubules *in vitro*, their cytotoxicity towards B16 melanoma cells, and their water solubility. N-Debenzoyl-N-(2-thienoyl)butitaxel (**8.4.4**) was found to be the most potent analog, possessing ca. 2-fold better tubulin-assembly properties and cytotoxicity against B16 melanoma cells compared with taxol; it was also ca. 25 times more water-soluble than taxol (*324*).

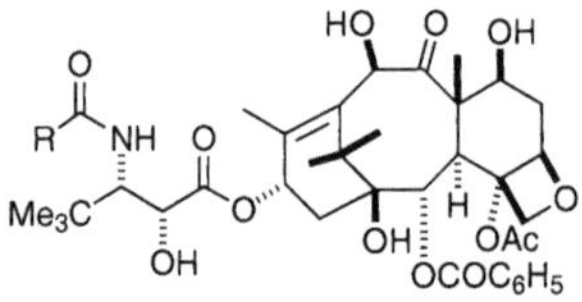

8.4.1 R = Me_3CO
8.4.2 R = Me_3CH_2CO
8.4.3 R = Me_3NH

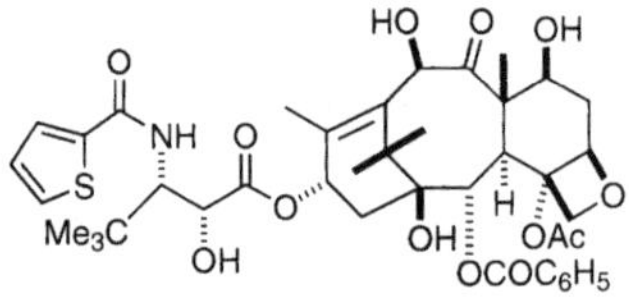

8.4.4

The Georg group also described the synthesis of various 3′-phenyl taxanes, prepared by the β-lactam route from appropriately substituted imines (*325*). The nine analogs prepared were almost all less active than taxol in both tubulin assembly and cytotoxicity assays, although the effects of the substituents on activity were relatively slight. A later paper describes the preparation of various heteroaromatic taxanes, with heteroaromatic moieties at the C-3′ and 3′-N positions; the analogs were also prepared by using the β-lactam coupling methodology, except for 3′-N substituted taxol analogs which were prepared from N-debenzoyltaxol via Schotten-Baumann acylation (*326*). 3′-Dephenyl-3′-(2-pyridyl)taxol, 3′-dephenyl-3′-(2-furyl)taxol, N-*t*-BOC-3′-dephenyl-3′-(2-furyl)taxol, 3′-dephenyl-3′-(2-furyl)-N-(hexanoyl)taxol, and N-debenzoyl-N-(3-furoyl)-taxol were found to be more cytotoxic than taxol to B16 melanoma cells.

Ojima and his colleagues have also made significant contributions in this area. Thus in addition to the 3′-(4-fluorophenyl)-10-acetyldocetaxel analogs reported in the previous section, they prepared analogs such as 3′-(4-fluorophenyl)-N-(4-fluorobenzoyl)taxol by the β-lactam method (*315*).

8.5. Taxol Analogs Modified Both on the Side Chain and the Ring System

The synthesis of analogs modified both on the side chain and the ring system is discussed in the section dealing with the major ring system modifications.

8.6. Synthesis of Taxol Analogs with Highly Modified Side Chains

A number of non-standard side chains have been prepared as part of various ongoing efforts to define the structure-activity relationships of taxol, and these are discussed in this section.

2′-Epitaxol **8.6.2** was prepared by formation of the oxazoline **8.6.1** from taxol, followed by hydrolysis (*327*).

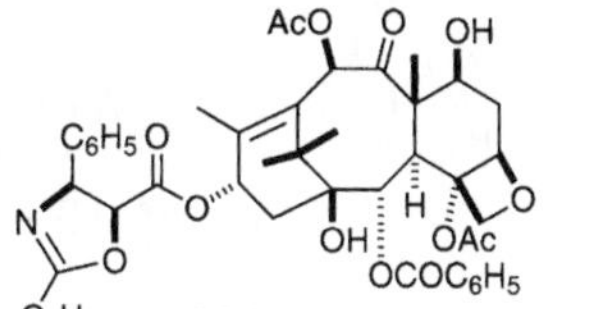

8.6.1

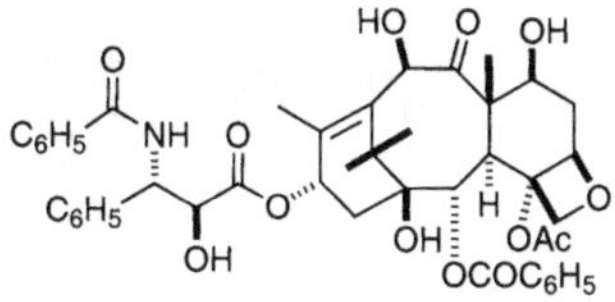

8.6.2

Kant and his collaborators reported the synthesis of 2′-methoxytaxol (**8.6.3**), 2′-desoxytaxol (**8.6.4**) and 2′-fluorotaxol (**8.6.5**). Treatment of taxol with silver oxide and methyl iodide in acetone at ambient temperature for 4 days afforded 2′-methoxytaxol; while 2′-desoxytaxol was prepared by using Barton's deoxygenation protocol. The synthesis of 2′-fluorotaxol was accomplished following the β-lactam methodology, using the appropriate fluorinated β-lactam to react with 7-(triethylsilyl)baccatin III. All of the analogs were at least 75-fold less cytotoxic than taxol (*328*).

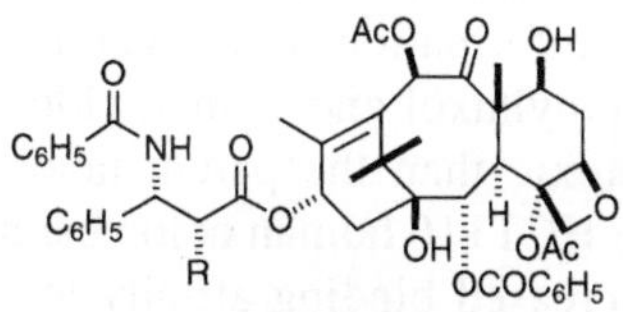

8.6.3 R = OMe
8.6.4 R = H
8.6.5 R = F

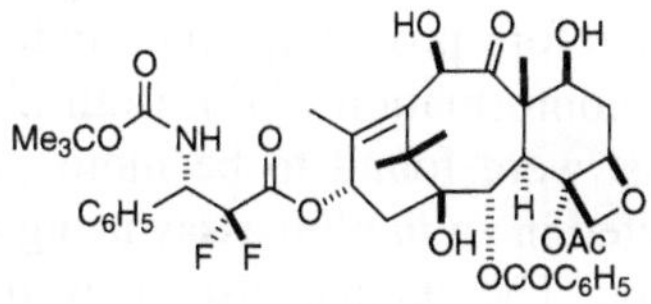

8.6.6

Uoto *et al.* synthesized a number of 2′-deoxy-2′-(difluoro)docetaxel derivatives. The parent compound **8.6.6** was prepared by coupling 7,10-(di-Troc)-10-deacetylbaccatin III with 3-(*t*-butoxycarbonylamino)-2,2-difluoro-3-phenylpropanoic acid in the presence of di-2-pyridylcarbonate and DMAP. In an *in vitro* study using P388 tumor cells, **8.6.6** had a GI_{50} value of 21.0 ng/mL compared with 30.4 ng/mL for taxol (*329*).

The synthesis of analogs with a homologated side chain was reported by Georg and her coworkers (*330*). The synthetic scheme began with phenylglycine **8.6.7**, which was converted to the allyl derivative **8.6.8** and thence to the oxazolidine **8.6.9**. This compound was converted to the taxol analogs **8.6.11** and **8.6.12** through the coupled product **8.6.10**. Both analogs were significantly less active than taxol.

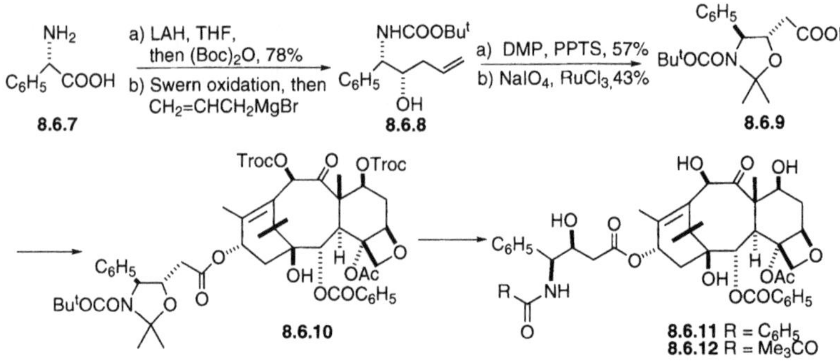

Methylated side chain analogs such as **8.6.13** have been prepared by Greene and his colleagues (*331*), by Kant and his colleagues (*332*) and by Ojima and his colleagues (*333*). The first group used niobium trichloride chemistry to prepare the side chain from an imine and pyruvic acid, while the latter two groups used β-lactam chemistry. Kant *et al.* prepared the β-lactam by diastereoselective addition of a Grignard reagent to azetidin-2,3-dione, while the Ojima *et al.* used a pathway involving a stereoselective alkylation of a 3-siloxy-β-lactam. Part of the logic for the introduction of a methyl group at the C-2′ position was that it would create some additional torsional strain (vs. H) for rotation around C-2′/C-3′, thus reducing the number of viable conformations, and therefore providing some indirect information on the nature of the active conformation (*332*). Both 2′-methyltaxol and 2′-methyldocetaxel analogs were found to be more cytotoxic than the parent taxol when evaluated in an *in vitro* assay using the HCT116 human colon carcinoma cell line. The analogs also showed increased binding affinity to microtubules compared to taxol. The enhanced potency due to the presence of the C-2′ methyl group may be a result of the postulated reduction in the degree of freedom of rotation at the C-2′/C-3′ bond, or simply of some additional hydrophobic binding interaction of the methyl group with the microtubule binding site.

Of particular interest is the conformationally restricted side chain analog **8.6.14** prepared by Appendino and his collaborators. This was converted into the taxol analog **8.6.15**, which had comparable cytotoxicity to taxol in several cell lines (*334*).

Chen *et al.* reported the synthesis of taxol analogs which were isomeric with docetaxel by virtue of having a 3′-*t*-butylaminocarbonyloxy group in place of the normal 3′-*t*-butoxycarbonylamino group

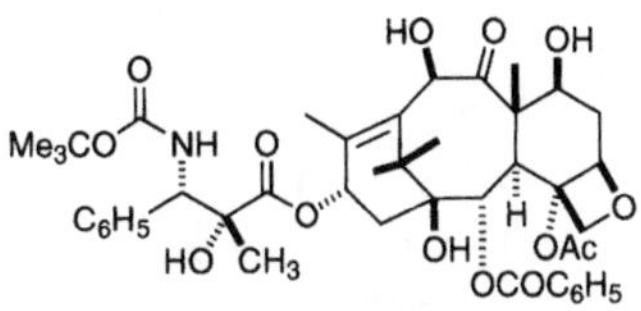

8.6.13

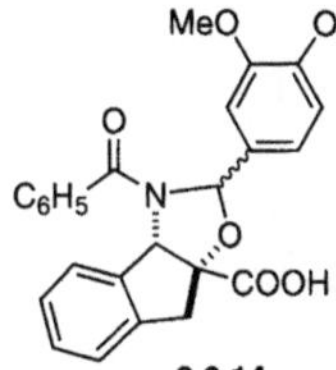

8.6.14

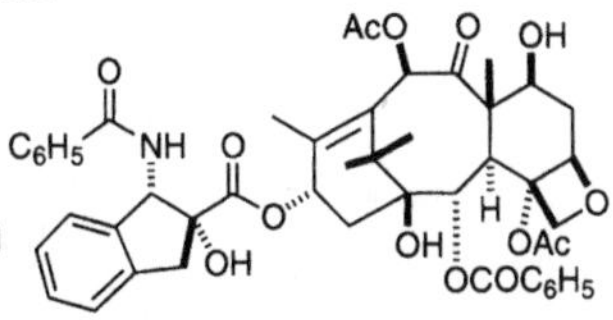

8.6.15

(*335*). Distereoselective dihydroxylation of the C-13 cinnamoyl derivative **8.6.16** was carried out using the Sharpless protocol, and provided the desired (2′*R*,3′*S*) diol as the major product in a 4:1 ratio. Conversion of the 2′-triethylsilyl derivative of the diol derived from **8.6.16** to the corresponding C-3′ carbonate was achieved using Hünig's base and *p*-nitrophenylchloroformate, and reaction of the carbonate with t-butylamine followed by deprotection provided the final target **8.6.17**. Deacylation at C-10 then gave the 10-deacetyl analog **8.6.18**. Analog **8.6.17** was 1.5-fold and 3.3-fold less potent than taxol in a tubulin polymerization assay and an *in vitro* cytotoxicity assay, respectively. In an *in vivo* experiment against the ip implanted M-109 lung carcinoma, **8.6.15** showed equivalent *in vivo* activity to taxol, with a T/C value of 166% (50 mg/kg/injection).

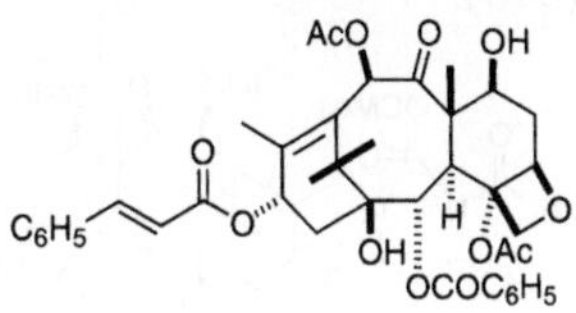

8.6.16

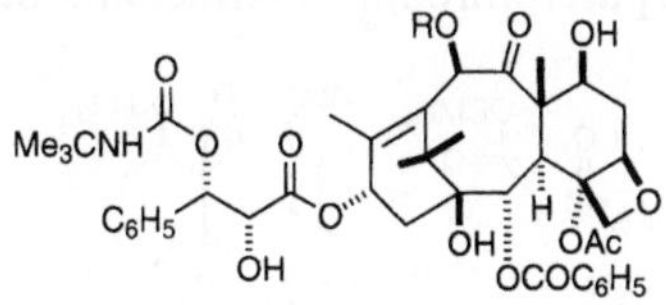

8.6.17 R = Ac
8.6.18 R = H

8.7. Side Chain Chemistry

A few reactions of the taxol side chain are of interest. The taxoid cephalomannine (**8.7.1**) co-occurs with taxol, and there is thus a potential for increasing the supply of taxol by using cephalomannine as a substrate. A simple conversion of cephalomannine to taxol was achieved by benzoylation and ozonolysis to the ketoamide **8.7.2**, followed by treatment with *o*-phenylenediamine under acidic conditions. The *o*-phenylenediamine induces a cleavage of the ketoamide to the amine, and intramolecular O→N acyl transfer gives rise to taxol in good overall yield (*336*).

Cephalomannine is not easily separated from taxol by chromatographic methods, but three chemical methods for accomplishing this

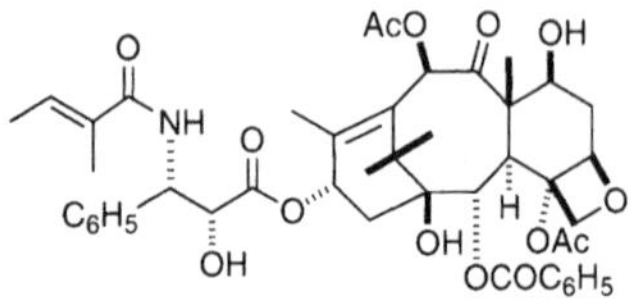

8.7.1

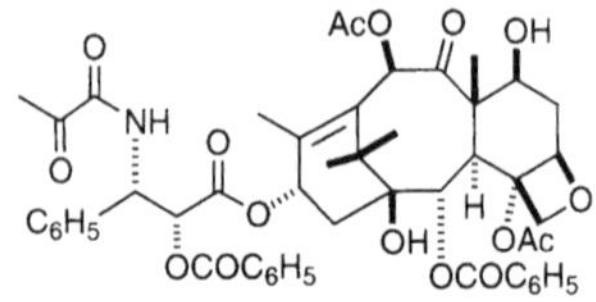

8.7.2

have been developed. All of these methods rely on the much greater chemical reactivity of the tigloyl double bond in the side-chain of cephalomannine as compared to the 11,12 double bond in ring A. The first method used selective osmylation of the tigloyl double bond followed by chromatography to effect the separation (*337*). The second method used ozonolysis and Girard's reagent to prepare a water-soluble derivative of cephalomannine (*338*), while the third method used the readily prepared dibromide derivatives of cephalomannine followed by chromatography to separate the two compounds (*339*).

Finally, selective N-debenzoylation of taxol has been achieved by reaction with di-t-butyl dicarbonate with 2′-(benzyloxycarbonyl)-7-(triethylsilyl)taxol to give the carbamate **8.7.3**, followed by treatment with base and deprotection. Using this chemistry, taxol can be converted to docetaxel in 39% yield, and various N-acyl analogs of taxol can also be prepared through intermediate **8.7.4** (*340*).

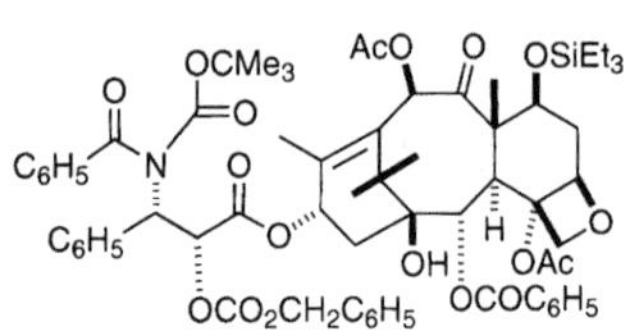

8.7.3

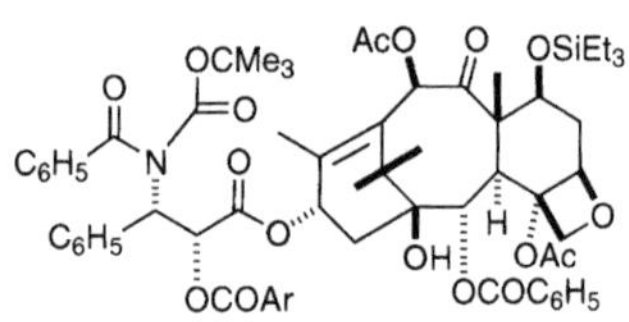

8.7.4

9. Taxol Metabolites

The metabolism of taxol and other taxoid drugs has been reviewed (*341–343*), and this section will thus primarily cover the synthesis of taxol metabolites. The major identified metabolites of taxol in the rat were found in the bile, and two of these were the phenolic metabolites **9.1** and **9.2**. The major taxol metabolite in man, which is excreted in the bile, is 6α-hydroxytaxol (**9.3**) (*344*), but the metabolites **9.3**, **9.4**, and **9.1** have all been identified in human feces (*345*). A study of taxol metabolites in rat bile by LC-MS identified nine metabolites, seven of which were hydroxylated compounds with the other two being a deacetyltaxol and a baccatin derivative (*346*). The metabolites were not completely characterized, but three were dihydroxylated derivatives with hydroxylation on the ring and on the side chain (two metabolites) or on the side chain only. The other four monohydroxylated metabolites were hydroxylated on the ring system (two compounds) or on the side chain (two compounds); the known compounds **9.1–9.3** were identified as three of the metabolites.

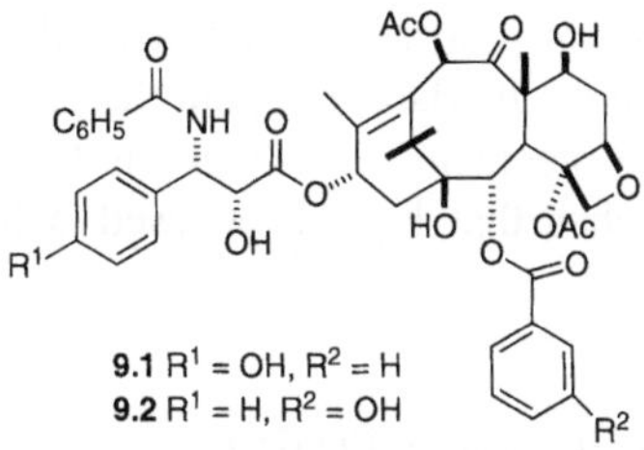

9.1 R^1 = OH, R^2 = H
9.2 R^1 = H, R^2 = OH

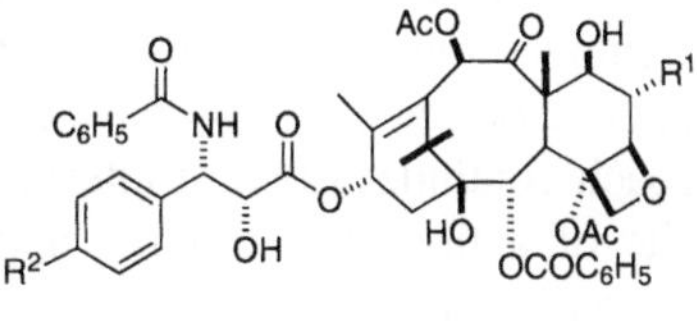

9.3 R^1 = OH, R^2 = H
9.4 R^1 = R^2 = OH

The metabolites **9.1** and **9.2** were both synthesized by Georg and her collaborators by standard methods, using an appropriate β-lactam to prepare **9.1** from baccatin III and converting taxol to a protected 2-debenzoyltaxol and then acylating it to prepare **9.2** (*347*). The 3′-metabolite **9.1** was about twice as active as taxol in a tubulin-assembly assay, but was sevenfold less cytotoxic than taxol to B-1 melanoma cells. The metabolite **9.2** was about 40 times less cytotoxic than taxol to L1210 leukemia cells. The synthesis of the major human metabolite **9.3** was described earlier in section 4.2.1.

The major human metabolites of docetaxel were found to be the side-chain oxidation products **9.5**, **9.6**, and **9.7** (*348*), and these were synthesized by Commerçon and his colleagues by appropriate conversions of an N-debenzoyl 10-deacetytaxol derivative (*349*).

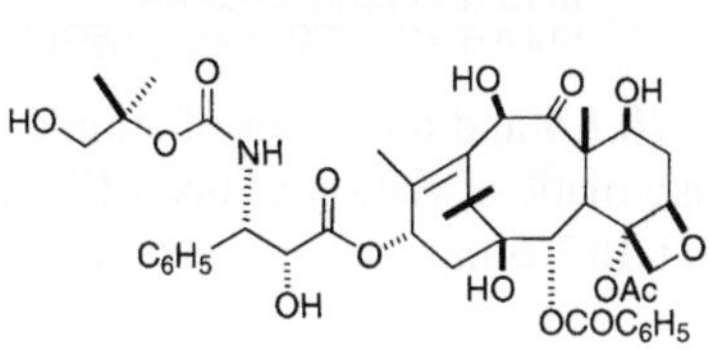

9.5

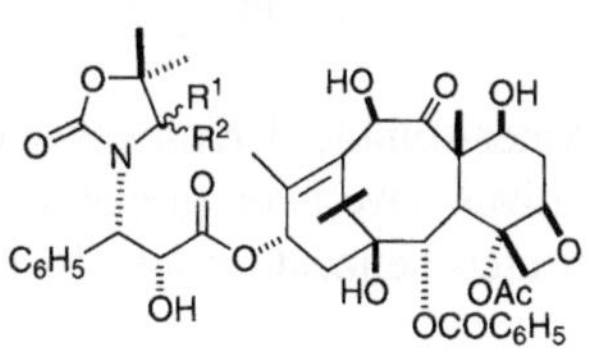

9.6 R^1 = R^2 = O
9.7 R^1, R^2 = H, OH

10. Taxol Analogs and Prodrugs

The limited water-solubility of taxol, coupled with its limited availability in Nature, almost doomed the development of taxol as an anticancer drug, and the fact that it was developed at all is a tribute the scientists at the National Cancer Institute, and especially to Dr. Matthew Suffness who championed its development when it was not at all clear that it would be the effective agent that it is now known to be.

Because of taxol's limited water-solubility, and because its clinical use initially required a 24-hour infusion, there was a great deal of interest in developing more water-soluble prodrugs. With experience, however, clinicians found that the infusion times could be reduced to 3 hours, and this removed much of the initial impetus for prodrug development. At this time, therefore, no prodrugs have entered the market place, but the prodrug approach continues to be of interest as a mechanism for targeting taxol to its site of action. The prodrug approaches that have been used will be discussed below.

10.1. Simple Ester Derivatives and Prodrugs

A number of different simple ester prodrugs have been prepared. In addition to the 2′-esters described in the previous review (*166, 350–352*), various other simple ester derivatives have been prepared such as the ester **10.1.1** (*353*) and various amino derivatives such as **10.1.2–10.1.3** (*354*); in this latter case the 3′-phenyl group was also replaced by a number of different aryl groups, although the best activity was reported for the analog **10.1.3** with Ar = Ph.

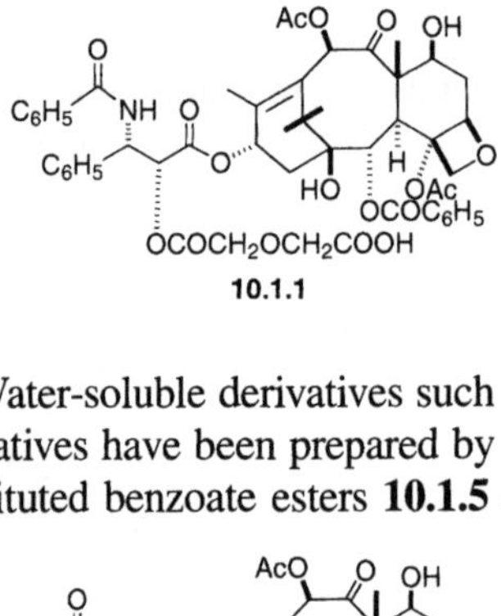

10.1.1

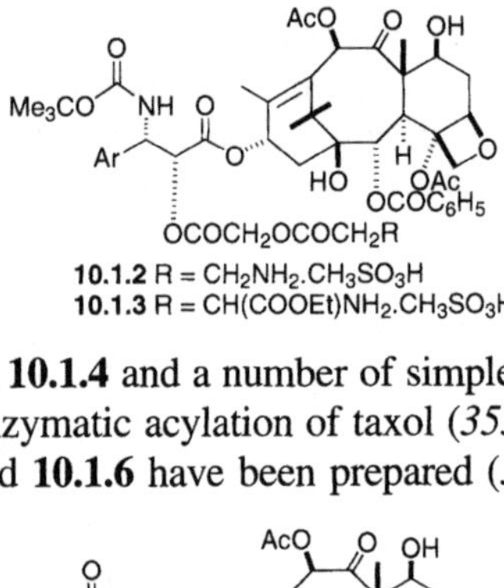

10.1.2 R = $CH_2NH_2.CH_3SO_3H$
10.1.3 R = $CH(COOEt)NH_2.CH_3SO_3H$

Water-soluble derivatives such as **10.1.4** and a number of simpler ester derivatives have been prepared by enzymatic acylation of taxol (*355*), and substituted benzoate esters **10.1.5** and **10.1.6** have been prepared (*356*).

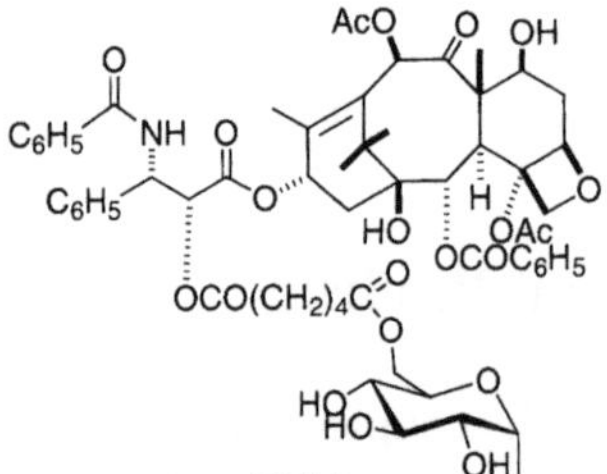

10.1.4

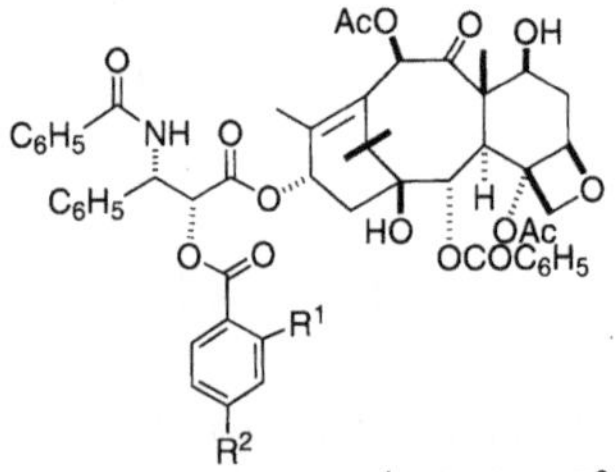

10.1.5 R^1 = COONa, R^2 = H
10.1.6 R^1 = H, R^2 = COONa

Both mono- and bis-malic acid esters of taxol have been made, and the sodium salt of the 2′-malyl derivative **10.1.7** showed enhanced activity and less toxicity as compared with taxol in the P-388 *in vivo* mouse leukemia model (*357*). Simple esters such as **10.1.8** have been prepared to enhance association with a liposomal lipid carrier in plasma (*358*), and the carbonate derivative **10.1.9** and related compounds were found to be as active as taxol in the M109 system (*359*).

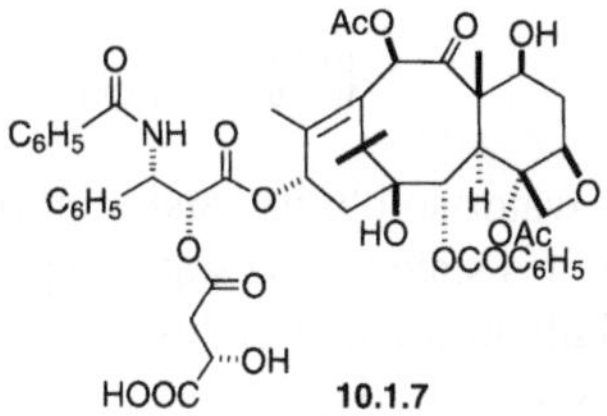

10.1.7

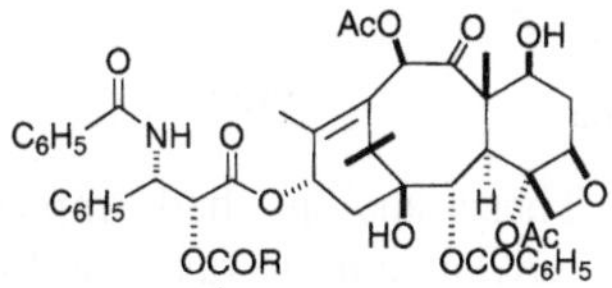

10.1.8 R = $(CH_2)_5CH_3$
10.1.9 R = OCH_2CH_3

10.2. Phosphate Ester and Other Prodrugs

Phosphate esters of various kinds have been developed by a number of groups, including the Bristol-Myers Squibb group, as useful prodrugs of taxol. The simple 2′- and 7-substituted analogs **10.2.1** and **10.2.2** were prepared by standard chemistry, but were found not to be good prodrugs because of their marginal *in vivo* antitumor activity against the M109 murine tumor model and because of their slow hydrolysis by phosphatase (*360*).

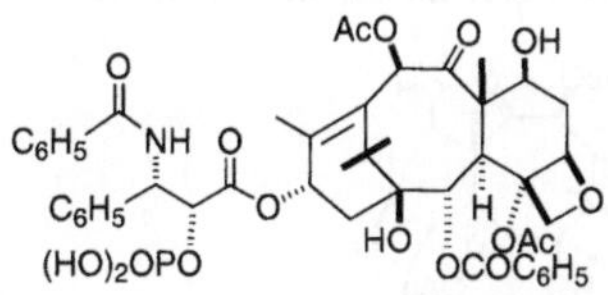

10.2.1

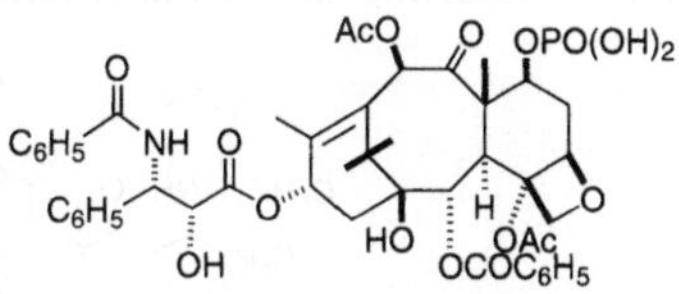

10.2.2

On the other hand, the derivatives **10.2.3** and **10.2.4**, prepared by logical extensions of previously used methodology, underwent much faster hydrolysis by phosphatase enzymes (which are present in more than normal amounts in cancer cells), and both were found to be much more soluble in water than taxol. The 7-substituted derivative **10.2.4** was found to be as active as taxol in the murine M109 tumor model (*361*), and the carbonate derivative **10.2.5** also had comparable activity to taxol in the M109 model (*362*).

Continued investigation by the Bristol-Myers Squibb group led to the development of the phosphonooxymethyl analogs **10.2.6**, **10.2.7**, and

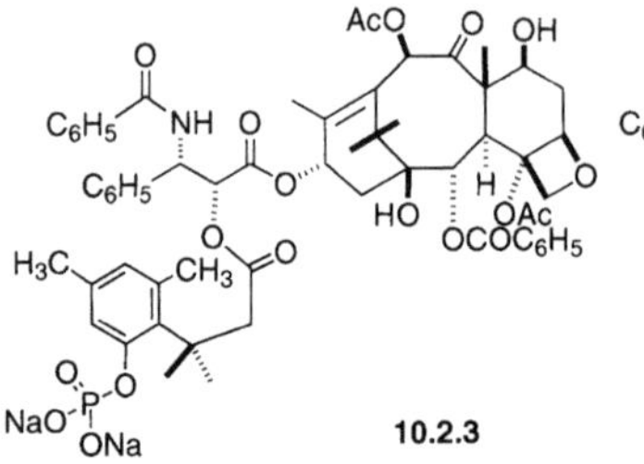

10.2.3

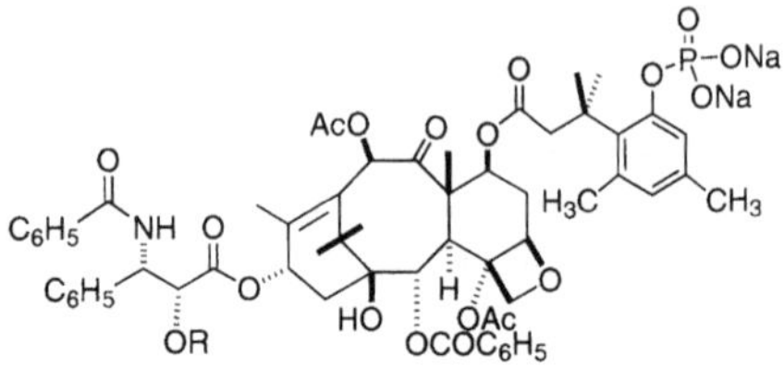

10.2.4 R = H
10.2.5 R = COOEt

10.2.8. These analogs had improved water-solubility and comparable M109 *in vivo* activity to taxol, and were thus stated to be suitable prodrugs of taxol; the carbonate **10.2.8** was especially promising in this regard (*363*), and a radioactive derivative of it was prepared (*364*).

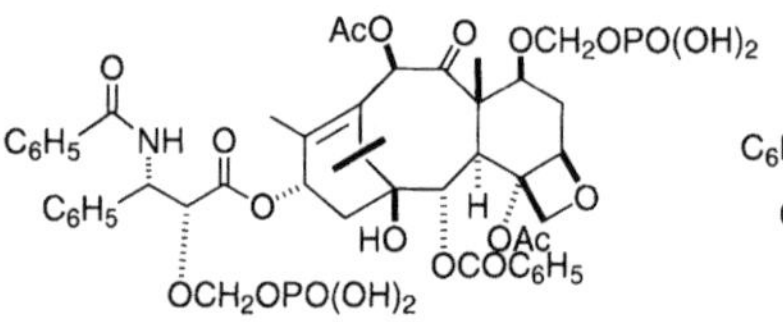

10.2.6

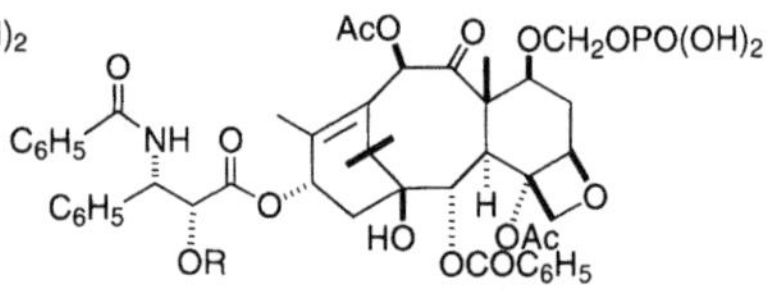

10.2.7 R = H
10.2.8 R = COOEt

Hwu *et al.* synthesized taxol-containing aminophosphates as protaxols. Treatment of taxol with $MeOPCl_2$ and collidine in THF and then with (monomethoxy)tritylated amino alcohols, I_2 and water produced zwitterionic taxol-containing aminophosphates such as **10.2.9** (*365*).

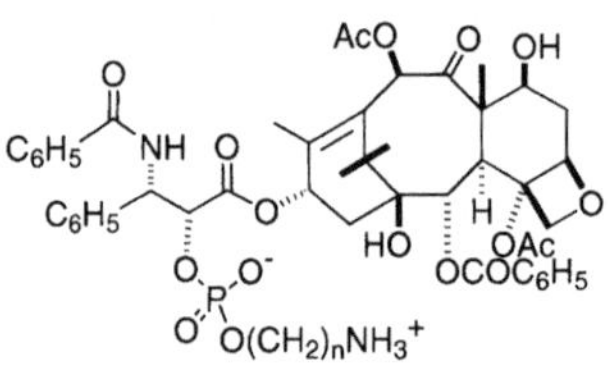

10.2.9

10.3. Taxol Analogs with Polymeric Acyl Substitutions

The concept of linking taxol to a water-soluble polymer to improve its solubility and possibly also its pharmacokinetics has been investigated by several groups. Thus Greenwald *et al.* prepared 2′-(polyethyleneglycol) esters of taxol as potential prodrugs (*366*, *367*), and similar compounds were also prepared by Li and his colleagues (*368*). The 2′-

bis-glycinyl(polyethyleneglycol) derivative **10.3.1** was prepared and shown to have improved activity and solubility as compared with taxol (*369*), and 2′-(polypyrrolecarboxamido-naphthalene) derivatives of taxol have been claimed to have increased bioavailability (*370*).

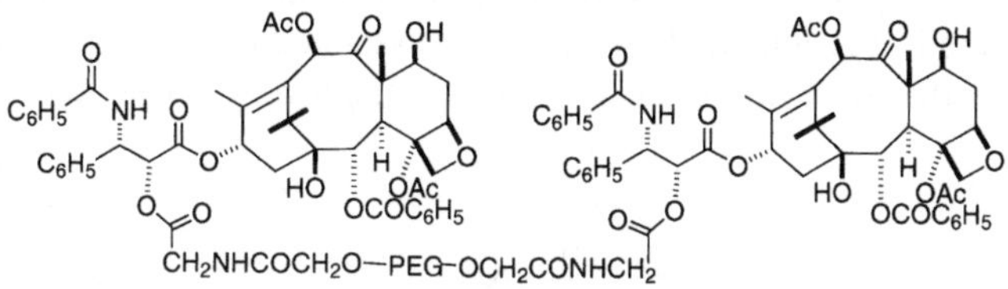

10.3.1

10.4. Targeted Analogs of Taxol

The concept of targeting taxol to a tumor is an attractive one and is beginning to be exploited by a few groups. Thus the cephalosporin-linked prodrug **10.4.1** was prepared by Rodrigues *et al.*; it is activated by β-lactamase (*371*). The β-glucuronyl carbamate-based prodrugs **10.4.2** and **10.4.3** were prepared by Scheeren and his colleagues as potential candidates for antibody-directed enzyme prodrug therapy (*372*), but were found to be two orders of magnitude less cytotoxic than taxol.

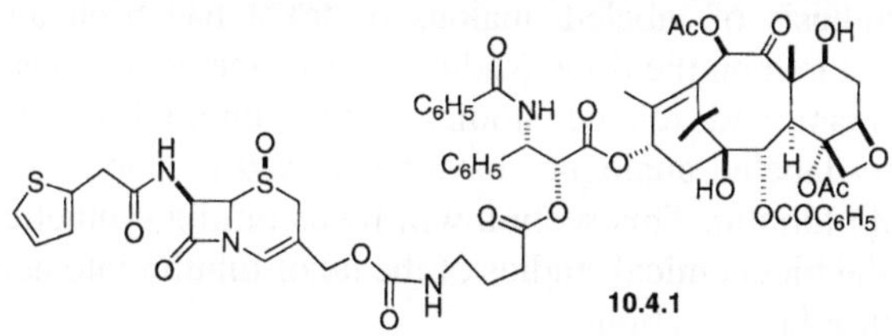

10.4.1

Safavy and his colleagues have developed an approach to "smart taxanes" in which taxol is conjugated through a 2′-succinate to solubilizing polyethylene glycol unit and thence to a bombesin/gastrin-releasing peptide receptor-recognizing peptide. This construct was prepared on the hypothesis that the soluble conjugate would be guided by the receptor-specific peptide to bind to to tumor cell surface receptors and then cleaved to deliver taxol. The cytotoxicity of the resulting compound to H1299 human nonsmall cell lung cancer cells was approximately tenfold greater than that of taxol for a 96 h exposure (*373*). A related approach was reported by Huang and his colleagues, who targeted taxol to tumor cells by linking it to octreotide, which is an octapeptide analog of the neuropeptide somatostatin. Fluorescein-labeled octreotide was shown to be internalized into the cytosol of human breast MCF-7 cells, and octreotide-linked taxol was found to be exclusively toxic to cells expressing somatostatin receptors (*374*). A number of 2′-carbonate-linked prodrugs were also prepared by Scheeren and his colleagues. The prodrug **10.4.4** was found to be the best of those investigated; it is activated by the tumor-associated enzyme plasmin, and is one of the most non-toxic prodrugs of taxol reported (*375*).

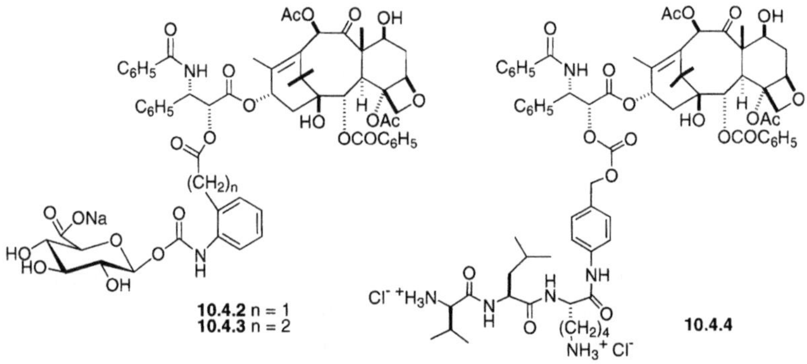

10.4.2 n = 1
10.4.3 n = 2

10.4.4

11. Labeled Taxol Analogs

The synthesis of labeled analogs of taxol has been an important aspect of research on the drug. Radioactively labeled analogs are needed for metabolism and related studies, while photoaffinity-labeled and fluorescently labeled analogues have been used to study the interaction of taxol with tubulin. This section will focus on the synthetic aspects of this work; the biochemical studies of the taxol-tubulin interaction will be discussed in a later section.

11.1. Isotopically Labeled Taxols

Tritiated taxol analogs were prepared by Kepler and his colleagues by two different routes (*376*). In one route 3″-bromotaxol was tritiated by exchange with carrier-free tritium over palladium, and in the other route 7-triethylsilyl-13-oxobaccatin III was reduced with [^{3}H]borane-THF to give a 13-[^{3}H]-7-triethylsilylbaccatin III, which was then coupled with side chain to give the corresponding tritiated taxol. A recent paper describes a method for selective tritiation of taxol and photoaffinity-labeled taxol analogs (*377*).

Several syntheses of carbon-14 labeled taxols have been reported. *N*-3′-[Carbonyl-^{14}C]-taxol **11.1.1** was prepared by coupling a [carbonyl-^{14}C]-*N*-benzoyl-β-lactam with 7-(triethylsilyl)baccatin III (*378*), while the same group prepared [3′-^{14}C]-taxol **11.1.2** by using a β-lactam prepared from carbonyl-^{14}C-benzaldehyde (*379*). [3′-^{14}C]-Docetaxel **11.1.3** was prepared by attachment of a labeled cinnamoyl side chain to 7,10-ditroc-10-deacetylbaccatin III, followed by hydroxyamination of the side chain (*380*). A slightly different approach was used to prepare [2′,3′-$^{14}C_2$]taxol, in that the required doubly labeled β-lactam synthon was prepared using a doubly labeled Oppolzer's bromoacylbornanesultam intermediate (*381*). 7-([carbonyl-^{14}C]-acetyl)taxol was prepared by simple acylation of 2′-(triethylsilyl)taxol with [carbonyl-^{14}C]-acetic anhydride and deprotection (*382*). The taxol analog PNU-105298 (**4.1.3.12**) was also prepared in both deuterium and carbon-14 labeled forms; in both cases the label was placed on the side chain (*383*).

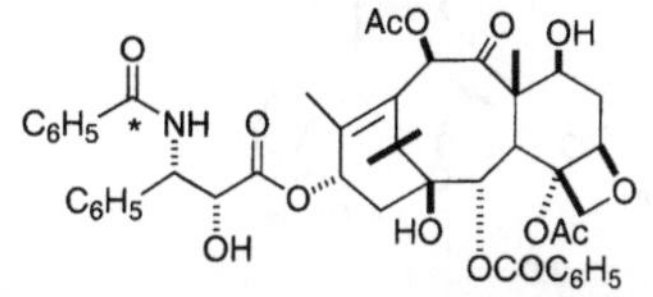

11.1.1 * = ^{14}C

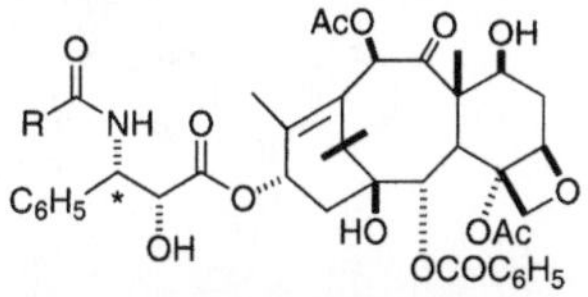

11.1.2 R = C_6H_5, * = ^{14}C
11.1.3 R = $OCMe_3$ * = ^{14}C

The [^{123}I]-labeled taxol derivative **11.1.5** was prepared by reacting the stannane **11.1.4** with sodium [^{123}I]-iodide (*384*).

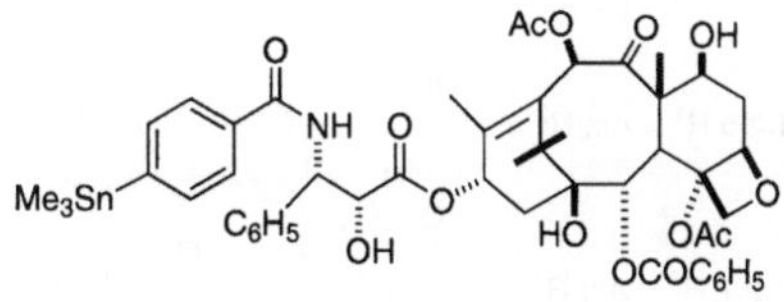

11.1.4

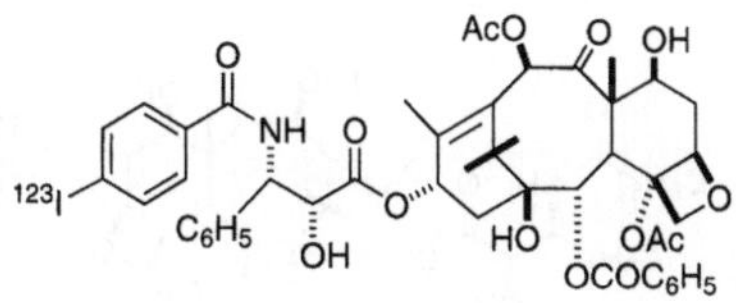

11.1.5

11.2. Photoaffinity Labeled Taxols

The syntheses of photoaffinity labeled analogs of taxol are discussed in this section, and the photoaffinity labeling studies with these analogs are described in section 15 below. Photoaffinity analogs of taxol have been prepared with the photoaffinity label on the side chain and on the 10-, 7-, and 2-positions of the ring system. The analogs **11.2.1** and **11.2.2** were prepared by Swindell (*385*), and the former was found to have the better microtubule binding ability and cytotoxicity; both analogs were prepared in tritiated as well as unlabeled forms, and the tritiated version of **11.2.1** was used by Horwitz in her photolabeling studies of tubulin (*386*). The analog **11.2.1** was also prepared independently by Georg and her collaborators (*387*), and similar analogs **11.2.3** and **11.2.4** were also prepared by the same group (*388*).

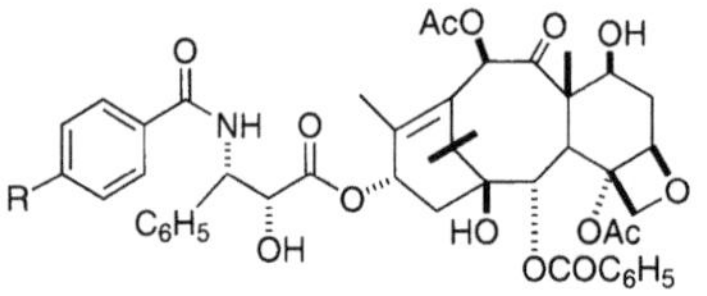

11.2.1 R = N_3
11.2.2 R = C_6H_5CO

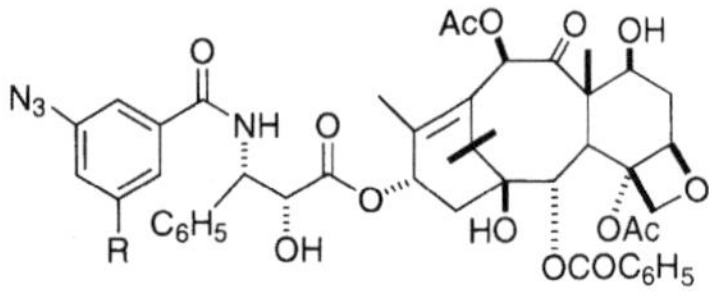

11.2.3 R = H
11.2.4 R = NO_2

The analog **11.2.6** was prepared by Ojima and his colleagues using tritiation of the corresponding $\alpha\beta$-unsaturated amide derivative to incorporate the tritium label (*389*).

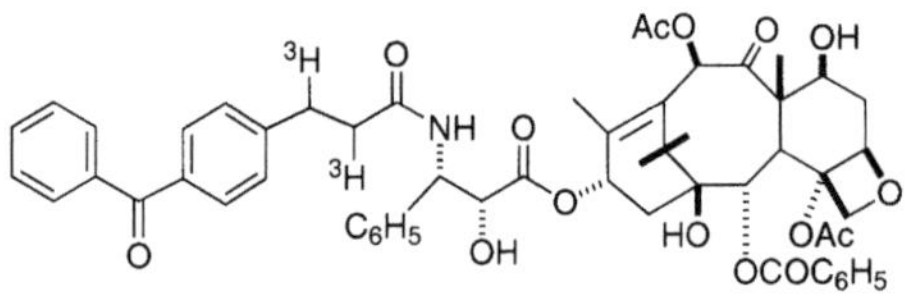

11.2.5

Taxol analogs **11.2.6** and **11.2.7** with photoaffinity labels at the 7- and 10-positions were prepared by similar methods to those used for the analog **11.2.6** (*390*). Analog **11.2.6** had the better cytotoxicity, with IC_{50} values approximately twice those of taxol in several cell lines.

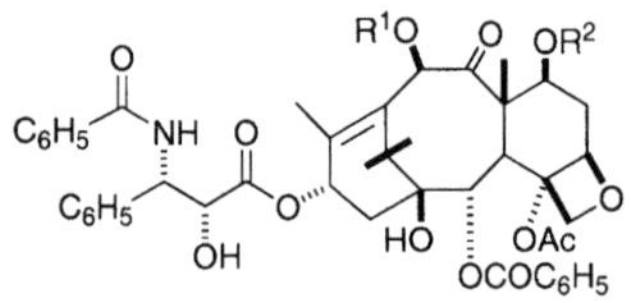

11.2.6 R^1 = Ac, R^2 =

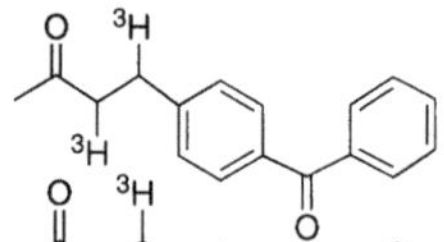

11.2.7 R^2 = Ac, R^1 =

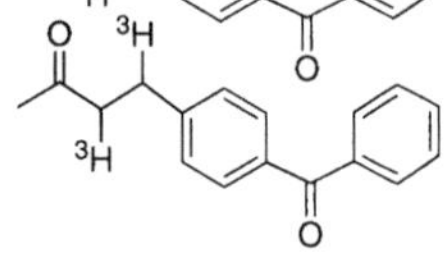

Finally, the interesting analog **11.2.8** was also prepared by Ojima; the biotin moiety was included to serve to as an affinity label so as to assist in the isolation of labeled fragments from photoaffinity studies (*391*).

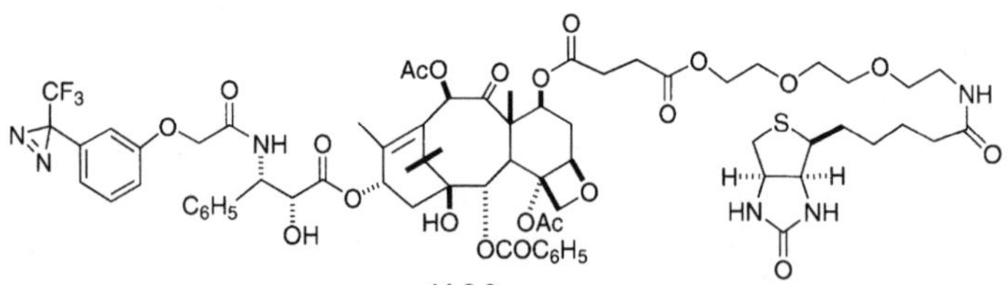

11.2.8

The related analog **11.2.9** was also prepared in both deuterium and tritium labeled forms, but its bioactivity was such that it was not the ideal derivative for photoaffinity labeling studies (*392*).

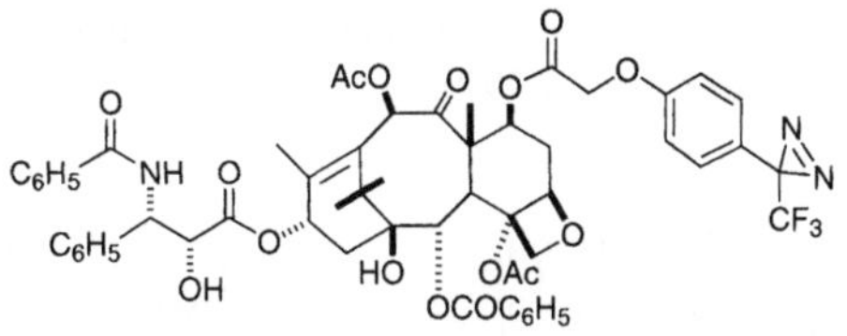

11.2.9

Taxol analogs labeled at the 2-position have been prepared by Kingston and by Georg. The tritiated 2-(*m*-azidobenzoyl) analog **11.2.10** was prepared by reductive debromination of the corresponding 3-nitro-4-bromobenzoyl derivative with sodium [^{3}H]-borohydride (*393*), and the corresponding non-radioactive analogs **11.2.11** and **11.2.12** were prepared by standard methods (*388*).

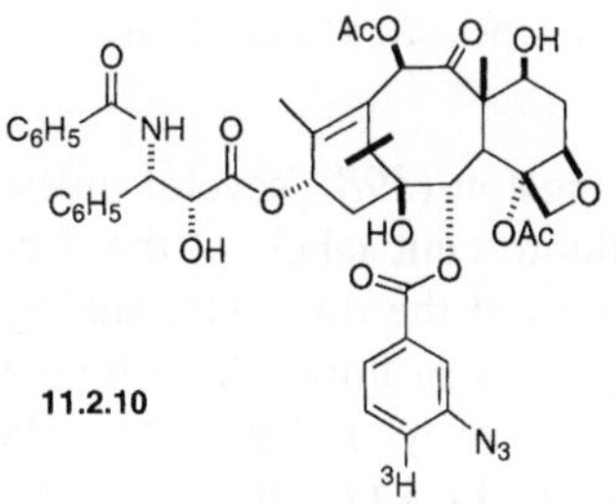

11.2.10

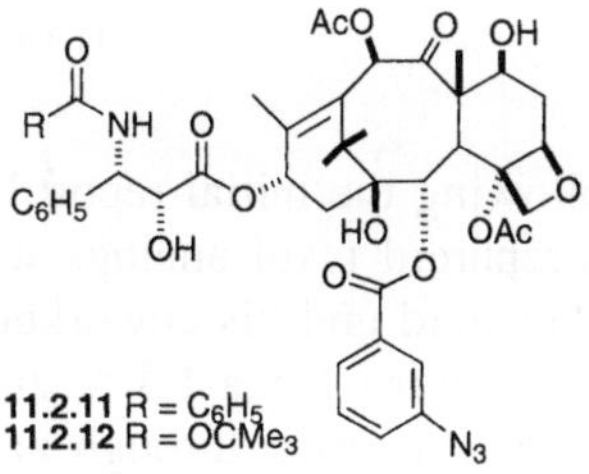

11.2.11 R = C_6H_5
11.2.12 R = $OCMe_3$

11.3. Fluorescent and Other Labeled Taxols

The synthesis of fluorescent analogs of taxol has yielded some valuable tools for studying its interaction with tubulin. Several different fluorescent analogs have been prepared, varying in their position of attachment of the fluorophore to the taxol molecule and in the nature of the fluorophore. Simple fluorophores, such as the aminobenzoyl group, provide the most "taxol-like" analogs, but suffer from relatively weak fluorescence activity. More strongly fluorescent fluorophores give better fluorescence, but at the expense (usually) of diminished bioactivity.

The taxol analog **11.3.1** with a *m*-dimethylaminobenzoyl group on the side chain has been prepared by Georg and her collaborators and used to probe the interaction of taxol and tubulin (*394*), and the corresponding desmethyl analog **11.3.2** has been prepared and used by Kingston and Bane and their coworkers (*395*). A later paper from the Georg group described the synthesis and study of the dimethylamino-substituted taxols **11.3.3–11.3.7** (*396*). Analog **11.3.8**, the desmethyl derivative of **11.3.5**, was prepared and investigated by Kingston and Bane and their collaborators (*397*).

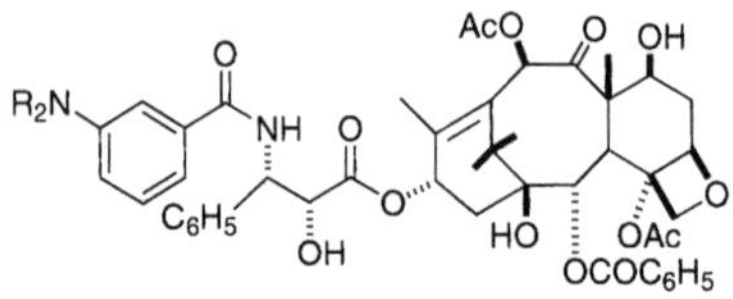

11.3.1 R = Me
11.3.2 R = H

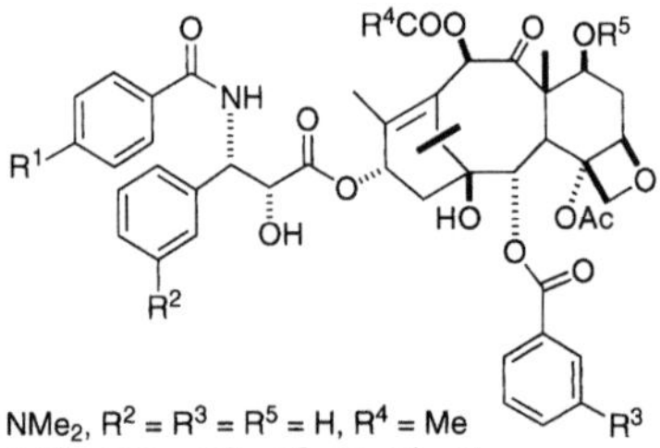

11.3.3 R^1 = NMe_2, R^2 = R^3 = R^5 = H, R^4 = Me
11.3.4 R^1 = H R^2 = NMe_2, R^3 = R^5 = H, R^4 = Me
11.3.5 R^1 = R^2 = H, R^3 = NMe_2, R^4 = Me, R^5 = H
11.3.6 R^1 = R^2 = R^3 = H, R^4 = Me, R^5 = *m*-N,N-dimethylbenzoyl
11.3.7 R^1 = R^2 = R^3 = R^5 = H, R^4 = *m*-N,N-dimethylphenyl
11.3.8 R^1 = R^2 = H, R^3 = NH_2, R^4 = Me, R^5 = H

Following the initial report by Kingston (*398*), several investigators have preparerd taxol analogs with fluorescent labels at the 7-position. Thus Guénard and his coworkers prepared the docetaxel analog **11.3.9** with the 7-nitro-2-oxa-1,3-diazol-4-yl (NBD) fluorophore linked to the C-7 hydroxyl group through an aminocaproyl linker (*399*). The same group also prepared analogs **11.3.10** and **11.3.11** with the NBD fluorophore linked to the side chain.

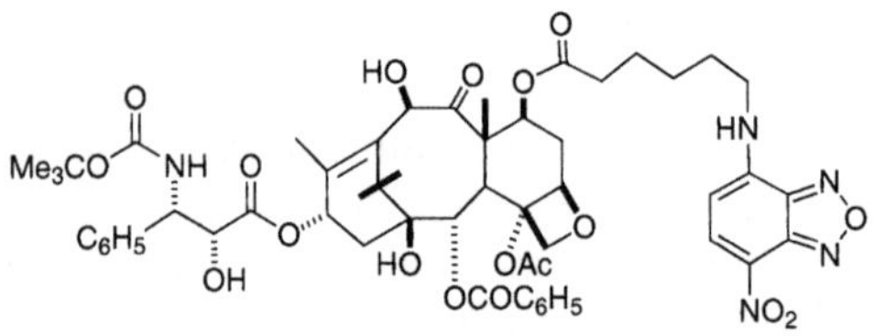

11.3.9

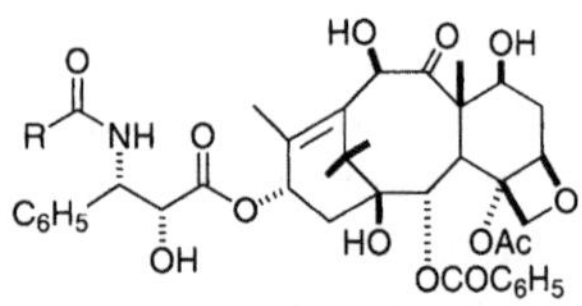

11.3.10 R = NBD-aminocaproyl
11.3.11 R = NBD-aminopropanoyl

The three fluorescent analogs **11.3.12–11.3.14** were prepared and studied by Amat-Guerri and Andreu (*400, 401*); analogs **11.3.12** (Flutax-1) and **11.3.13** (Flutax-2) were found to be useful for the study of the microtubule cytoskeleton, as discussed later (section 15.2).

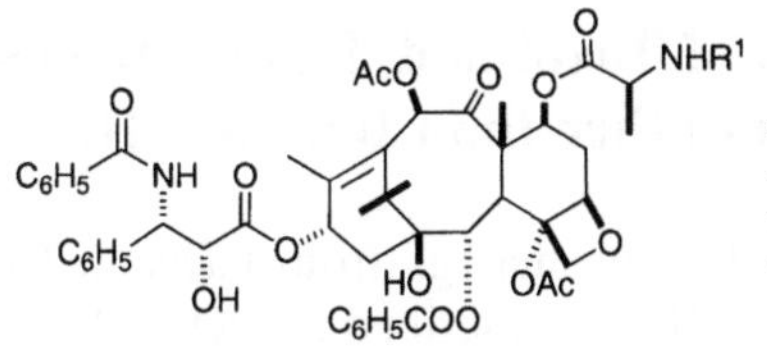

11.3.12 R^1 = A, R^2 = H
11.3.13 R^1 = A, R^2 = F
11.3.14 R = B
11.3.15 R = C

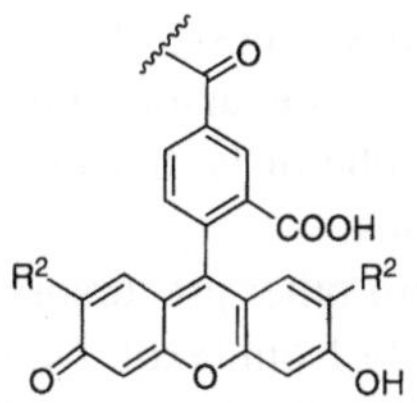

A

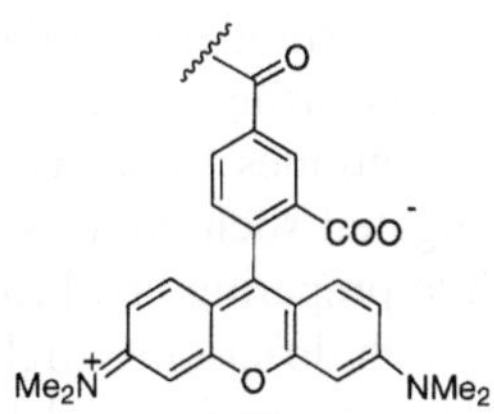

B

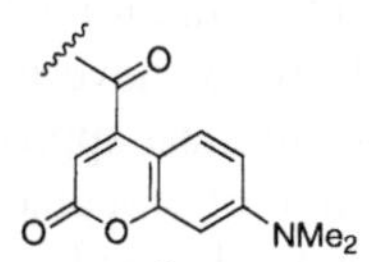

C

The four fluorescent taxols **11.3.16–11.3.19** were prepared by Nicolaou and his colleagues (*402*), and **11.3.18** was found to be the most suitable derivative for optical microscopy studies of the cell.

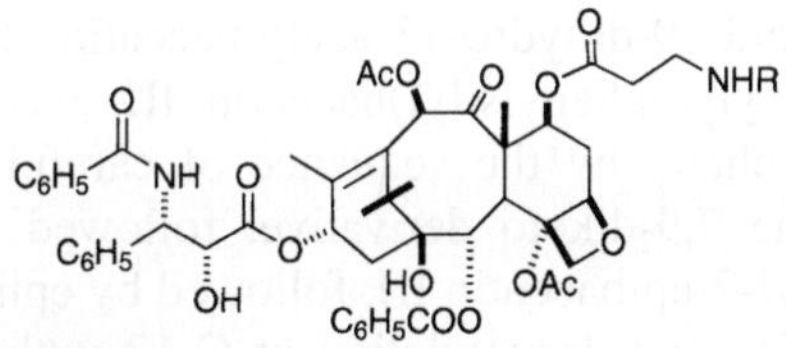

11.3.16 R = A
11.3.17 R = D
11.3.18 R = E
11.3.19 R = F

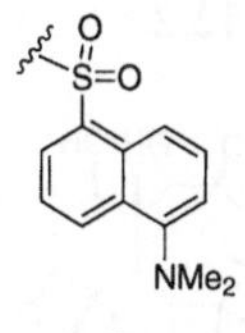

D

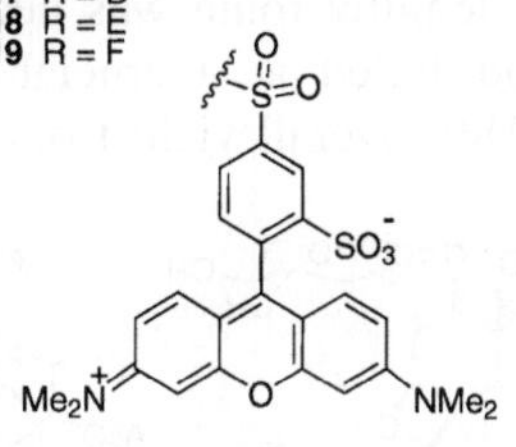

E

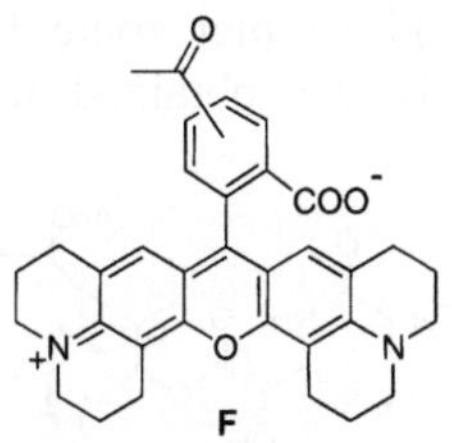

F

Similar analogs to those described above were prepared by Lai and coworkers (*382*). These investigators used a 7-(aminocaproyl) linker to couple lissamine rhodamine B (structure **E** above) or fluorescein isothiocyanate to taxol, and confirmed that both derivatives can be used to carry out intracellular fluorescence mapping.

Two nitroxylated taxol analogs, 7-(4-carboxy-2,2,6,6-tetramethyl-1-piperidinyloxy)-taxol and the corresponding 2′-analog were prepared and used to obtain information about the orientation of taxol at its binding site on tubulin using EPR. It was concluded that taxol is strongly immobilized after polymerization of tubulin, but that it is not constrained in the dimer structure (*403*).

12. The Synthesis of Taxol and Taxol Analogs from Precursors other than Baccatin III

The synthesis of taxol and taxol analogs from baccatin III and its precursor 10-deacetyl baccatin III was described in section 8 above, and this approach constitutes the current commercial synthesis of taxol and docetaxel. The concept of converting other taxoid precursors into taxol or bioactive taxol analogs continues to intrigue chemists, however, and several groups have investigated such conversions.

The simplest approach to preparing taxol from other precursors is by conversion of precursors which differ only slightly from taxol. Naturally occurring glycoside precursors such as 10-deacetyltaxol-7-xyloside (**12.1**) cannot be hydrolyzed to taxol without hydrolyzing the side chain, but they have been converted to taxol by the sequence of periodate cleavage, followed by acetylation, followed by reaction with phenylhydrazine; the overall yield was however only about 16% (*404*). The abundant taxoid 9-dihydro-13-acetylbaccatin III (**12.2**) was converted to 7-(*t*-butylydiphenylsilyl)baccatin III, ready for coupling with the taxol side chain, by the sequence of careful oxidation with Jones' reagent to the 7,9-diketo derivative, followed by borohydride reduction to 13-acetyl-7-epibaccatin III, followed by epimerization at C-7 to the normal series and deacetylation at C-13 with borohydride in phosphate buffer. This rather lengthy route was necessary because an earlier and simpler route had failed at a crucial deprotection step; baccatin III was obtained in 18% overall yield from **12.2** (*405*).

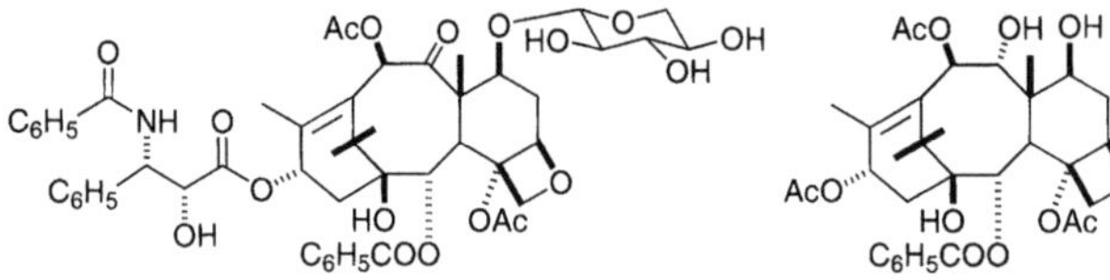

12.1 **12.2**

Three groups have investigated the conversion of taxine into 7-deoxytaxol analogs. The first group to publish in this area was that of Scheeren and his colleagues, who converted crude taxine (containing **12.3** and other compounds) into the 4-deacetyl-7-deoxytaxol analog **12.8** in some 14 steps (*406*). The sequence began by conversion of crude **12.3** into the protected and deaminated analog **12.4** in five steps. Saponification of **12.4**, hydroxylation of the C-4(20) double bond, protection and mesylation gave the mesylate **12.5**. Deprotection of **12.5** with fluoride ion led to the formation of oxetane **12.6**, which was reduced and acylated at C-13 to give **12.7**. Deprotection and oxidation of the benzylidene acetal with t-butylhydroperoxide gave the final product **12.8**, which was not acetylated at C-4. Other analogs similar to **12.8** were also prepared by this group, but none of them had any significant cytotoxic activity, consistent with the lack of the C-4 acetyl group.

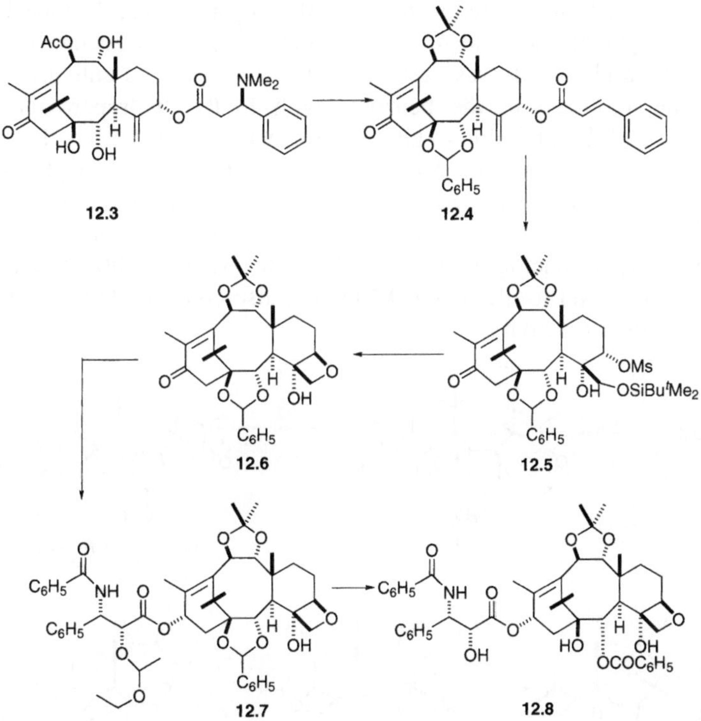

12.3 **12.4** **12.6** **12.5** **12.7** **12.8**

A similar approach was described by Ahond and Poupat and their colleagues, who prepared the analogs **12.9** and **12.10** from a mixture of taxine B and isotaxine B (*407*, *408*). Neither compound was cytotoxic, but **12.10** showed weak inhibition of microtubule disassembly.

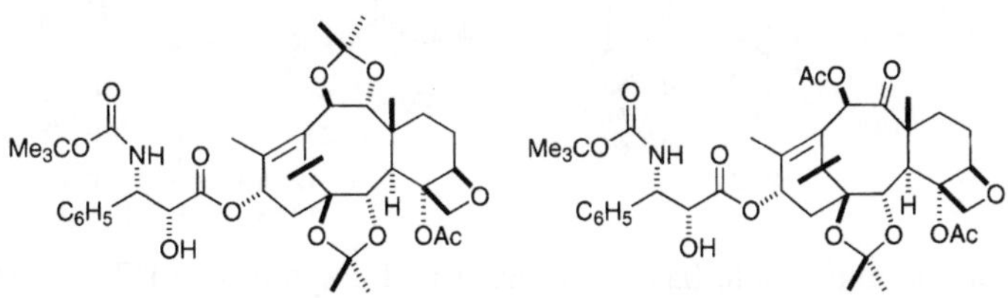

12.9 **12.10**

The same group were able to prepare analogs with an intact 2-benzoyl group in a later paper (*409*). In this approach the advanced precursor **12.5** used by Scheeren and his coworkers (*406*) was carefully hydrolyzed and then resilylated to give the diol **12.11**. Acetylation of **12.11** gave a mixture of acetates **12.12** and **12.13**, which was oxidized to a mixture of the ketone **12.14** and its regioisomer with the carbonyl group at C-10; the desired isomer was the major product. Ketone **12.14** was desilylated and treated with tetrabutylammonium acetate to give oxetane **12.15**, which was acetylated at C-4 with acetic anhydride and pyridine, oxidized to the benzoate with *t*-butyl hydroperoxide and reduced with borohydride to give 7-deoxybaccatin III (**12.16**). The final step of coupling with the docetaxel side chain gave 7-deoxydocetaxel (**12.17**) in 1.7% overall yield. Modifications of the protecting groups used gave the related compounds **12.18** and **12.19**. Compound **12.17** was the most active analog prepared, with a cytotoxicity slightly better than that of docetaxel itself; analog **12.18** was also comparable to docetaxel in its cytotoxic activity.

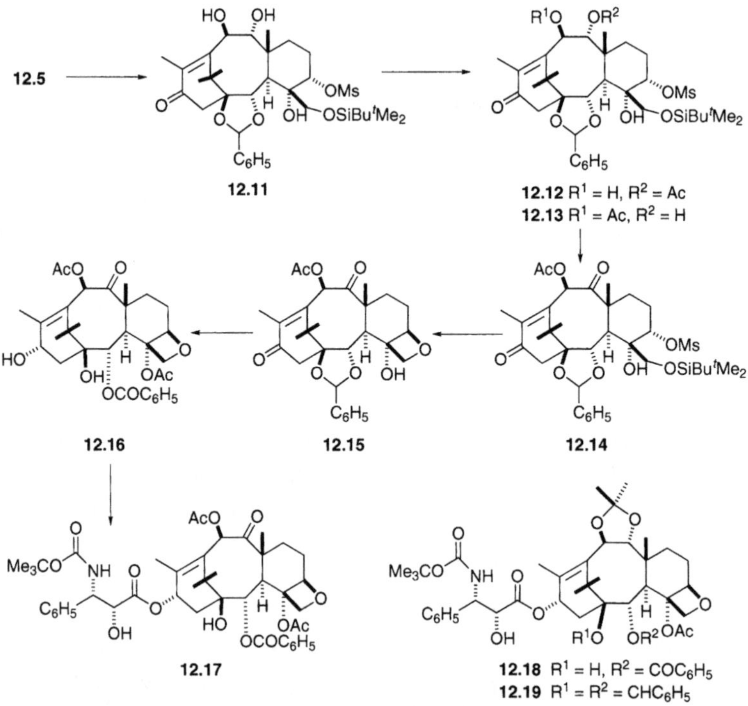

Matovic and Saicic have also described a synthesis of 7-deoxytaxol from taxine (*410, 411*). Taxine **12.3** was converted to the triol **12.20** by standard methods, and this was protected as the cyclic 1,2-carbonate and oxidized to **12.21**. Removal of the cinnamoyl side chain was effected by osmylation followed by mild base to give the triol **12.22** and, in a nice touch, the diol **12.23**; the latter compound was obtained in optically pure form and could be converted to the taxol side chain by known chemistry. Formation of the oxetane **12.24** from **12.22** followed the same pathway used by Scheeren and his coworkers (*406*), and ring-opening of the carbonate followed by reduction at C-13 gave 7-deoxybaccatin III (**12.25**). The synthesis was completed by attachment of the side chain to give 7-deoxytaxol (**12.26**).

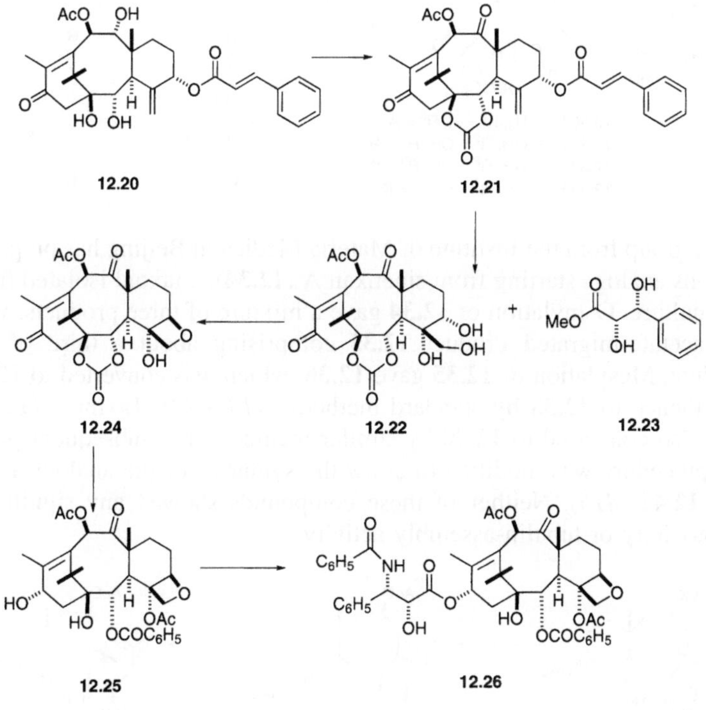

Cheng and his colleagues have prepared the 7-deoxypaclitaxel analog **12.30** starting from 5-cinnamoyltriacetyltaxicin I (**12.27**), the major taxoid of Japanese yew (*412*). Their synthesis paralleled those described above, and proceeded through the key intermediates **12.28** and **12.29** to the 9-(nucleosidosuccinoyl)-7-deoxytaxol analogs **12.30–12.33**. None of the compounds were as effective at promoting microtubule assemply as taxol, but analogs **12.31** and **12.32** both had improved cytotoxicity as compared to taxol in two cell lines.

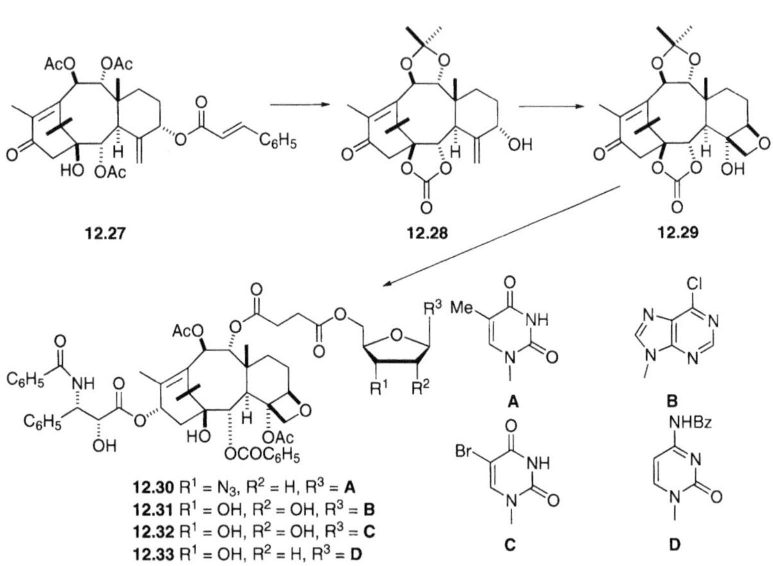

12.27 12.28 12.29

12.30 $R^1 = N_3$, $R^2 = H$, $R^3 = $ **A**
12.31 $R^1 = OH$, $R^2 = OH$, $R^3 = $ **B**
12.32 $R^1 = OH$, $R^2 = OH$, $R^3 = $ **C**
12.33 $R^1 = OH$, $R^2 = H$, $R^3 = $ **D**

A group from the Institute of Materia Medica in Beijing has prepared various analogs starting from sinenxan A (**12.34**), a taxoid isolated from cell culture. Osmylation of **12.34** gave a mixture of three products, with the acetate-migrated product **12.35** comprising about a third of the product. Mesylation of **12.35** gave **12.36**, which was converted to **12.37** and thence to **12.38** by standard methods (*413*, *414*). Taxinine (**12.39**) was also converted to **12.40** by similar methods. In a subsequent paper the procedure was modified to allow the synthesis of the analogs **12.41** and **12.42** (*415*). Neither of these compounds showed any significant cytotoxicity or tubulin-assembly activity.

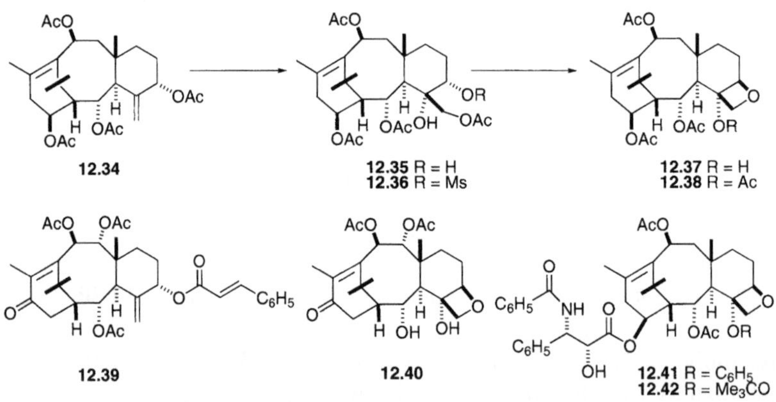

12.34; 12.35 R = H, 12.36 R = Ms; 12.37 R = H, 12.38 R = Ac

12.39; 12.40; 12.41 $R = C_6H_5$, 12.42 $R = Me_3CO$

13. The Synthesis of Simplified and Unusual Taxol Analogs

Several workers have investigated the synthesis of simplified taxol analogs in the hope of finding compounds with equal or better activity than taxol but with much simpler (and thus synthetically accessible) structures. So far these hopes have not been realized, but some interesting structures have been prepared and some interesting bioactivities have been discovered, so the studies have been worthwhile.

13.1. Simplified Analogs

Some of the earliest work in this area was reported by Blechert, who prepared the diastereomeric diols **13.1.1** (*416*) and **13.1.2** (*417*). Both compounds were reported as being able to inhibit the depolymerization of tubulin, but the level of this activity was not reported. An even more simplified set of analogs was prepared by Fuji and Kohno and their collaborators; these investigators prepared the analogs **13.1.3–13.1.8**, in which the alkyl chain comprised two to seven carbons (*418*); none of these compounds showed any promising tubulin inhibitory activity.

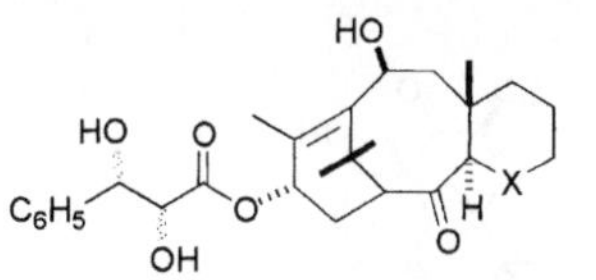

13.1.1 X = CH_2
13.1.2 X = O

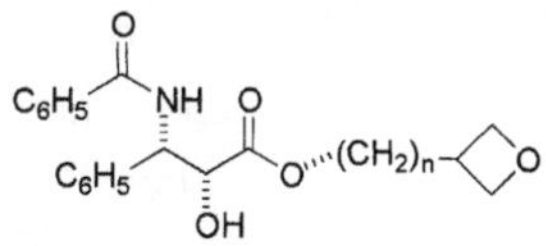

13.1.3 - 13.1.8 n = 2 -7

The macrolactam **13.1.9** was prepared by Botta and Corelli and their collaborators, based on the observation that an analog lacking the taxol side chain had moderate antiproliferative activity (*419*). Regrettably **13.1.9**, with the taxol side chain in place, did not show any enhanced activity.

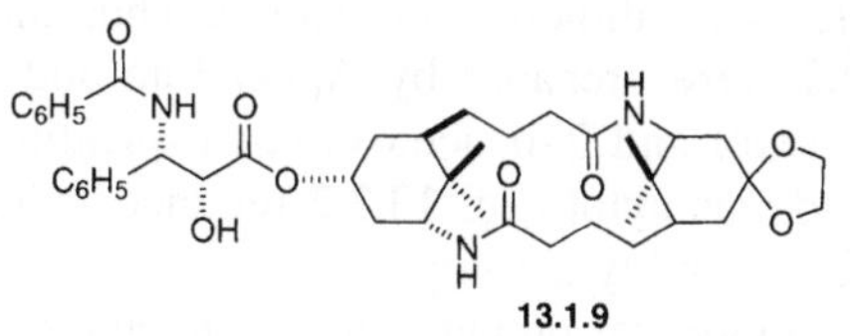

13.1.9

Two analogs which do show interesting activities have been reported recently. The C-aromatic analog **13.1.10** had the same activity as verapamil as an MDR-reversal agent (*420*). The borneol derivative **13.1.11** was prepared, together with several related analogs, as the result of a finding that a rearrangement product in a synthetic approach to taxol had weak microtubule depolymerization activity after it was esterified with the taxol side chain. The synthesis of several analogs led to analog **13.1.11**, which had better microtubule-stabilizing activity than taxol (*421*). Interestingly, although this compound was more active than taxol in the tubulin assay, it was 2–3 orders of magnitude less cytotoxic than taxol, indicating that a separation had been achieved between tubulin-assembly activity and cytotoxic activity.

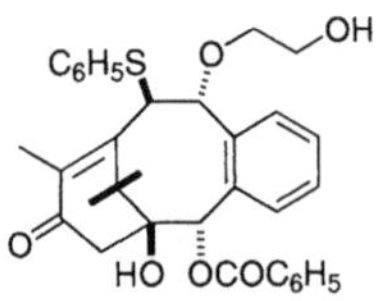

13.1.10

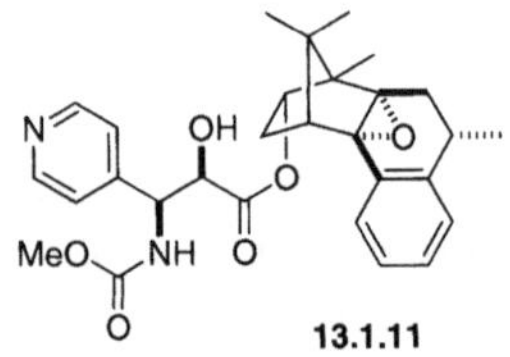

13.1.11

The taxol analog **13.1.12** was prepared from deaminoacyltaxine A isolated from *Taxus baccata* (*422*). The analog was more cytotoxic than deaminoacyltaxine A, but was significantly less cytotoxic than taxol.

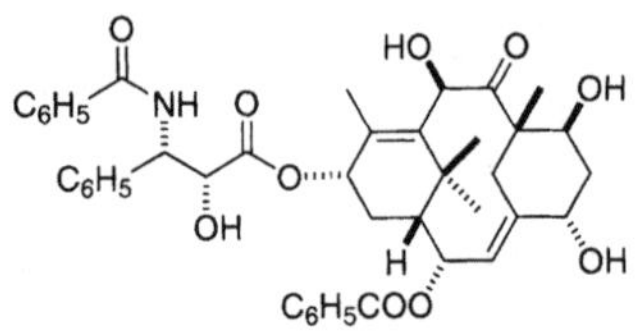

13.1.12

13.2. Dimeric Analogs

Several taxol analogs have been prepared in which taxol is linked to another molecule as a "dimeric" derivative. Thus the dimeric taxanes **13.2.1** and **13.2.2** were prepared by Appendino and Danieli and their colleagues from taxol and 2′-deacetoxyaustrospicatine (*423*). Although neither compound was cytotoxic, **13.2.2** retained approximately a third of taxol's tubulin-assembly activity.

A true taxoid dimer was prepared by Cheng and coworkers by Diels-Alder cyclization of taxinine B and taxacin I derivatives. Thus oxidation of 20-hydroxy-4,5-ene-7-oxotaxinine derived from natural taxinine B

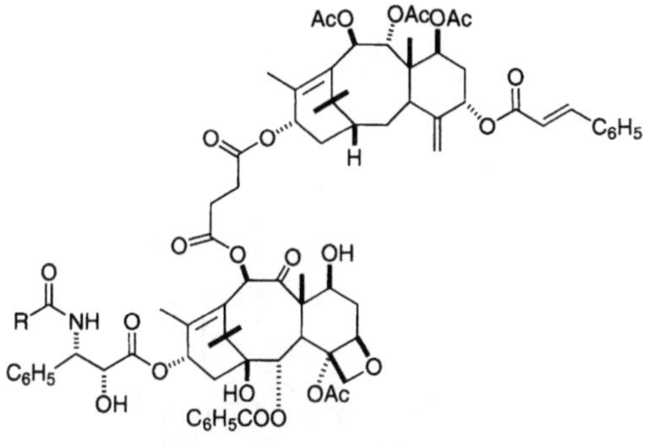

13.2.1 R = C_6H_5
13.2.2 R = Me_3CO

gave the aldehyde **13.2.3**, which underwent Diels-Alder cycloaddition to give the dimeric compound **13.2.4** (*424*). Similar treatment of the aldehyde **13.2.5** gave the dimeric compound **13.2.6** as the major isomer. Cycloadducts were also reported for taxoid derivatives with *N*-methylmaleimide, but no biological data were reported for any of these new compounds.

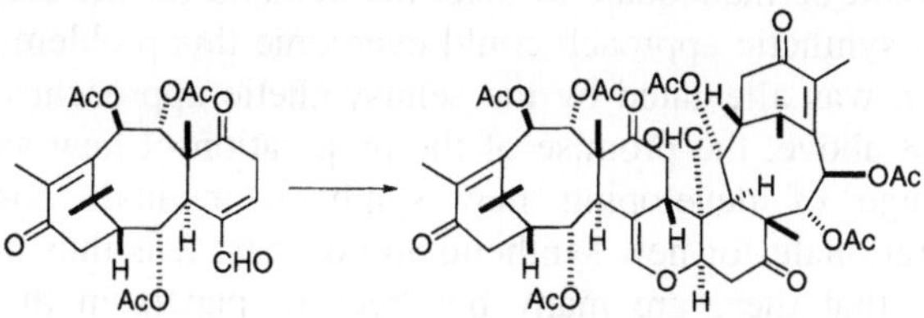

13.2.3 **13.2.4**

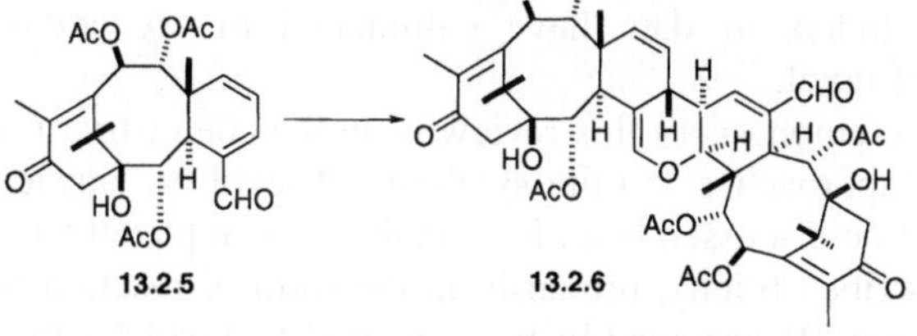

13.2.5 **13.2.6**

A different type of dimeric product was prepared by Wandless and his colleagues, who coupled taxol with daunorubicin to prepare analogs of the general structure **13.2.7**, where X is one of several alkyl and aryl linker units. None of the dimeric compounds were as cytotoxic as the individual monomeric compounds (*425*).

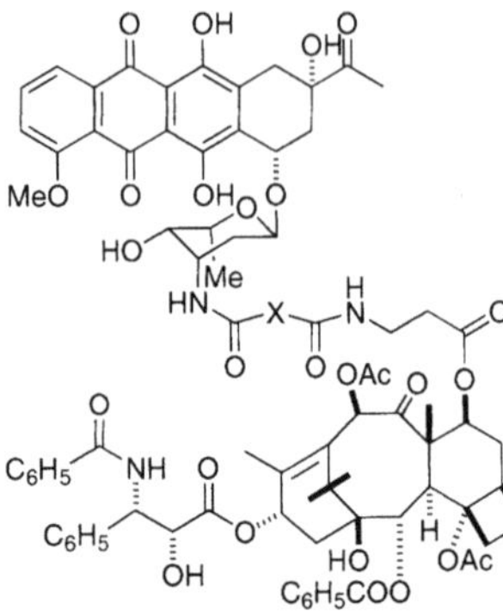

13.2.7

14. The Synthesis of Taxol

The total synthesis of taxol represents one of the greatest challenges to synthetic chemistry among molecules with molecular weights less than 1000 Da, and it thus provides an attractive target for total synthesis. An added incentive was the fear in the early 1990's that natural supplies of taxol would be inadequate to meet the demand for the drug, and the hope that a synthetic approach could overcome this problem. Although this concern was alleviated by the semisynthetic approaches described in section 8 above, the promise of the preparation of new analogs and the challenge of developing new synthetic methods continues to provide a rationale for new synthetic approaches. It is thus not surprising to find that there are many hundreds of papers in the literature describing approaches to the synthesis of taxol, and that its synthesis has attracted the attention of some of the world's best synthetic chemists. Studies to date have culminated in six completed total syntheses of taxol.

Since the emphasis of this review is on the chemistry of taxol itself, the many "approaches" to the synthesis of taxol are outside its scope and will not be discussed here. Instead the six completed total syntheses will be described briefly, primarily in the form of reaction flow charts. Since baccatin III can readily be converted to taxol by the chemistry described in Section 8.1, the total synthesis of baccatin III constitutes a formal synthesis of taxol, and thus the synthetic routes described will stop at baccatin III, although the authors did in each case convert this to taxol.

The synthesis of baccatin III presents a formidable challenge, because the molecule contains an unusual and distorted ABCD ring system, and also because of the large number of stereocenters which need to be controlled. The presence of the four ring systems means that there is considerable flexibility in the order in which the rings are synthesized, and the six reported syntheses use five different general approaches to this problem.

The six reported syntheses are described in the following sections in order of their completion. The Holton and Nicolaou syntheses were published within days of each other, with the Nicolaou synthesis appearing first. The Holton synthesis is described first on the basis of the earlier submission date of the manuscript describing the work.

14.1. The Holton Synthesis

The Holton synthesis was published in the form of two communications in early 1994 (*426*, *427*), but at this point the full papers describing this work have yet to appear. The synthesis is of the form AB→ABC→ABCD (426). Building on earlier studies on the synthesis of taxusin (*428*) and his discovery of the "epoxy alcohol fragmentation," the taxusin intermediate **14.1.1** was converted to the AB system **14.1.2**. The C-ring was then elaborated through the intermediates **14.1.3–14.1.10**. Some interesting chemistry emerged during these steps: thus the conversion of **14.1.5** to **14.1.6** involves an unusual rearrangement related to the Chan rearrangement, and the intermediate **14.1.7** underwent preferential deprotonation with LTMP at C-1 (baccatin numbering) rather than at the C-3 positions which would normally be expected to be more acidic.

Elaboration of the D-ring was achieved through the key intermediate **14.1.13**, which was converted to an oxetane derivative by the steps of mesylation, osmylation, and treatment with base; similar, although not identical pathways were used by all the other workers in this area. The synthesis was completed by conversion of the 1,2-carbonate protective group to the C-2 benzoate (another sequence used by most of the other investigators), and insertion of the C-9 oxygenation to give the protected baccatin III derivative **14.1.17.**

The overall yield of taxol from **14.1.1** in this synthesis is 4–5%. Compound **14.1.1** is readily available from the commercial intermediate (−)-patchino, but unfortunately it is the enantiomer of the structure required for taxol. Compound **14.1.1** with the correct absolute stereochemistry can be prepared from (−)-borneol, but the additional steps required obviously detract from the elegance of the approach.

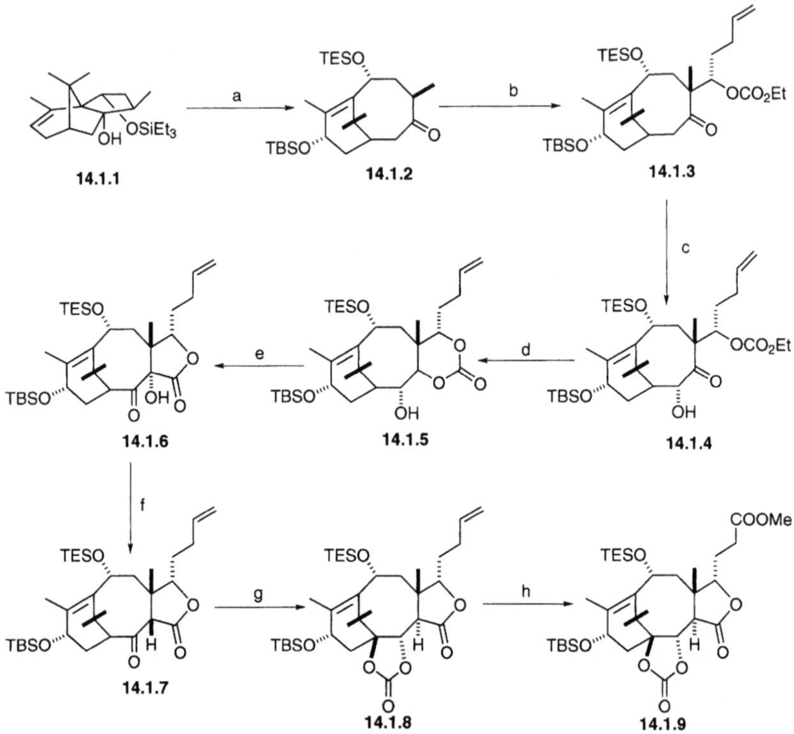

(a) Peracetic acid, then Ti(O*i*Pr)$_4$, then TBDMSiOTf/pyridine, 93%; (b) HN(*i*Pr)$_2$, THF, MeMgBr, then 4-pentenal, then Cl_2CO/py, then EtOH, 75%; (c) LDA, (+)-camphorsulfonyl oxazidiridine, 85%; (d) RedAl, toluene, - 78 °C, then warm, then Cl_2CO, pyridine, CH_2Cl_2, 97%; (e)Swern oxidation, then LTMP, 85%; (f) SmI_2, then SiO_2, 91%; (g) LTMP, - 10 °C, camphorsulfonyl oxazidiridine, then RedAl, THF, - 78 °C, basic workup, then Cl_2CO, py , 77%; (h) O_3, then $KMnO_4$, KH_2PO_4, then CH_2N_2, 93%.

14.2. The Nicolaou Synthesis

The Nicolaou group published a preliminary account of their synthesis in early 1994 (*429*) and followed this up with a series of full papers in 1995 (*430–433*). The synthesis is of the form A+C→A-C→ABC→ABCD, and it is thus a convergent synthesis as opposed to the linear approach adopted by Holton. The C ring was prepared by a Diels-Alder reaction between diene **14.2.2** and dienophile **14.2.1**, using phenylboronic acid as a tether to hold the reactants together. Elaboration of the adduct **14.2.3** then gave the protected ring C synthon **14.2.8**.

The A-ring was also prepared by a Diels-Alder reaction this time between diene **14.2.9** and dienophile **14.2.10** to give **14.2.11**. Adduct

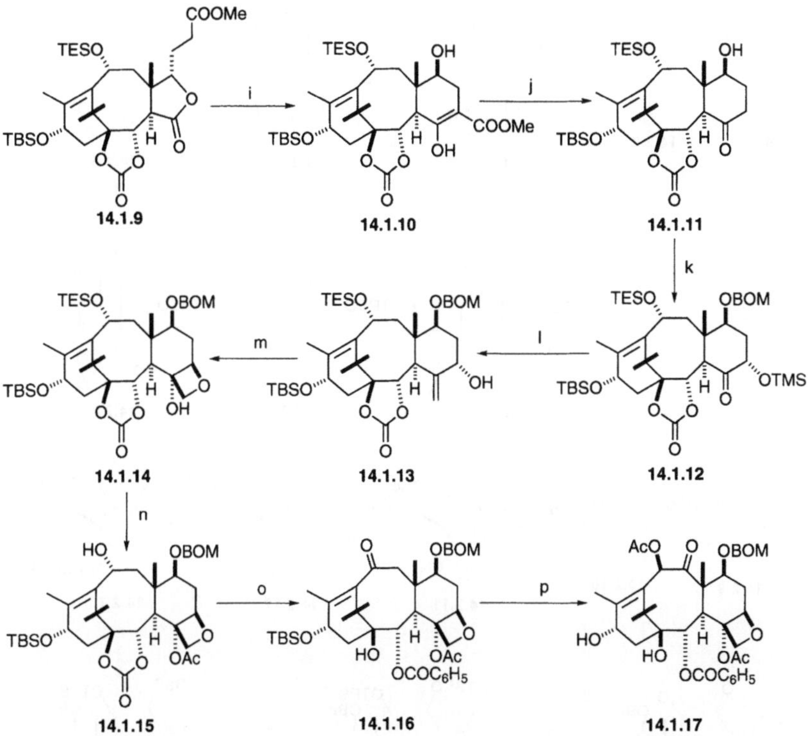

(i) LDA, THF, - 78 °C, then AcOH, THF, 93%; (j) pTsOH, 2-methoxypropene, then PhSK, DMF, 86 °C, acidic workup, 92%; (k) Benzyloxymethyl chloride, EtN(*i*Pr$_2$), CH_2Cl_2, Bu_4NI, then LDA, THF, TMSCl, -78 °C, then *m*CPBA, hexane, 25 °C, 86%; (l) MeMgBr, CH_2Cl_2, then Burgess' reagent, acidic workup, 60%; (m) MsCl/py, then OsO_4, then DBU, toluene, 105 °C, 48%; (n) Ac_2O, py, DMAP, 24 h, then HF/py, CH_3CN, 70%; (o) PhLi, THF, -78 °C, then TPAP, NMO, CH_2Cl_2, molecular sieves, 85%; (p) KO^{t}Bu, then benzeneseleninic anhydride, 0 °C, then KO^{t}Bu, then Ac_2O, py, DMAP, then HF/py, 90 °C, 12 h, 94%.

14.2.11 was hydrolyzed and protected to give **14.2.12**, which was then converted to the sulfonylhydrazone **14.2.13**.

The first of the two key coupling reactions linked **14.2.8** and **14.2.13** through a Shapiro reaction to give **14.2.14**. The C-1 hydroxyl group was introduced through an epoxide intermediate, and protection of the C-1 and C-2 hydroxyl groups as their cyclic carbonate gave **14.2.16** and set the stage for the second key coupling reaction. The planned McMurry coupling of **14.2.16** to **14.2.17** proved troublesome, and even after careful experimentation the best yield of **14.2.17** obtained was only 25% using titanium trichloride under controlled conditions.

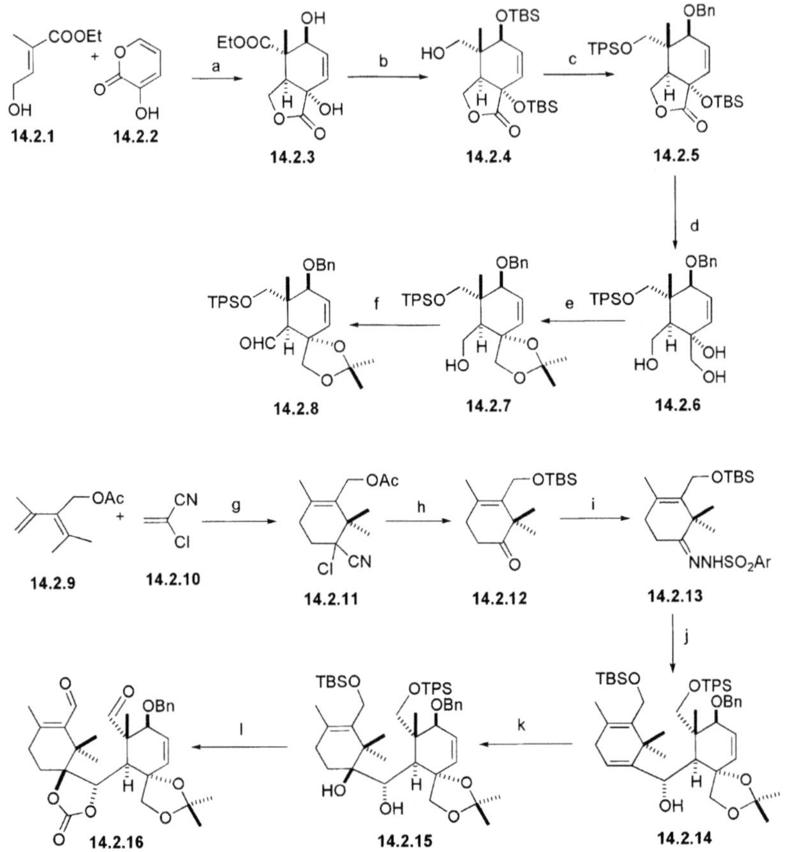

(a) $PhB(OH)_2$, PhH, 90 °C, 48 h, then 2,2-dimethylpropane-1,3-diol, 25 °C, 30 min, 61%; (b) Bu^tMe_2SiOTf, 2,6-lutidine, DMAP, CH_2Cl_2 0 °C, then $LiAlH_4$, Et_2O, 85%; (c) CSA, MeOH, CH_2Cl_2, then Bu^tPh_2SiCl, imidazole, DMF, 25 °C, then KH, Et_2O, n-Bu_4NI, BnBr, 25 °C, 72%; (d) $LiAlH_4$, Et_2O, 80%; (e) 2,2-dimethoxypropane, CSA, CH_2Cl_2, 25 °C, 82%; (f) TPAP, NMO, CH_3CN, 25 °C, 95%. (g) 135 °C, 96 h, 85%; (h) KOH, Bu^tOH, 70 °C, then Bu^tMe_2SiOTf, DMAP, ca. 85%; (i) 2,4,6-triisopropylbenzenesulfonylhydrazide, MeOH, 25 °C, 70%; (j) n-BuLi, THF, -78 °C to 25 °C, cool to 0 °C and add **14.2.7**, 0.5 h, 82%; (k) $VO(acac)_2$, t-BuOOH, molecular sieve, PhH, 12 h, then $LiAlH_4$, Et_2O, 25 °C, 7 h, 66%; (l) KH, HMPA/Et_2O, $COCl_2$, 25 °C, 2 h, then TBAF, THF, 25 °C, 7 h, then TPAP, NMO, CH_3CN/CH_2Cl_2, 25°C, 2h, 31%.

With the ABC ring system and much of the required functionality in place, intermediate **14.2.17** was efficiently converted to baccatin III by the sequence **14.2.17→14.2.24**.

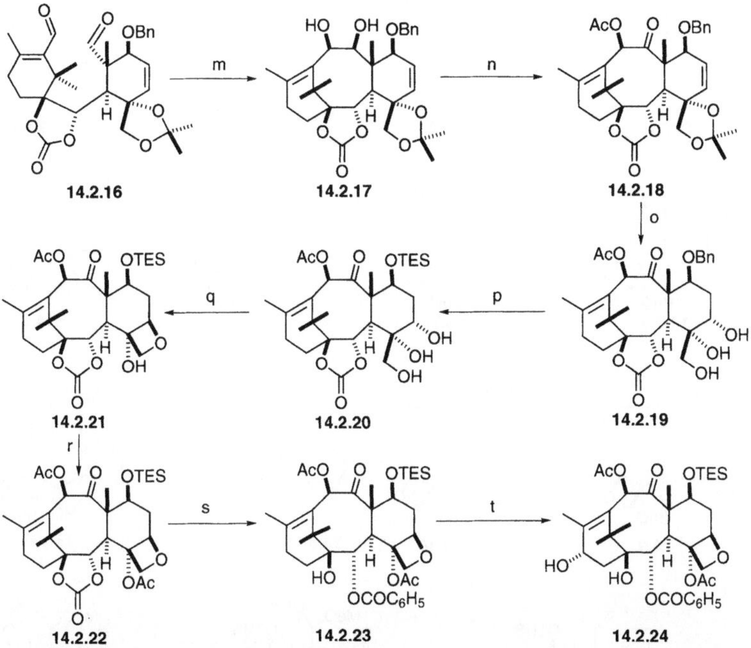

(m) $(TiCl_3)_2$-$(DME)_3$, Zn-Cu, DME, 70 °C, 1 h, 23%; (n) Ac_2O (1.5 eq.), DMAP, CH_2Cl_2, rt, 2 h, then TPAP, NMO, CH_3CN, rt, 2h, 88%; (o) BH_3-THF, THF, 0 °C, 2 h, then H_2O_2, aq. $NaHCO_3$, 0.5 h, then conc. HCl, MeOH, H_2O, rt, 5 h, 44%; (p) Ac_2O, DMAP, CH_2Cl_2, then H_2, 10% $Pd(OH)_2$(C), EtOAc, rt, 0.5 h, then Et_3SiCl, py, rt, 12 h, then K_2CO_3, MeOH, 0 °C, 15 min, 75%; (q) Me_3SiCl, py, CH_2Cl_2, 0 °C, 15 min, then Tf_2O, *i*-Pr_2NEt, CH_2Cl_2, 25 °C, 4 h, then CSA, MeOH, rt, 10 min, then silica gel, CH_2Cl_2, rt, 4 h, 40%; (r) Ac_2O, DMAP, CH_2Cl_2, rt, 4 h, 94%; (s) PhLi, THF, -78 °C, 10 min, 80%; (t) PCC, NaOAc, Celite, PhH, reflux, 1 h, then $NaBH_4$, MeOH, rt, 5 h, 63%.

14.3. The Danishefsky Synthesis

The Danishefsky synthesis was published in 1996 (*434*) and it is the only synthesis to date in which the oxetane D-ring is formed first and maintained throughout the synthesis. The synthesis is of the form C→CD→A-CD→ABCD. A key to this strategy was the protection of the C-4 hydroxyl group as its benzyl ether rather than as an acetate. This protection arrangement avoided the neighboring group participation by acetate which is a large part of the reason for the oxetane ring's lability in taxol, as described in section 5.1.

The synthesis started with the well-known Wieland-Miescher ketone **14.3.1**, which was converted in several steps to the triol **14.3.5** and thence to the oxetane **14.3.7**. Cleavage of the A-ring of **14.3.7** and adjustment of oxidation levels gave the aldehyde **14.3.10**, which was

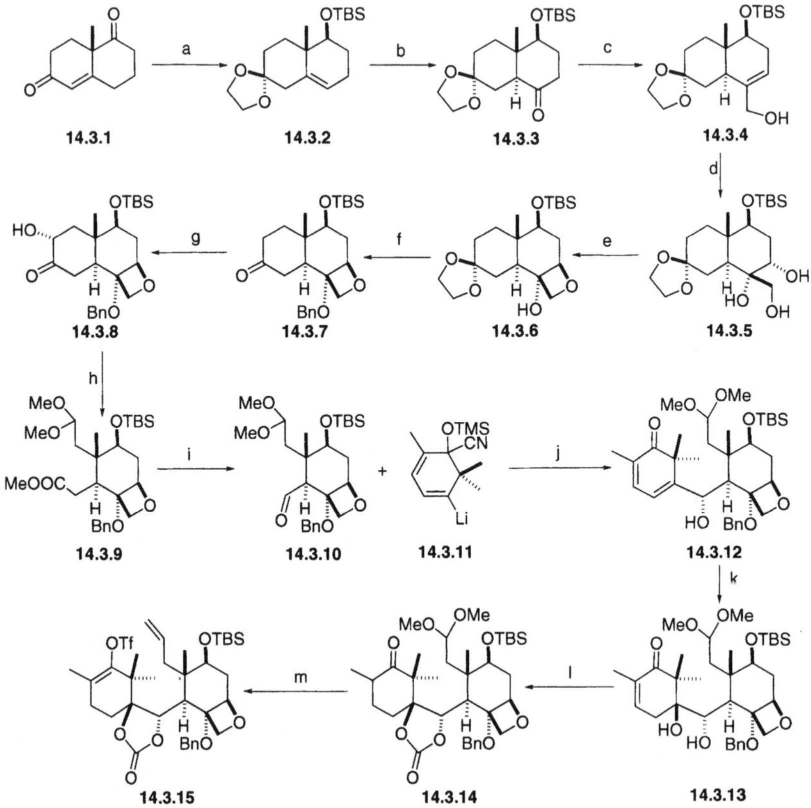

(a) $NaBH_4$, EtOH, 0 °C, then Ac_2O, py, DMAP, CH_2Cl_2, 0 °C, then $(HOCH_2)_2$, PhH, naphthalene sulfonic acid, reflux, then NaOMe, MeOH, THF, then TBSOTf, 2,6-lutidine, CH_2Cl_2, 0 °C, 64%; (b) BH_3.THF, THF, 0 °C to rt, then H_2O_2, NaOH, H_2O, then PDC, CH_2Cl_2, 0 °C to rt, then NaOMe, MeOH, 62%; (c) $Me_3S^+I^-$, KHMDS, THF, 0 °C, then $Al(O\textit{i}Pr)_3$, PhMe, reflux, 98%; (d) OsO_4, NMO, acetone, H_2O, 45%; (e) TMSCl, py, CH_2Cl_2, -78 °C to rt, then Tf_2O, -78 °C to rt, then $(HOCH_2)_2$, 40 °C, 66%; (f) BnBr, NaH, TBAI, THF, 0 °C to rt, then TsOH, acetone, H_2O, 70 °C, 82%; (g) TMSOTf, Et_3N, CH_2Cl_2, -78 °C, then 3,3-dimethyldioxirane, CH_2Cl_2, 0 °C, then CSA, acetone, rt, 89%; (h) $Pb(OAc)_4$, MeOH, PhH, 0 °C, then MeOH, PPTS, 70 °C, 94%; (i) $LiAlH_4$, THF, 0 °C, then *o*-$NO_2C_6H_4SeCN$, PBu_3, THF, rt, then 30% H_2O_2, THF, rt, then O_3, CH_2Cl_2, -78 °C, then PPh_3, 62%; (j) THF, -78 °C, then TBAF, THF, -78 °C, 74%; (k) *m*CPBA, CH_2Cl_2, rt, then H_2, Pd/C, -5 °C, EtOH, 52%; (l) CDI, NaH, DMF, then L-Selectride, THF, -78 °C, 75%; (m) $PhNTf_2$, KHMDS, THF, -78 °C, then PPTS, acetone, H_2O, then $Ph_3P{=}CH_2$, THF, -78 °C to 0 °C, 72%

coupled with the organolithium derivative **14.3.11** to give the A-CD synthon **14.3.12**.

The key cyclization to give tetracyclic product was achieved by the Heck reaction. The vinyl triflate **14.3.15** underwent cyclization in the presence of $Pd(PPh_3)_4$ to give the product **14.3.16** in 49% yield. Subsequent elaboration of **14.3.16** gave the protected baccatin III

14.3.23, with the introduction of the C-9 ketone using the procedure developed by Holton and the conversion of **14.3.21** to baccatin III following Nicolaou's procedure.

The Danishevsky synthesis, as noted earlier, is unique in its early incorporation of the oxetane ring. It also makes effective use of the starting Wieland-Miescher ketone, since this is available in enantiomerically pure form and all the stereochemistry of baccatin III derives from the chiral center of this ketone.

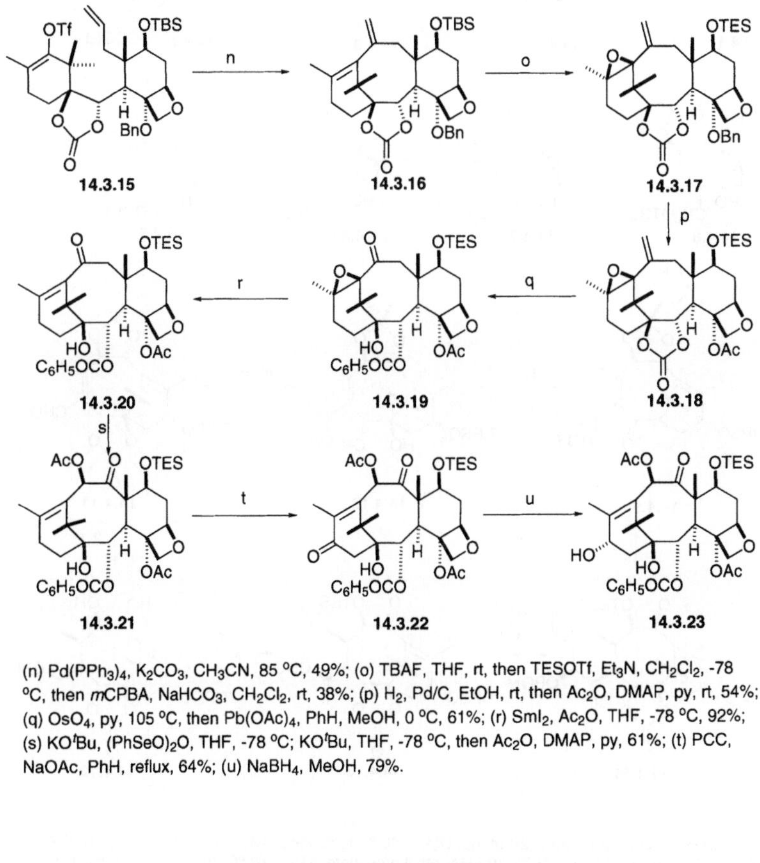

(n) $Pd(PPh_3)_4$, K_2CO_3, CH_3CN, 85 °C, 49%; (o) TBAF, THF, rt, then TESOTf, Et_3N, CH_2Cl_2, -78 °C, then *m*CPBA, $NaHCO_3$, CH_2Cl_2, rt, 38%; (p) H_2, Pd/C, EtOH, rt, then Ac_2O, DMAP, py, rt, 54%; (q) OsO_4, py, 105 °C, then $Pb(OAc)_4$, PhH, MeOH, 0 °C, 61%; (r) SmI_2, Ac_2O, THF, -78 °C, 92%; (s) KO^{t}Bu, $(PhSeO)_2O$, THF, -78 °C; KO^{t}Bu, THF, -78 °C, then Ac_2O, DMAP, py, 61%; (t) PCC, NaOAc, PhH, reflux, 64%; (u) $NaBH_4$, MeOH, 79%.

14.4. The Wender Synthesis

The Wender synthesis was reported in two communcations in 1997 (*435*, *436*), but the full papers have yet to appear. The synthesis is linear of the form A→AB→ABC→ABCD, but it is very different from the formally similar Holton approach.

The synthesis depends on some key rearrangement chemistry of verbenone (**14.4.1**), the oxidation product of the abundant natural product pinene. Verbenone provided 10 of the 20 carbons of the baccatin III ring system and also its chirality. Verbenone (**14.4.1**) was converted to the chrysanthenone derivative **14.4.3** by photorearrangement of

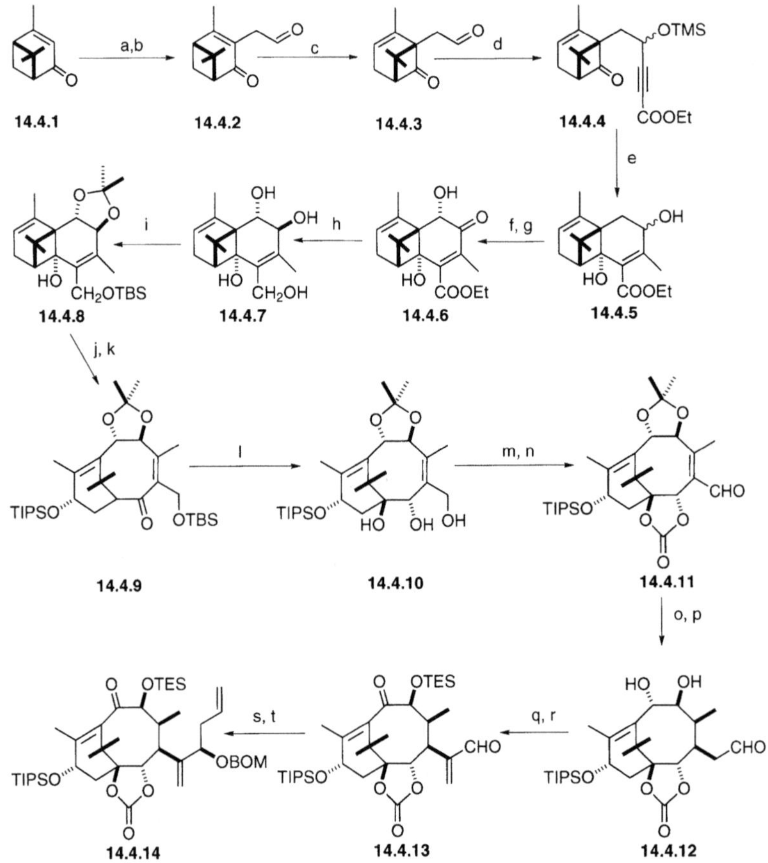

(a) KO*t*-Bu, 1-bromo-3-methyl-2-butene, DME, -78 °C to rt, 79% at 41% conversion; (b) O_3, CH_2Cl_2, MeOH, 85%; (c) *hv*, MeOH, 85%; (d) LDA, ethyl propiolate, THF, -78 °C, then TMSCl, 89%; (e) Me_2CuLi, Et_2O, -78 °C to rt, then AcOH, H_2O, 97%; (f) $RuCl_2(PPh_3)_3$, NMO, acetone, 97%; (g) KHMDS, Davis' oxaziridine, THF, -78 °C to -20 °C, 97% at 57% conversion; (h) $LiAlH_4$, Et_2O, 74%; (i) TBSCl, imid., then PPTS, 2-methoxypropene, rt, 91%; (j) *m*-CPBA, Na_2CO_3, CH_2Cl_2; (k) DABCO (cat.),CH_2Cl_2, heat, then TIPSOTf, 2,6-lutidine, -78 °C, 85%; (l) KO*t*-Bu, O_2, $P(OEt)_3$, THF, -40 °C, then NH_4Cl, MeOH, rt, then $NaBH_4$, 91%; (m) H_2, Crabtree's catalyst, CH_2Cl_2, rt, then TMSCl, py, -78 °C, then triphosgene, 0 °C, 98%; (n) PCC, 4 A molecular sieves, CH_2Cl_2, 100%; (o) $Ph_3PCHOMe$, THF, -78 °C, 91%; (p) 1 N HCl, NaI, dioxane, 94% at 90% conversion; (q) TESCl, py, CH_2Cl_2, -30 °C, 92%; (r) Dess-Martin periodinane, CH_2Cl_2, then Et_3N, Echenmoser's salt, 97%; (s) allyl-MgBr, $ZnCl_2$, THF, -78 °C, 89%; (t) BOMCl, $(i\text{-}Pr)_2NEt$, 55 °C.

14.4.2, and **14.4.3** was then converted to the tricyclic intermediate **14.4.8**. Oxidative cleavage of **14.4.8** then gave the AB synthon **14.4.9** in good yield.

The C-ring of baccatin III was elaborated through the C-3 position, which was used as a base on which to build aldehyde side chain of **14.4.16**. Aldol cyclization and protection then gave **14.4.17**, which was converted to baccatin III (**14.4.23**) or 10-deacetylbaccatin III (**14.4.22**) by the reactions shown. The overall synthesis, at 37 steps from verbenone, is claimed to be the shortest reported synthesis of taxol.

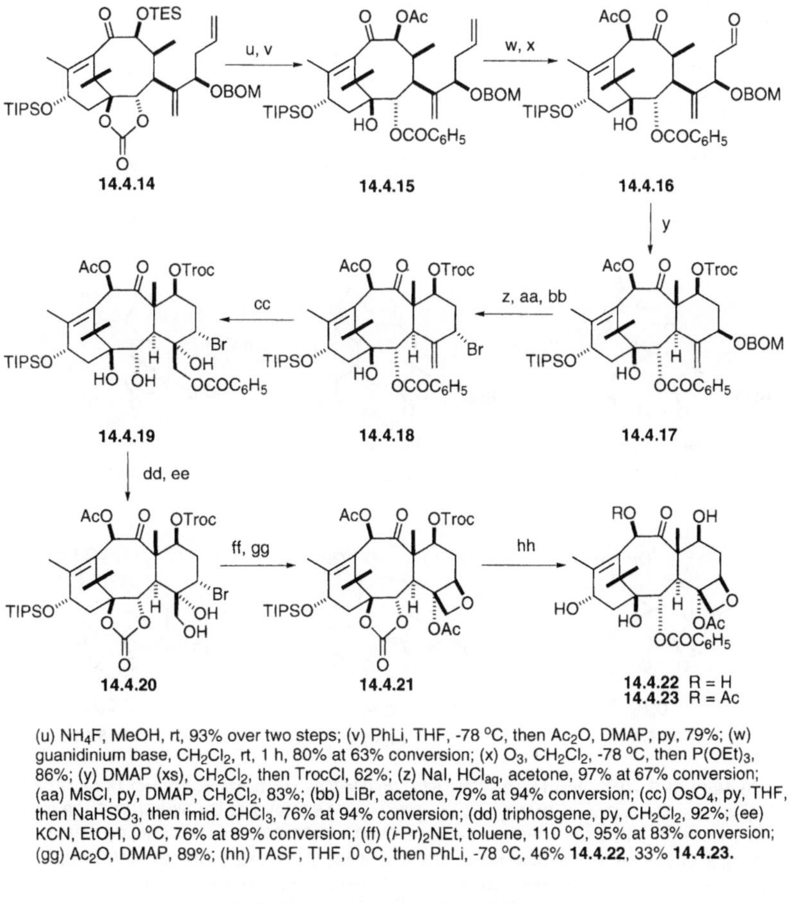

(u) NH_4F, MeOH, rt, 93% over two steps; (v) PhLi, THF, -78 °C, then Ac_2O, DMAP, py, 79%; (w) guanidinium base, CH_2Cl_2, rt, 1 h, 80% at 63% conversion; (x) O_3, CH_2Cl_2, -78 °C, then $P(OEt)_3$, 86%; (y) DMAP (xs), CH_2Cl_2, then TrocCl, 62%; (z) NaI, HCl_{aq}, acetone, 97% at 67% conversion; (aa) MsCl, py, DMAP, CH_2Cl_2, 83%; (bb) LiBr, acetone, 79% at 94% conversion; (cc) OsO_4, py, THF, then $NaHSO_3$, then imid. $CHCl_3$, 76% at 94% conversion; (dd) triphosgene, py, CH_2Cl_2, 92%; (ee) KCN, EtOH, 0 °C, 76% at 89% conversion; (ff) $(i\text{-}Pr)_2NEt$, toluene, 110 °C, 95% at 83% conversion; (gg) Ac_2O, DMAP, 89%; (hh) TASF, THF, 0 °C, then PhLi, -78 °C, 46% **14.4.22**, 33% **14.4.23**.

14.5. The Kuwajima Synthesis

The Kuwajima synthesis was reported in a communication in 1998 (*437*) with the full report appearing in 2000 (*438*). The approach used is an A+C→A-C→ABC→ABCD approach. The A-ring synthon **14.5.1** was prepared from propargyl alcohol in about 16 steps with the stereochemistry being provided by an asymmetric dihydroxylation reaction. Coupling of **14.5.1** with the C-ring synthon **14.5.2** gave **14.5.3**, which underwent Lewis acid-catalyzed cyclization in an original reaction to the tricyclic intermediate **14.5.4**. The C-18 methyl group was introduced via the cyclopropane intermediate **14.5.7**. The synthesis was completed by addition of the oxetane ring and the C-9 oxygen function. One difference from previous approaches was the protection of the C-1, C-2 hydroxyl groups as a cyclic benzylidene acetal rather than as a cyclic

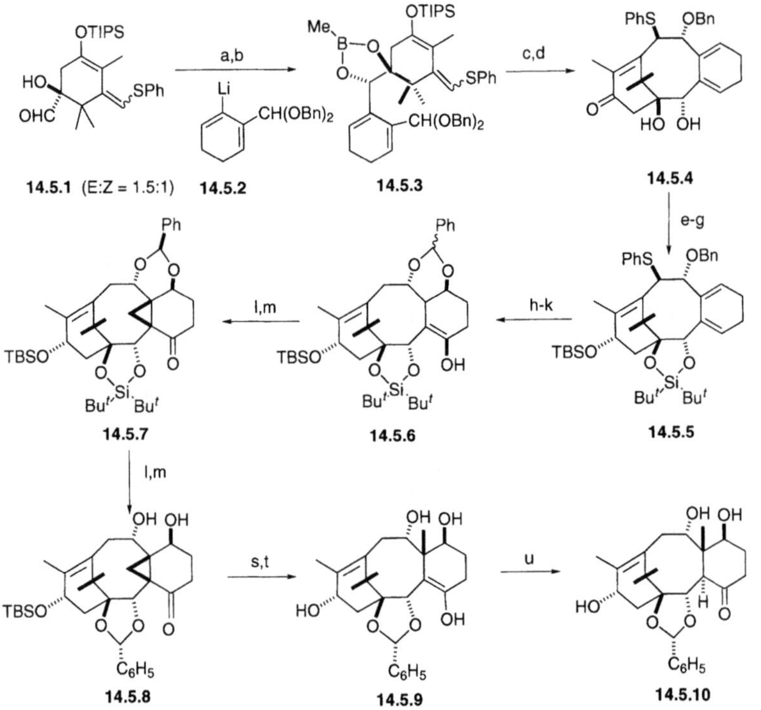

(a) ^{t}BuMgCl, THF, -78°C; then **14.5.2**, -78°C, 68%; (b) $(MeBO)_3$, pyr, benzene, 77%; (c) $TiCl_2(O^{i}Pr)2$, CH_2Cl_2, -78-0°C; (d) Pinacol, DMAP, benzene, 59% from **14.5.3**; (e) BuLi, $^{t}Bu_2Si(H)Cl$, THF, -78-0°C; (f) DIBAL, CH_2Cl_2, -78°C; (g) TBSOTf, 2,6-lutidine, CH_2Cl_2, -23°C, 62% from **14.5.4**; (h) O_2, TPP,hv, CH_2Cl_2; (i) Bu_3SnH, AIBN, benzene, reflux, 80% from **14.5.5**; (j) Pd/C, H_2, EtOH. (k)$PhCH(OMe)_2$, CSA, CH_2Cl_2, -23°C, 79% in 2 steps (a:b = ca. 1:4; (l) Et_2Zn, $ClCH_2I$, toluene, 0 °C, 66%; (m) Dess-Martin, CH_2Cl_2, 77%; (n) $Pd(OH)_2$, H_2, EtOH, 84%; (o) Triphosgene, pyr, CH_2Cl_2, -45 °C; (p) TBAF, AcOH, THF, 95% in 2 steps; (q) $PhCH(OMe)_2$, PPTS, benzene, reflux, 86%; (r) K_2CO_3, MeOH-THF, quant.; (s) SmI_2, THF-MeOH-HMPA, quant.; (t) TBAF, BHT, THF. (u) NaOMe, BHT, degassed MeOH, 45% in 2 steps (ca 50% of **14.5.9** was recovered); 65% by repeating this isomerization procedure twice.

carbonate; the acetal was, however, replaced with the cyclic carbonate late in the synthesis. An additional feature of interest is the derivation of all the stereochemistry from the C-1 position of synthon **14.5.1**, derived ultimately from asymmetric dihydroxylation.

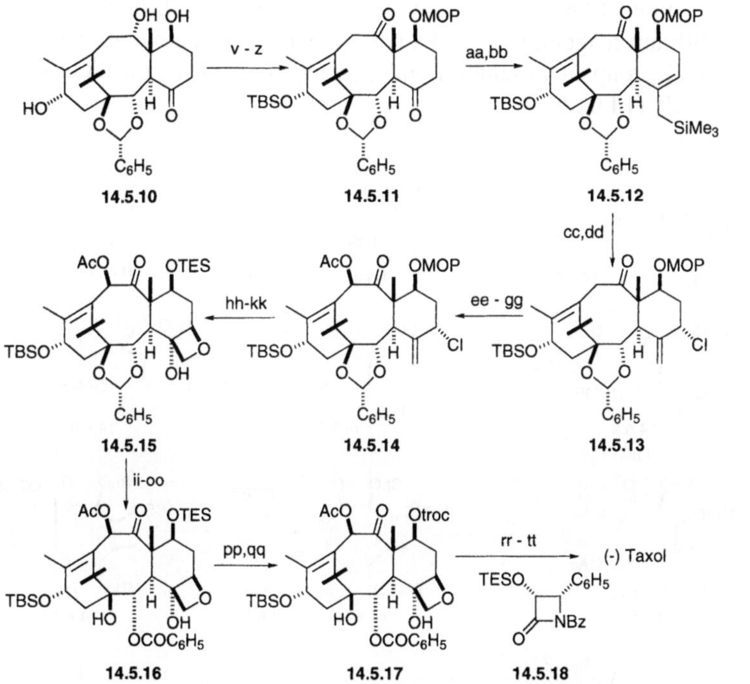

(v) $PhB(OH)_2$,CH_2Cl_2; (w) TBSOTf, 2,6-lutidine, CH_2Cl_2, -45 °C; (x) H_2O_2, $NaHCO_3$, AcOEt, 70% from **14.5.10**; (y) Dess-Martin, CH_2Cl_2, 92%; (z) 2-Methoxypropene, PPTS, CH_2Cl_2, 97%; (aa) KHMDS, $PhNTf_2$, -78 °C, 89%; (bb) $Pd(PPh_3)_4$, $TMSCH_2MgCl$, Et_2O, 91%; (cc) NCS, MeOH, 88%; (dd) 2-methoxypropene, PPTS, CH_2Cl_2, 89%; (ee) LDA, MoOPH, THF, -23 °C, 80%. (ff) Ac_2O, DMAP, CH_2Cl_2, 92%; (gg) DBN, toluene, reflux, 68% at 92% conversion; (hh) OsO_4, pyr, Et_2O, 86%; (ii) DBU, toluene, reflux, 86%; (jj) PPTS, MeOH; (kk) TESCl, imidazole, DMF, 97% in 2 steps; (ll) $Pd(OH)_2$, H_2, EtOH, 97%; (mm) triphosgene, pyr, CH_2Cl_2, -78 -0 °C, 94%; (nn) Ac_2O, DMAP, CH_2Cl_2, 66%; (oo) PhLi, THF, -78 °C, 83%; (pp) HF.pyr, THF, 88%; (qq) TrocCl, pyr, CH_2Cl_2, 94%; (rr) TASF, THF, 80%; (ss) LHMDS, **14.5.18**, THF, -78 -0 °C, 77% at 90% conversion; (tt) Zn, AcOH-H_2O, 84%.

14.6. The Mukaiyama Synthesis

The Mukaiyama synthesis was published in full in 1999 (*278*), after various preliminary reports (*439*, *440*, *441*).This synthesis was unique in that the B-ring was formed first, leading to a B→BC→ABC→ABCD approach. The aldehyde **14.6.3** was prepared by an asymmetric aldol reaction, and this was elaborated to **14.6.5** and thence to **14.6.8**. Cyclization of **14.6.8** and elimination gave the key ring-B synthon **14.6.10**. Ring C was constructed by conjugate addition of a vinyllithium reagent derived from **14.6.11**, followed by an aldol reaction, and gave the BC unit **14.6.15**. Ring A was then added through another organolithium reagent, followed a few steps later by a pinacol reduction. Functional group manipulation and oxetane ring formation then led to baccatin III (**14.6.26**).

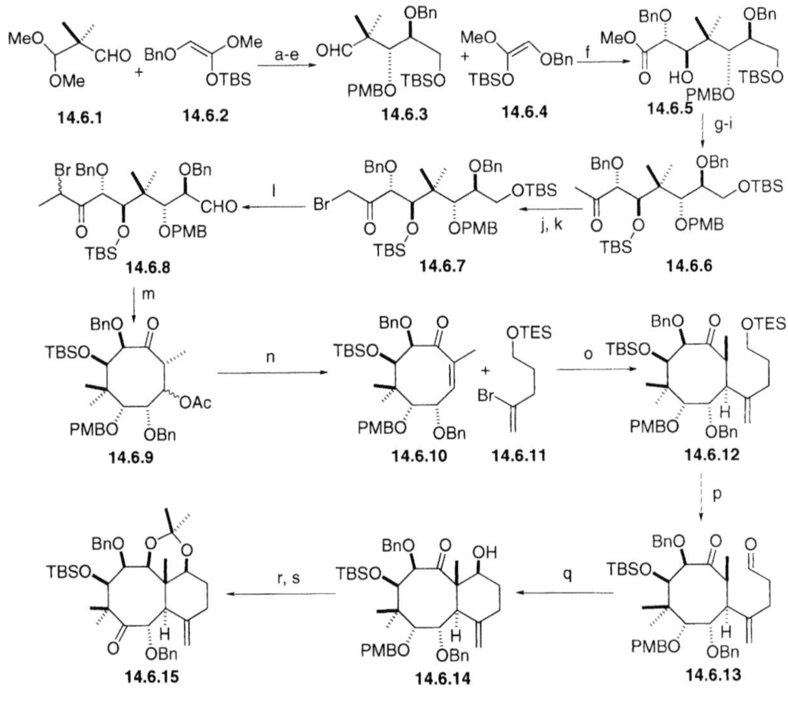

(a) $Sn(OTf)_2$, chiral diamine, $n$$Bu_2Sn(OAc)_2$, CH_2Cl_2, -23 °C, 68%, *anti/syn* = 80/20; (b) $PMBOC(CCl_3){=}NH$, TfOH, Et_2O, 0 °C, 95%, *anti/syn* = 80/20; (c) $LiAlH_4$, THF, 0 °C, 86%; (d)TBSCl, Imidazole, CH_2Cl_2, rt (93%); (e) AcOH, H_2O, THF, rt, 87% (f) $MgBr_2$.OEt, toluene, -15 °C, 87% at 88% conversion; (g) TBSOTf, 2,6-lutidine, CH_2Cl_2, 0 °C, 100%; (h) DIBAL, toluene, -78 °C, then $(COCl)_2$, DMSO, Et_3N, CH_2Cl_2, -78 °C to rt, 94%; (i) MeMgBr, Et_2O, -78 °C, then $(COCl)_2$, DMSO, Et_3N, CH_2Cl_2, -78 °C to rt 97%; (j) LHMDS, TMSCl, THF, -78 °C (k) NBS, THF, 0 °C, 100% from **14.6.6**; (l) LHMDS, MeI, HMPA, THF, -78 °C, then 1N HCl, THF, rt, then $(COCl)_2$, DMSO, Et_3N, CH_2Cl_2, -78 °C to rt 79%; (m) SmI_2, THF, -78 °C, 70%; then Ac_2O, DMAP, py, rt 87%; (n) DBU, PhH, 60 °C, 91%; (o) *t*BuLi, CuCN, Et_2O, -23 °C, 99% at 93% conversion; (p) 0.5 N HCl, THF, 0 °C, then TPAP, NMO, 4 molecular sieves, CH_2Cl_2, 0 °C, 89%; (q) NaOMe, MeOH, THF, 0 °C, then NaOMe, THF, rt, 86% at 85% conversion; (r) AlH_3, toluene, -78 °C, then $Me_2C(OMe)_2$, CSA, CH_2Cl_2, rt, 94%; (s) DDQ, H_2O, CH_2Cl_2, rt, then PDC, CH_2Cl_2, rt, 91% on 96% conversion.

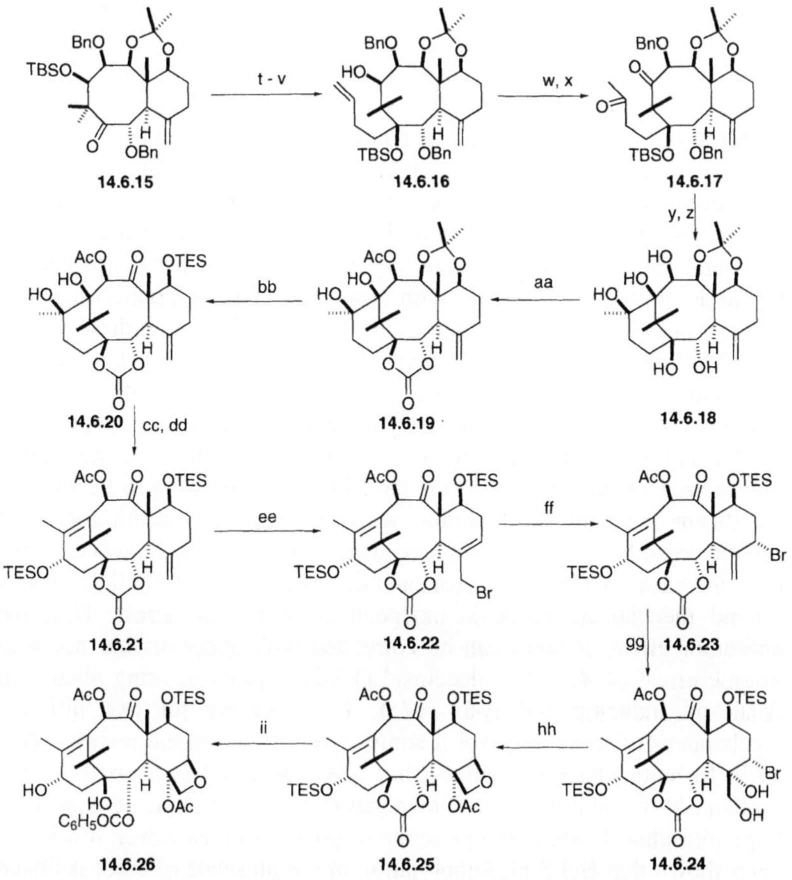

(t) Homoallyl iodide, *s*BuLi, *c*-hexane, PhH, -23 °C to 0 °C, then TBAF, THF, 50 °C, 96%; (u)*t*BuMeSi(OTf)$_2$, py, 0 °C, 100%; (v) MeLi, HMPA, THF, -45 °C, 96%; (w) TPAP, NMO, 4 molecular sieves, CH_2Cl_2, CH_3CN, rt, 85%; (x) $PdCl_2$, H_2O, DMF, rt, 91%; (y) $TiCl_2$, $LiAlH_4$, THF, 40 °C, 50%; (z) Na, liq. NH_3, -78 °C to -45 °C, then TBAF, THF, rt, 93%; (aa) $(CCl_3)_2CO$, py, CH_2Cl_2, -45 °C, 100%; (bb) 3N HCl, THF, 60 °C, then TESCl, py, rt, then TPAP, NMO, 4 molecular sieves, CH_2Cl_2, rt, 63%; (cc) TCDI, DMAP, toluene, 100 °C, then $P(OMe)_3$, 110 °C, 53%; (dd) PCC, NaOAc, Celite, PhH, 95 °C, 78%, then K-Selectride, THF, -23 °C, 87%, then TESOTf, py, -23 °C, 98%; (ee) CuBr, $PhCO_3$*t*Bu, CH_3CN, -23 °C, 62% of **14.6.22**, 15% of **14.6.23**; (ff) CuBr, CH_3CN, 50 °C, 25% of **14.6.22**, 69% of **14.6.23**; (gg) OsO_4, py, THF, rt, 96% at 96% conversion; (hh) DBU, py, toluene, 50 °C, 81% at 52% conversion; (ii) PhLi, THF, -78 °C, then HF-py, THF, rt, 90%.

15. The Interaction of Taxol with Tubulin

The bioactivity and mechanism of action of taxol has been extensively investigated, but since this review is targeted to a chemical audience this section will focus primarily on chemical aspects of its mechanism. As noted in the introduction to this review, taxol was discovered on the basis of the antileukemic and cytotoxic activities of *Taxus brevifolia* extracts (*6*); taxol itself was found to be both strongly cytotoxic to mammalian cancer cells and also antileukemic (*1*). A major breakthrough in understanding its mechanism of action at the molecular level came in 1979 with the report by Horwitz of its ability to promote the assembly of tubulin into microtubules (2). The action of taxol on microtubule assembly has been reviewed on numerous occasions (*442*, *443*) and this summary can thus be abbreviated. In brief, taxol binds to the assembled microtubule with a stoichiometry of approximately 1 mole of taxol to 1 mole of tubulin dimer and stabilizes it to dissociation. This binding occurs in the absence of any cofactors, and the resulting disruption of the equilibrium between tubulin and microtubules also disrupts cell division and ultimately leads to cell death by apoptosis.

The binding of taxol to tubulin polymers and the associated interruption of the cell cycle was thought for a long time to be its only significant mechanism of action, with apoptotic cell death following mitotic arrest by an unknown mechanism. In recent years, however, it has been increasingly clear that taxol can bring about apoptotic cell death by a second mechanism which is independent of mitotic arrest. Thus the antitumor effect of taxol can be correlated with apoptosis but not with mitotic arrest (*444*), while docetaxel (**1.1.2**) appears to bring about cell death by inducing cell lysis (*445*). The evidence for two different mechanisms of action of taxol is summarized in two recent reviews (*446*, *447*). Recently the protein Bcl-2 has been identified as a taxol-binding protein (*448*), and it has been shown that Bcl-2 undergoes dose-dependent hyperphosphorylation in the presence of taxol (*449*). However, it has also been shown that Bcl-2 phosphorylation in the presence of taxol is linked to the latter's tubulin-assembly activity, since activation of Raf-1 kinase (a key enzyme for phosphorylation of Bcl-2) requires the interaction of taxol with tubulin. It has thus been proposed that taxol's tubulin-assembly activity leads to Raf-1 activation and Bcl-2 phosphorylation, leading to apoptosis (*450*). Regardless of the eventual outcome of this debate, the binding of taxol to tubulin is clearly biologically significant, and has been studied extensively by several methods. A detailed understanding of this binding has become much more achievable in recent years thanks to the work of Downing and his collaborators who have reported the structure of tubulin at a resolution of 3.7Å using electron crystallography on crystalline sheets formed in the presence of zinc (*451*, *452*).

It should also be pointed out that taxol and its analogs have been found to have other effects besides their anticancer activity. Thus Georg and her colleagues have reported taxol's activity in protecting against

β-amyloid toxicity (*453*), and both Ojima and his colleagues (*147*) and Kuwajima and his colleagues (*454, 455*) have reported on various semisynthetic taxanes with activity as reversal agents for multidrug resistance in cancer cells; these studies have been reviewed (*456, 457*). The ability of taxol and its analogs to induce nitric oxide and tumor necrosis factor has been described (*458, 459, 460*), and the effects of changes in the A-ring and the C-10 acyl group on this activity have been studied (*461*). Taxol has also been shown to induce the production of cytokines and to induce tyrosine phosphorylation of proteins (*462, 463, 464*); recently it has been shown to mediate serine phosphorylation of the 66-kDa SHc isoform, a protein implicated in a signaling cascade leading to Ras activation (*465*).

15.1. Photoaffinity Labeling Studies

The location of the binding site of taxol on tubulin has been determined by photoaffinity labeling using several differently labeled taxol analogs. An early study showed that taxol labeled β-tubulin as opposed to α-tubulin (*466*), and this result was confirmed with the azidophenyl taxol analog **15.1.1** (*467*).

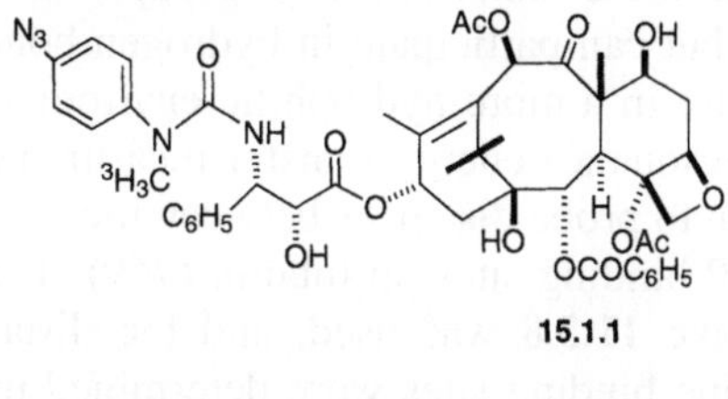

15.1.1

More detailed studies showed that 3′-(*p*-azidobenzamido)taxol (**11.2.1**) photolabeled β-tubulin preferentially (*387*) and that the site of labeling was in the N-terminal 31 amino acid unit (*386*). On the other hand, 2-(*m*-azidobenzoyl)taxol (**11.2.10**) labeled a peptide containing amino acids 217–231 of β-tubulin (*393*). The photoaffinity probe **11.2.5** was also found to bind (most probably) to β-tubulin (*389*), while the probe **11.2.6** was shown to bind to Arg^{282} in β-tubulin (*468*).

15.2. Fluorescence Spectroscopic Studies

Fluorescence spectroscopy has several useful applications in studying the binding site, and it has the significant advantage that it is performed on systems in solution, thus avoiding potential problems with changes in shape on preparing solid samples for analysis. The 3′-*N*-(dimethylaminobenzoyl) derivative **11.3.1** was shown to bind to tubulin in a dimeric state as well as to polymerized tubulin, but the binding affinity to dimeric tubulin was a relatively weak 49 μM, as compared with an estimated 10 nM for binding to polymerized tubulin (*394*).

The 2-(*m*-aminobenzoyl)taxol derivative **11.3.8** gave interesting results, since it was found that its absorption and emission spectra, and specifically the difference between them, were solvent dependent. Using this information it was possible to ascertain that the fluorophore binding site on the microtubule is in an environment of intermediate polarity, and it was also shown that tubulin has two binding sites for taxol, one high affinity site and one low affinity site (*397*).

The Georg and Himes groups carried out a detailed investigation of the fluorescence of the derivatives **11.3.3–11.3.7** (*396*). In this study they observed differences between the 2- and *N*-substituted derivatives **11.3.5** and **11.3.3** on the one hand and the 7- and 10-substituted derivatives **11.3.6** and **11.3.7** on the other hand. The 2- and *N*-substituted derivatives were found to bind to microtubules with an increase in fluorescence intensity over that observed with tubulin, while binding of the 7- and 10-substituted derivatives resulted in a large quenching of fluorescence. It was concluded that the 2- and *N*-benzoyl groups bind in a hydrophobic pocket of tubulin but can participate in hydrogen bonding, while the 7- and 10-positions are in a more hydrophilic environment.

Fluorescence resonance energy transfer measurements were used by Bane and Kingston to probe distances between the taxol, colchicine, and exchangeable GTP binding sites on tubulin (*469*). The 2-(*m*-aminobenzoyl)taxol derivative **11.3.8** was used, and the distances between the taxol and colchicine binding sites were determined using two different acceptor ligands in the colchicine site; an average distance of 17 Å was found in each case. Results with a fluorescent ligand bound in the GTP binding site were less clearcut, and it was suggested that compound **11.3.8** serves as an energy transfer donor for two GTP sites, one on the same subunit and one on an adjacent subunit.

A later paper from the same group describes results with the 3′-*N*-(*m*-aminobenzoyl)taxol **11.3.2** (*395*). It was found that **11.3.2** bound to two types of site on microtubules with binding affinities of 61 nM and 3.3 μM. It bound to a single site on microtubules assembled from GDP-tubulin with a dissociation constant of 2.5 μM, and it bound to a single site on microtubules assembled from the GTP analog GMPCPP with a dissociation constant of 15 nM. It was thus proposed that although all the subunits of the microtubule at the steady state are the same "GTP-tubulin-taxol", they are formed through two different pathways: either from taxol binding to a tubulin subunit before GTP hydrolysis (a high affinity binding) or taxol binding to a tubulin subunit after GTP hydrolysis (a low affinity binding).

A recent paper describes the use of Flutax-1 and -2 (**11.3.12** and **11.3.13**) as probes for the study of the binding of taxol to microtubules. A binding mechanism consisting of a fast bimolecular reaction followed by at least two slow monomolecular rearrangements was developed. The results also suggested that the taxol binding site is directly accessible, in opposition to some other models of microtubules (*470*).

15.3. Nuclear Magnetic Resonance Studies

A number of investigators have studied the NMR spectra of taxol in various solvents in an attempt to determine the solution conformation. Work in this area in nonpolar solvents has been reported by Baker (*471*), Hilton *et al.* (*472*), Falzone *et al.* (*473*), Dubois *et al.* (*474*), Williams *et al.* (*475, 476*), Balasubramanian *et al.* (*477*), Cachau *et al.* (*478*), Ojima *et al.* (*479*), and Milanesio *et al.* (*480*), and work in polar solvents has been reported by Dubois *et al.* (474), Williams *et al.* (475), Ojima *et al.* (*479, 481*), Vander Velde *et al.* (*482*), Paloma *et al.* (*483*), and Jimenez-Barbero *et al.* (*484*). Different types of conformations in polar and non-polar solvents are reported by these groups; in non-polar solvents such as chloroform taxol seems to exist primarily in a set of "open" conformations in which the side chain is oriented away from the 2-benzoyl group, but in polar aqueous solvents it adopts a set of "hydrophobically collapsed" conformations in which the 3′-phenyl group is oriented towards the 2-benzoyl group. The tricyclohexyltaxol analog prepared by Georg and her coworkers did not undergo hydrophobic collapse in polar solvents, however, suggesting that the presence of aromatic rings is important to this process (*119*). Over-interpretation of these results in terms of a "binding conformation" of taxol to tubulin is, however, dangerous, since it has been pointed out by Snyder (*485*) that taxol exists in chloroform (and presumably also in polar solvents) as a population of different conformations. Eight optimized conformations were found in chloroform by the NAMFIS methodology, with the most important one contributing 35% to the total. The point was made that attempts to express the taxol conformation as a single or "virtual" conformation are incomplete, and that the presence in solution of energetically similar conformations means that the molecule does not need to "preorganize" to bind to tubulin. It was also suggested that taxol's relatively weak association with tubulin may be due in part to the presence of a large number of nonproductive conformers.

NMR studies have also been carried out on active cyclohexyl analogs of taxol (*119*), on the taxol analogs 2′-deoxytaxol and 2′-methoxytaxol, which were shown to adopt conformations similar to those of taxol (*486*), and on D-secotaxol analogs (*208*) as noted earlier in section 5.4.

Under ideal circumstances it would be wonderful to be able to study the conformation of taxol bound to microtubules by NMR, using the nuclear Overhauser effect to determine internuclear distances. Unfortunately this simple experiment cannot be performed because of the insolubility of microtubules, but NMR studies can be made using the technique of solid state magic angle spinning to enable data to be obtained on polymerized tubulin. Two such studies have been reported to date. In one study fluorine-containing taxol analogs were used to obtain an F-F distance of 6.5 Å in the difluorotaxol **15.3.1** (*487, 488*). In the second study, internuclear distances between ^{13}C and F were obtained by the

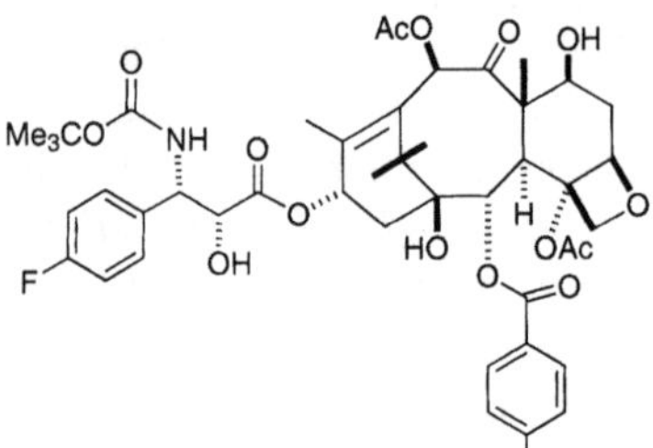

15.3.1

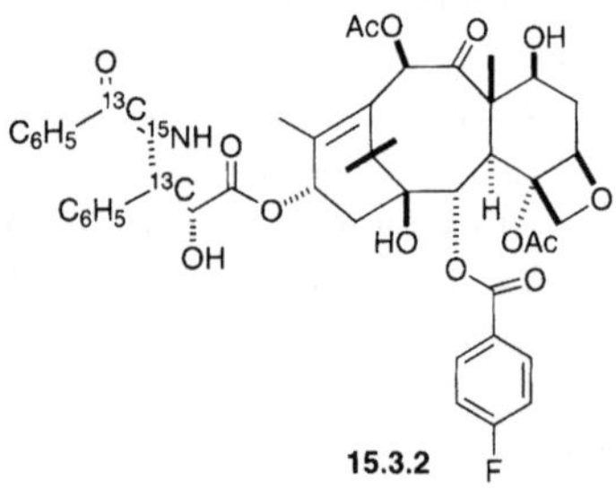

15.3.2

REDOR technique on the quadruply labeled analog **15.3.2** (*489*), and these results coupled with fluorescence data led to the proposal of structure **15.3.3** as the most probable conformation of taxol on tubulin.

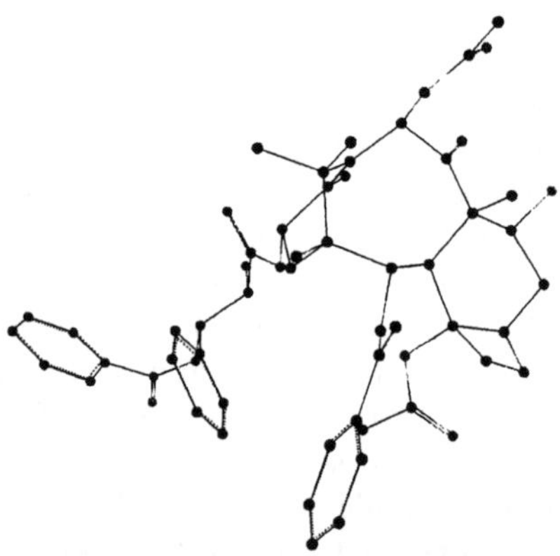

15.4. The Taxol Pharmacophore

Several attempts have been made to define the key pharmacophore of taxol. Much of the chemical work reported in the previous sections was carried out with a view to determining structure-activity relationships for taxol and its analogs and thus to defining the pharmacophore in chemical terms. Some of this work can be summarized as in structure **15.4.1** below, where the key structure-activity relationships of taxol are shown.

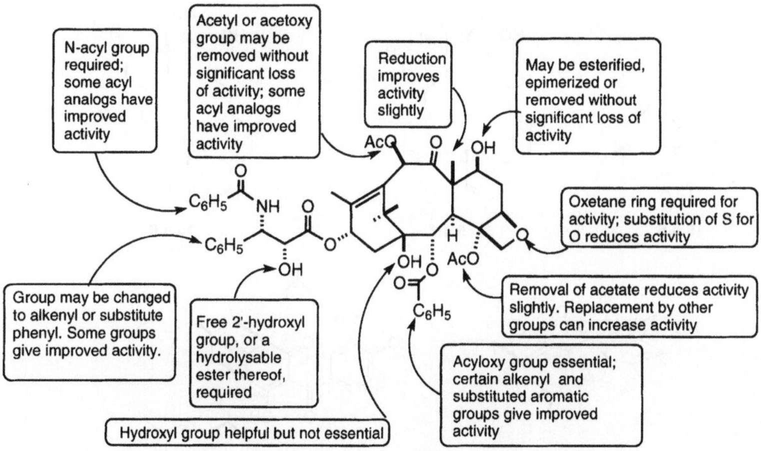

Although taxol was for many years unique in terms of its mechanism of action, in recent years a number of other natural products have been found to promote the assembly of tubulin into microtubules in the same way as taxol. The most important and familiar compounds of this class are the epothilones A (**15.4.2**) and B (**15.4.3**) (*490, 491*), discodermolide (**15.4.4**) (*492*), and eleutherobin (**15.4.5**) (*493*), but other compounds with this activity have also been discovered.

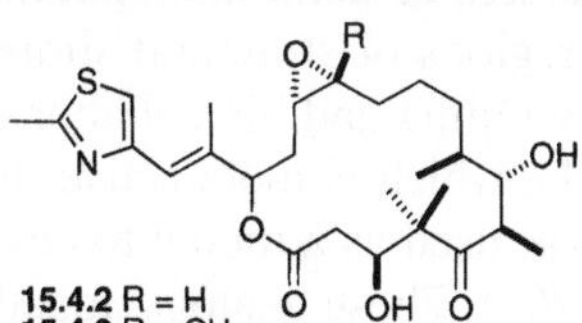

15.4.2 R = H
15.4.3 R = CH_3

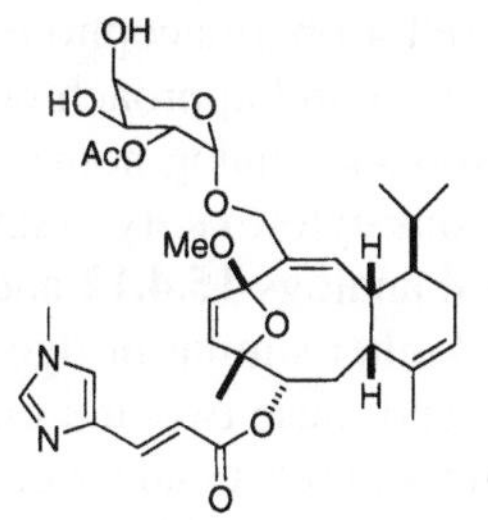

15.4.5

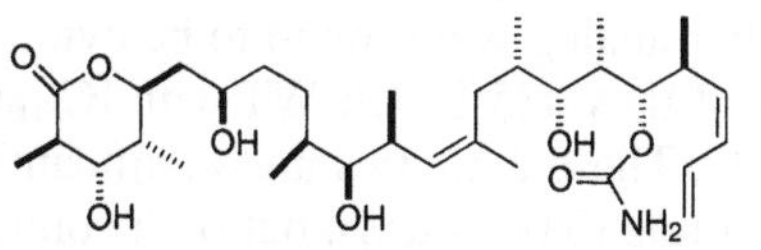

15.4.4

These include rhazinilam (**15.4.6**) (*494*), which inhibits the disassembly of microtubules but has a different mechanism of action than taxol, laulimalide (**15.4.7**) and isolaulimalide (**15.4.8**) (*496*), WS9885B (**15.4.9**) (*497*), and polyisoprenylated benzophenones such as guttiferone E (**15.4.10**) (*497*). The naturally occurring 3(2H)-furanone derivative geiparvin has been found to counteract the microtubule-assembly effects of taxol, suggesting that it is a competitive inhibitor at the taxol-binding site of tubulin (*498*).

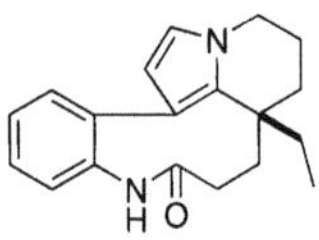
15.4.6

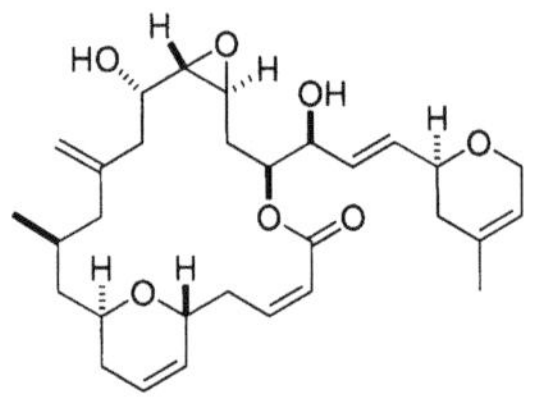
15.4.7

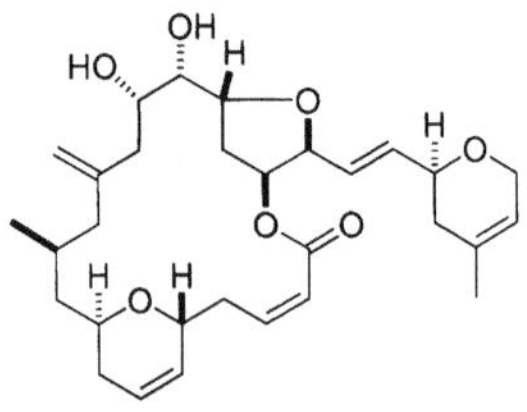
15.4.8

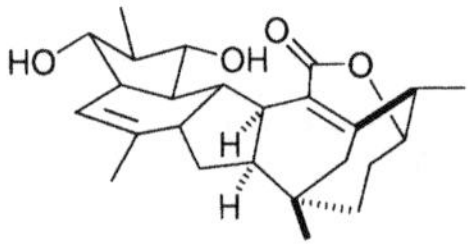
15.4.9

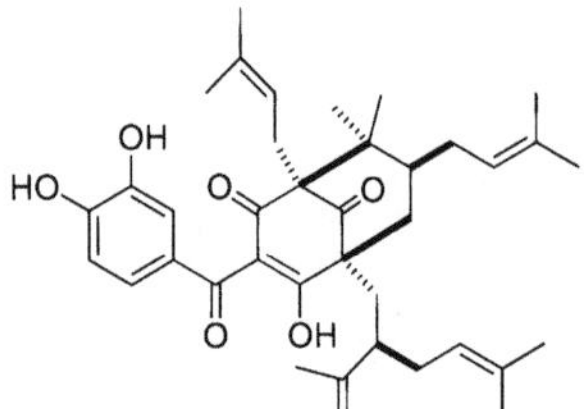
15.4.10

The discovery of the tubulin-assembly properties of these natural products have raised the obvious questions as to whether they bind to the same binding site as taxol and whether there is thus a common pharmacophore that can be deduced for all these compounds. These questions have been addressed in various ways. The first attempt at developing a common pharmacophore was by Winkler and Axelsen (*499*), who simply compared the structures of taxol and epothilone and proposed a pharmacophore based on regions of structural similarity.

In a second approach developed by Ojima and his collaborators, the non-aromatic analog nonataxel (**15.4.11**), which is more active than taxol in various cytotoxicity assays (*135*), was used as a model to develop the bridged analogs **15.4.12** and **15.4.13** (*481*). These analogs, together with several other similar bridged analogs, were prepared by olefin metathesis (*500*), and these two and one other analog were found to be cytotoxic to the human breast cancer cell line MDA-435/LCC6-WT with IC_{50} values of 0.48 and 0.39 μM, respectively. These activities are significantly less than that of taxol in the same cell line (0.0031 μM), but both compounds also showed tubulin-assembly activity 19% and 37% as great as that of

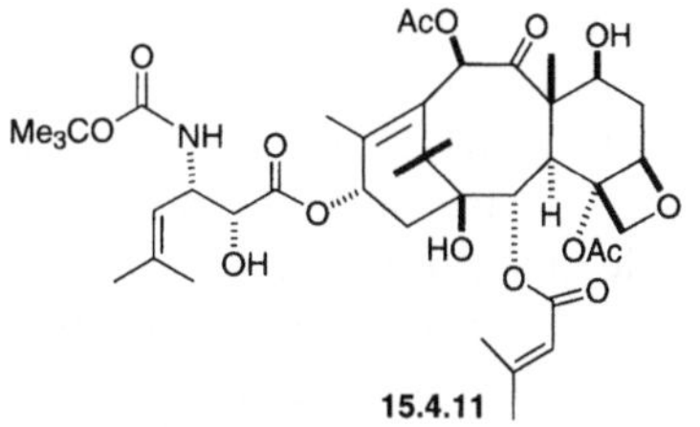

15.4.11

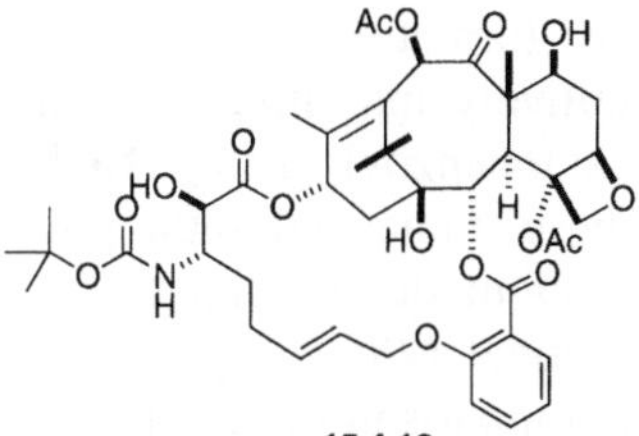

15.4.12

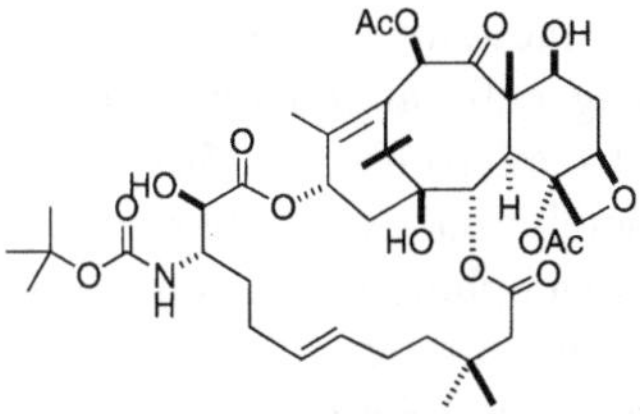

15.4.13

taxol, and so they are presumably binding to the same binding site as taxol. These data were used to support a model of the pharmacophore in which the aryl sector of epothilone overlaps the C-13 acyl side chain of taxol and in which the baccatin portion of the taxol molecule is relatively non-essential.

Two other bridged analogs have been prepared, but in both cases they proved to be inactive. Compounds **15.4.14** (two isomeric substances) were prepared by Georg *et al.* (*501*), and compound **15.4.15** was prepared by Yuan and Kingston (*502*). Compounds **15.4.14** proved to be inactive in a tubulin-assembly assay, and compound **15.4.15** was inactive in tubulin-assembly and cytotoxicity assays. These results indicate that the constraints on designing an effective taxol pharmacophore are fairly rigid, and in particular that that shorter bridges and bridges under ring C cannot be tolerated.

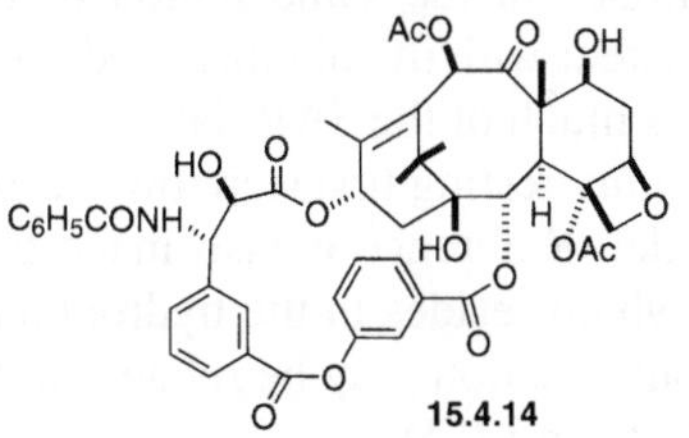

15.4.14

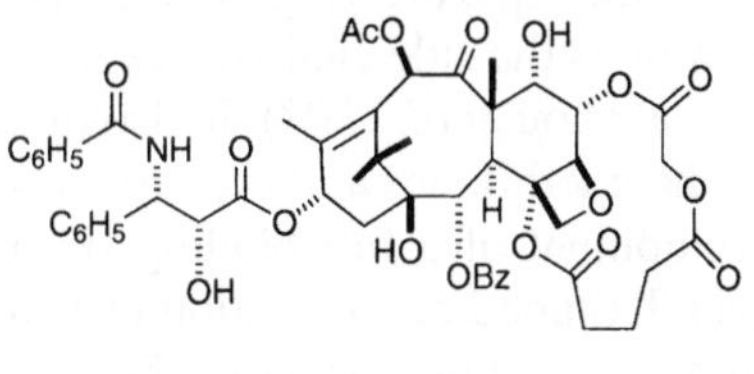

15.4.15

A third approach to the taxol pharmacophore was developed by Giannakakou *et al.* (*503*), who compared the effects of taxol and various epothilone analogs on the polymerization of native tubulin and of modified tubulins carrying β-tubulin mutations near the taxol-binding site. Two possible common overlaps of the epothilones and taxol were found; binding mode I in which the C-2 benzoyl group of taxol overlapped with much of the 1-methyl-2-thiazolyl side chain of the epothilones, while in binding mode II the thiazole portion of the epothilones overlapped with the side chain of taxol.

A fourth study by the Horwitz and Kingston groups reported that the side chain of taxol is not as essential for activity as was previously thought, since 2-(*m*-azidobenzoyl)baccatin III (**15.4.16**) is significantly active, inducing microtubule bundle formation in cultured cells, arresting cells at mitosis, and competitively inhibiting the binding of taxol to microtubules (*504*); the *p*-azidobenzoyl analog **15.4.17** was completely inactive in these assays. These studies led to the proposal of a pharmacophore model similar to binding mode I of Giannakakou *et al.* (*503*), and different from the Ojima model (*481*). A perceptive evaluation of the Giannakakou work and comparison of it with the Ojima and Horwitz-Kingston results has been provided by Day (*505*).

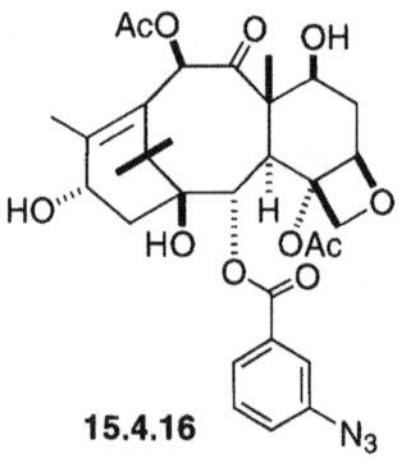

15.4.16

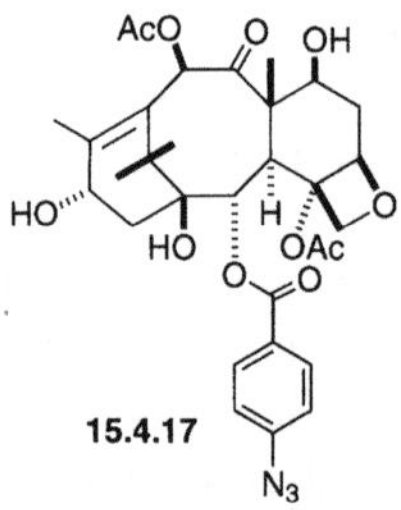

15.4.17

A different approach was taken by Snyder (*506*), who developed a minireceptor model for the binding of taxol and epothilone to the microtubule based on an analysis of tubulin-assembly data, with activities expressed as approximate K_i's and ΔG values. This model places the thiazole ring of the epothilones in the same region of the receptor as the side chain of taxol, consistent with the binding mode II of Giannakakou *et al.* (*503*). It also predicts much of the SAR data for taxol and the epothilones and appears to be an interesting first step towards the development of a full binding site model. Reference is also made to a novel T-shaped conformation of taxol which resides in the hydrophobic β-tubulin binding pocket (*507*); this conformation may be related to the "open" conformation of taxol in chloroform (*485*).

Three studies on the SAR of taxol and its analogs using comparitive molecular field analysis (CoMFA) have been published (*508*, *509*, *510*); all of the analyses indicated that the approach is a useful one which has predictive power for the development of new analogs. Finally, it has been noted that several of the electronegative atoms of taxol lie in a plane in one conformation of the side chain, and it is suggested that this observation may also assist in analog design (*511*).

In summary, we are at an interesting point in our understanding of the taxol pharmacophore. At least two rather different hypotheses have been suggested for the pharmacophore of taxol, and it will be interesting to see how future work clarifies this situation. If the initial results described above are supplemented with new results from the use of physical methods such as solid state NMR and additional computations, it is reasonable to expect that a final model of the taxol pharmacophore could be developed and used in a predictive way to create new and improved taxol analogs.

Addendum

A number of important papers have appeared since the submission of the manuscript in December 2000, and the key results from these are summarized below. The arrangement of this addendum follows that of the main text.

A1. Introduction

Two additional reviews on taxol chemistry and biology have been published. One covers its history and gives an overview of its chemistry (*512*), and the other surveys recent aspects of its chemistry and structure-activity relationships (*513*). A recent book recounts the history of the development of taxol as a drug, with an emphasis on the difficulties of the supply problem and the controversial agreement with Bristol-Myers Squibb to commercialize it (*514*).

Reviews of the clinical use of taxol continue to appear regularly. Although a discussion of the clinical use of taxol is outside the scope of this review, a useful summary of the use of taxanes in breast cancer and ovarian cancer has been published (*515*). Not all taxanes are beneficial, and a timely review of the toxicity of yew alkaloids has also appeared (*516*).

A2. A-Ring Chemistry

The synthesis of 14-bromo derivatives of baccatin III has been achieved by conversion of baccatin III to the 13-oxo-1,2-carbonate derivative **A2.1** followed by bromination to give **A2.2** (*517*). 14β-Hydroxy-13-oxo-baccatin III (**A2.4**) was formed by treatment of the enol ether **A2.3** with mCPBA. Neither **A2.2** nor **A2.4** could be reduced to the 13α-hydroxy derivatives, however, so the corresponding taxol analogs could not be prepared (*517*).

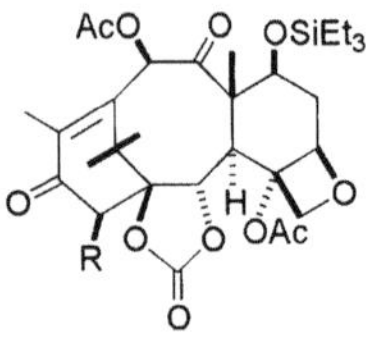

A2.1 R = H
A2.2 R = Br

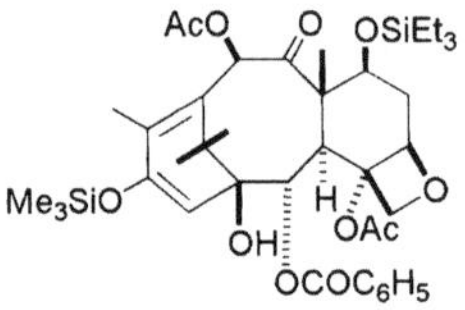

A2.3

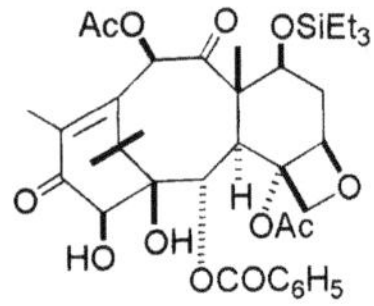

A2.4

A3. B-Ring Chemistry

Additional chemical modifications have been reported at C-10, C-9, and C-2, with most of the work centered on the accessible C-10 position.

A3.1. Modifications at C-10

Two different groups have reported the synthesis of taxol analogs with a basic group at C-10. In an extension of work reported in section *3.1.3* (*79*, *80*), Soga and his collaborators modified their C-10 analogs by preparing various ethers at C-7. The methyl ether **A3.1.1** showed the best cytotoxicity, and was active against the B16 melanoma in mice both intravenously and orally (*518*).

A different approach was taken by Park, who prepared 10-*O*-nicotinyltaxol (**A3.1.2**) and related compounds by conventional methods (*519*); unfortunately no bioactivity data was reported for these compounds.

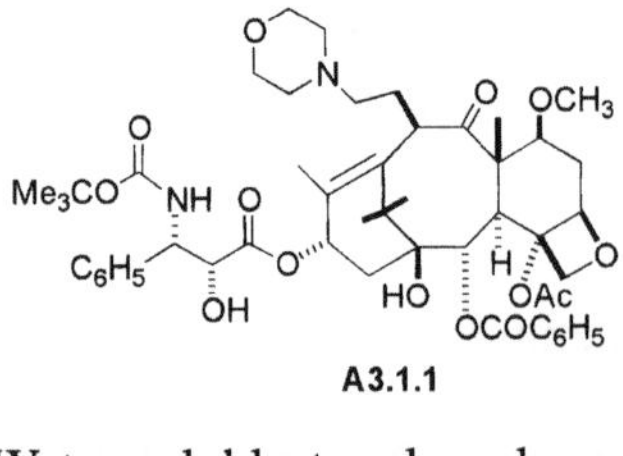

A3.1.1

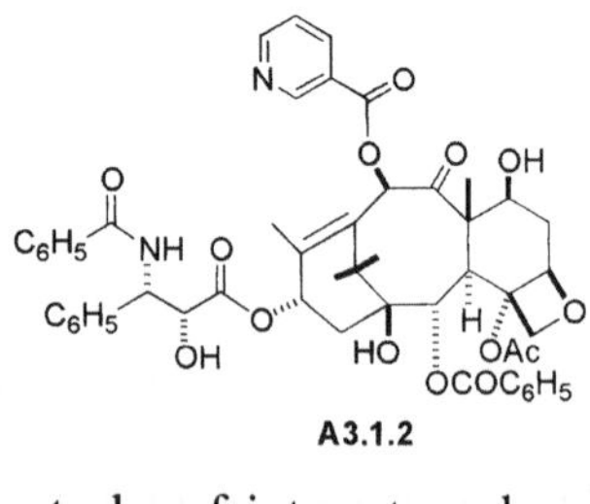

A3.1.2

Water-soluble taxol analogs continue to be of interest, and various C-10 glycosyl derivatives have been prepared by Mandai and his colleagues. The best results were obtained with simple α- or β-glucose analogs and α- or β-galactose analogs such as the α-glucosyl analog **A3.1.3**; these analogs were all more water-soluble and more cytotoxic than taxol (*520*).

A different approach to modification at C-10 was taken by Walker and his colleagues, who converted taxol into its 10-oxo derivative and then treated this with dimethylsulfonium ylide to give a spiro-epoxide which was further converted to the taxol analogs **A3.1.4** and **A3.1.5** (*521*). Both **A3.1.4** and **A3.1.5** were more active than taxol in tubulin-assembly and cytotoxicity assays, but only **A3.1.5** retained its activity against the taxol-resistant cell line HCT VM46.

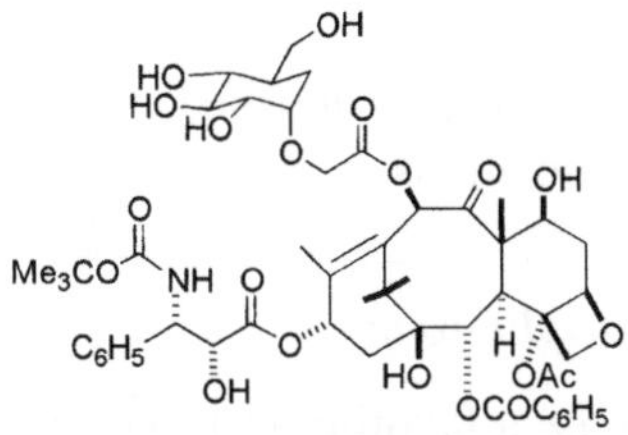

A3.1.3

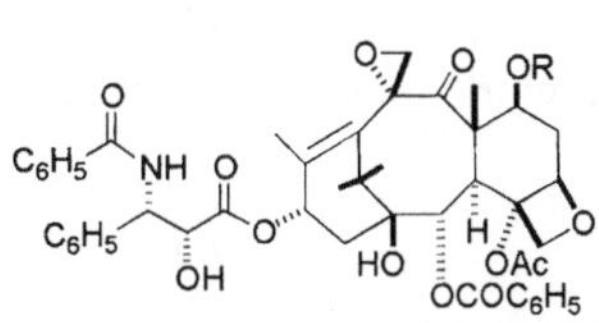

A3.1.4 R = H
A3.1.5 R = CH_2OCH_3

A final interesting finding is that baccatin III forms the crystalline complexes **A3.1.6** and **A3.1.7** with imidazole and isopropanol, respectively. These complexes allow for a simple purification of baccatin III from complex mixtures of taxanes (*522*).

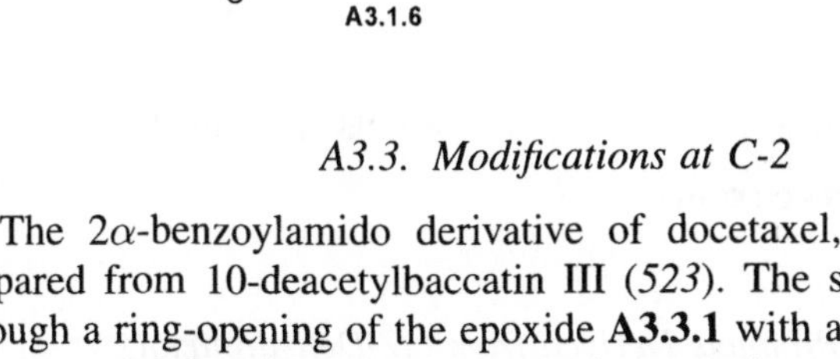

A3.1.6

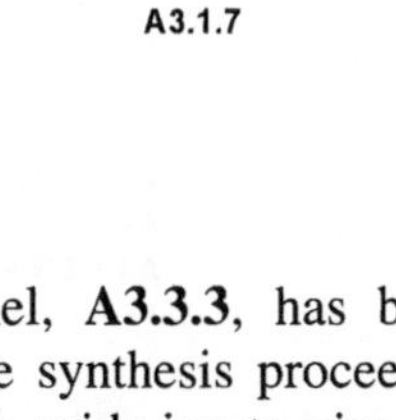

A3.1.7

A3.3. Modifications at C-2

The 2α-benzoylamido derivative of docetaxel, **A3.3.3**, has been prepared from 10-deacetylbaccatin III (*523*). The synthesis proceeded through a ring-opening of the epoxide **A3.3.1** with azide ion to give the α-azide **A3.3.2**, followed by standard conversions to the final product **A3.3.3**. The α-amide **A3.3.3** was 16–20 times less cytotoxic than taxol in A-549 and KB cell lines, but had comparable cytotoxicity to taxol in the A2780 cell line. This result is in contrast to that for the 2α-analog of taxol, which was inactive in both cytotoxicity and tubulin-assembly assays (*121*).

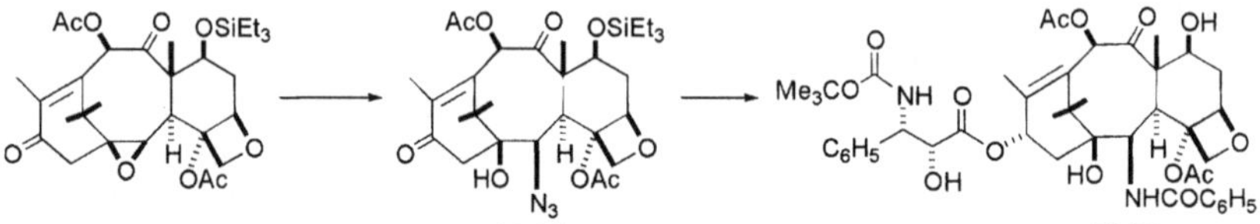

A3.3.1 **A3.3.2** **A3.3.3**

A4. C-Ring Chemistry

Several studies of C-ring chemistry have been published, with most of the work focused on the C-7 and C-4 positions.

A4.1. The C-7 Hydroxyl Group

The C-7 hydroxyl group continues to be a favorite position for derivatization. Mastalerz and his colleagues prepared 7β-thiol derivatives such as **A4.1.1–A4.1.3** by displacement of the 7β-triflate with potassium thioacetate followed by epimerization and alkylation. The analog **A4.1.2** was five times more cytotoxic than taxol to the HCT-116 cell line (*524*).

The Bristol-Myers group also prepared three taxol-chlorambucil derivatives at the C-7 position; of these the analog **A4.1.4** proved to have the best *in vivo* activity against the mouse M109 lung carcinoma, with a T/C value of 224 compared to 194–246 for taxol (*525*).

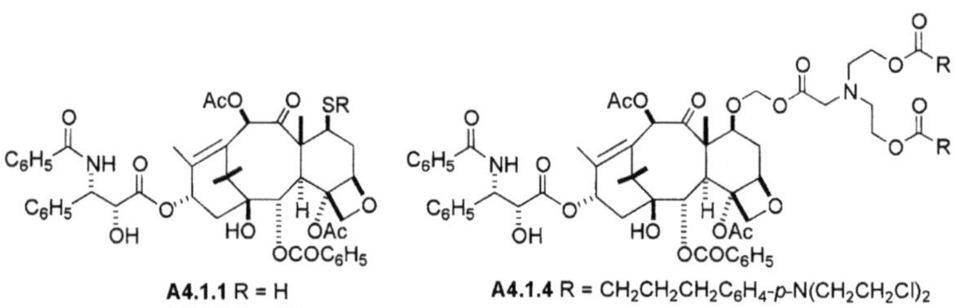

A4.1.1 R = H
A4.1.2 R = CH_3
A4.1.3 R = CH_2OCH_3

A4.1.4 R = $CH_2CH_2CH_2C_6H_4$-*p*-$N(CH_2CH_2Cl)_2$

A4.2. Products Derived from 6,7-Dehydrotaxol

The primary human metabolite of taxol is 6α-hydroxytaxol, so analogs which have the 6α-position blocked should have a longer half-life in a clinical setting. The synthesis of the 6α-halotaxols **A4.2.1**, **A4.2.2**, and **A4.2.3** has been achieved from the diol **4.2.1.1** (*526*). The analogs were comparable to taxol in tubulin-assembly and cytotoxicity assays, and also had similar *in vivo* activity in the M109 murine lung cancer model. They also failed to yield any detectable metabolites when treated with a human liver S9 fraction capable of hydroxylating taxol.

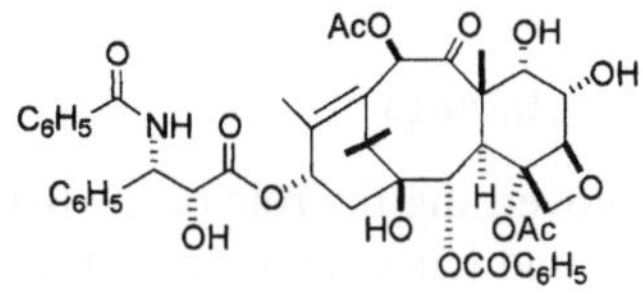

4.2.1.1

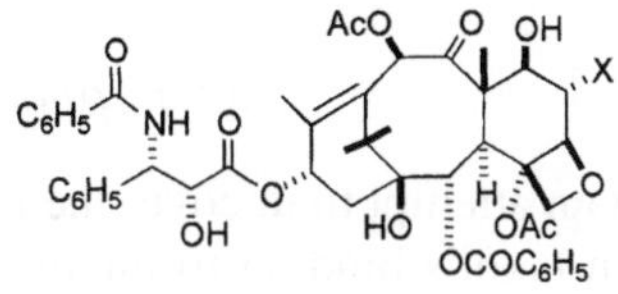

A4.2.1 R = F
A4.2.2 R = Cl
A4.2.3 R = Br

A4.3. Modifications at C-4

Efforts continue to prepare biologically active taxol analogs with modified groups at C-4. In one approach the alkaloid taxine was converted in several steps to the analogs **A4.3.1** and **A4.3.2**; both compounds were, however, virtually inactive against all cell lines tested, with IC_{50} values 1,000–10,000 times greater than those of taxol (*527*).

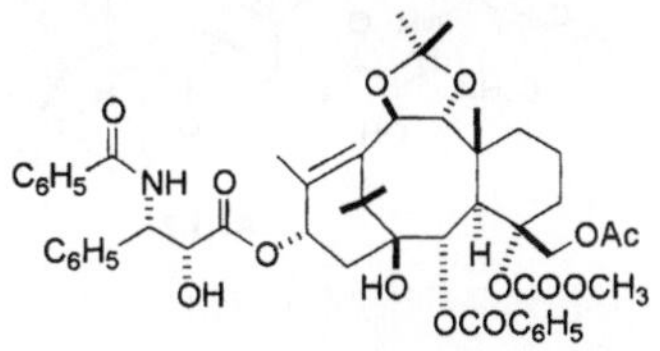

A4.3.1

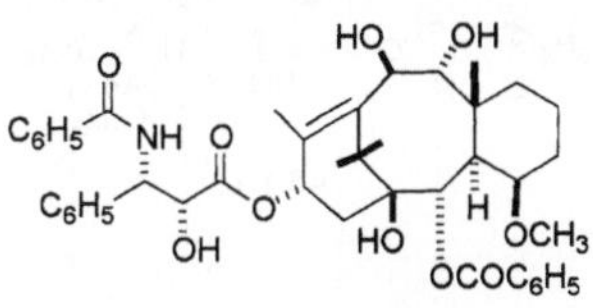

A4.3.2

In an alternative approach the naturally occurring alkaloid 2-deacetoxytaxinine J was treated with *m*-chloroperoxybenzoic acid to give a mixture of the compounds **A4.3.3**, **A4.3.4**, and **A4.3.5** (*528*). The product ratios could be adjusted by varying the temperature, with **A4.3.4** being the major product at 0–10°C. This was the only product to show any cytotoxicity, with an IC_{50} value 80 times that of taxol against the TFK-1 cell line.

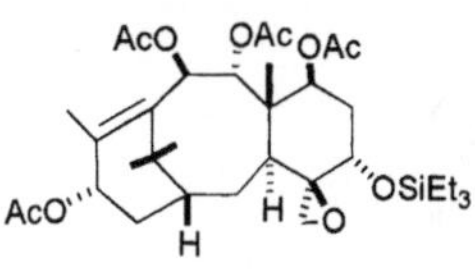

A4.3.3

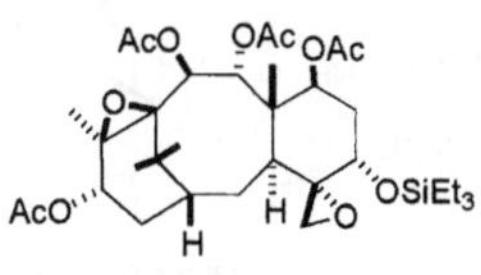

A4.3.4

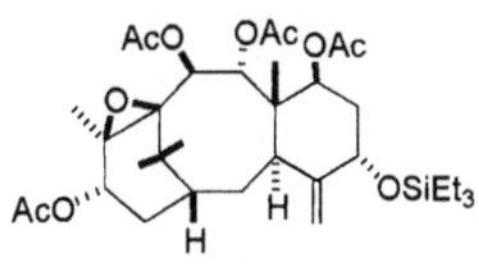

A4.3.5

A5. D-Ring Chemistry

The D-ring of taxol has been a difficult site to manipulate, but has been of interest to many investigators because of its apparent key role in the bioactivity of taxol. Several new studies in this area have thus appeared.

A5.1. Ring Opened Analogs

In an attempt to uncover the factors governing the role of the oxetane ring in taxol's binding to tubulin, Georg and her collaborators prepared several D-seco analogs using previously reported chemistry for the key step (*529*). Analogs such as **A5.1.1** and **A5.1.2** were prepared: the key difference between these analogs and similar ones prepared earlier is the fact that these were acetylated at C-4. In spite of this difference, none of the analogs prepared were cytotoxic or showed any tubulin-assembly activity, contrary to predictions from the minireceptor model. As a result of this work, the minireceptor model was modified to take steric bulk at C-20 into account; the modified model then accurately predicted the inactivity of the D-seco analogs.

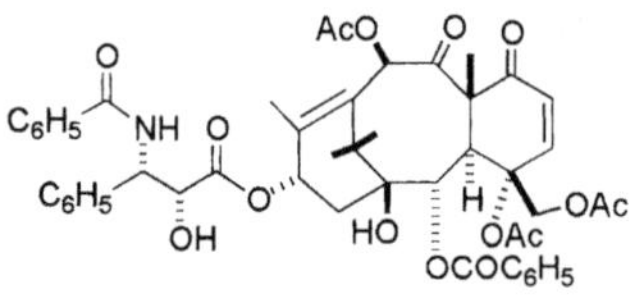

A5.1.1

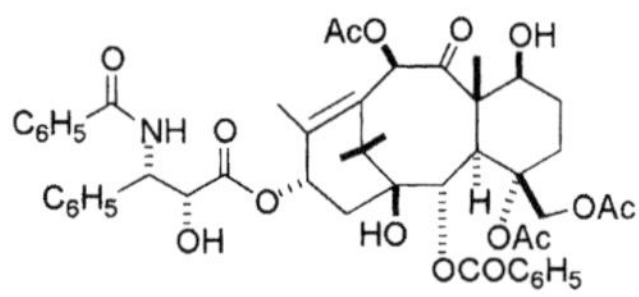

A5.1.2

A5.3. D-Ring Analogs

The synthesis of additional novel thia derivatives of docetaxel previously reported in an abstract (*206*) has now been reported in a full paper (*530*). The route employed incorporated an ingenious modification that allowed the preparation of 4-acetyl analogs. The thia analogs **A5.3.1** and **A5.3.2** were prepared, and were found to be much less cytotoxic than taxol.

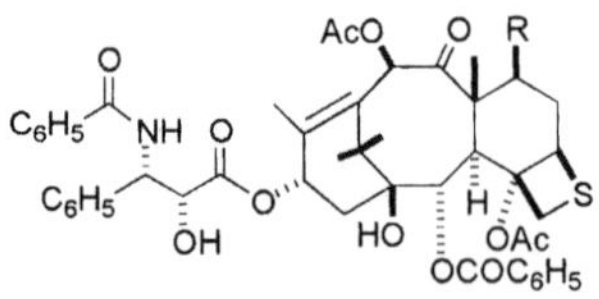

A5.3.1 R = OH
A5.3.2 R = H

References, pp. 193–225

The presumed importance of the oxetane ring to taxol's activity continues to prompt synthetic studies, and the Oritani group has developed a method for the selective hydroboration of taxa-4(20)-enes to the corresponding primary alcohols (*531*). Yields up to 84% and high diastereoselectivity were observed for the conversion of a taxadiene such as **A5.3.3** to the alcohol **A5.3.4**. This alcohol was then converted to the baccatin VI analog **A5.3.5** to demonstrate the utility of the reaction.

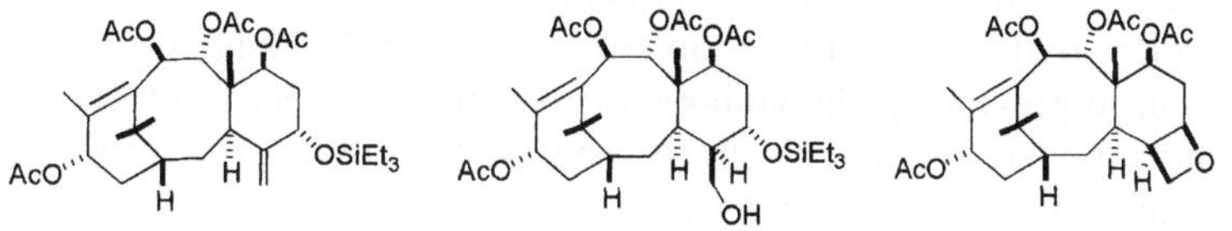

A5.3.3 A5.3.4 A5.3.5

A6. Rearrangements and Related Reactions

A study of the biotransformation of taxanes by the Zamir group gave the interesting finding that the taxane **A6.1** gave the rearranged taxanes **A6.2** and **A6.3** on treatment with the fungus *Absidia coerula*. The non-rearranged product **A6.4** was also formed (*532*).

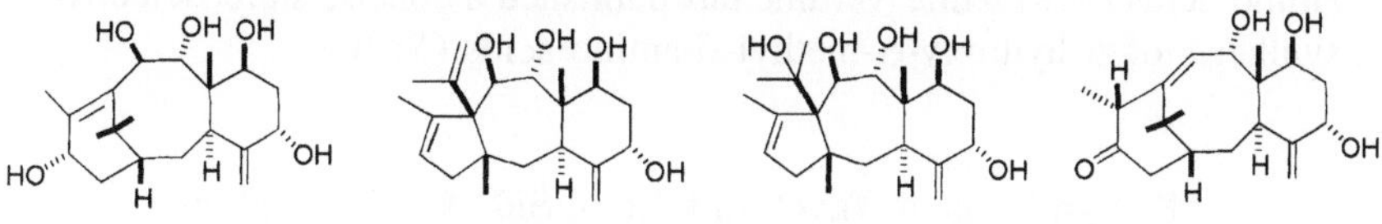

A6.1 A6.2 A6.3 A6.4

A7. The Side Chain

New synthetic approaches to the taxol side chain and related compounds continue to be developed. Two different precursors to the side chain of the clinical candidate ABT-271 (**A7.1**) (*533*) have been prepared on a large scale. The β-lactam **A7.2** and the *trans*-oxazolidine acid **A7.3** were both prepared in about 40% yield from L-leucine and subsequently converted to **A7.1** (*534*).

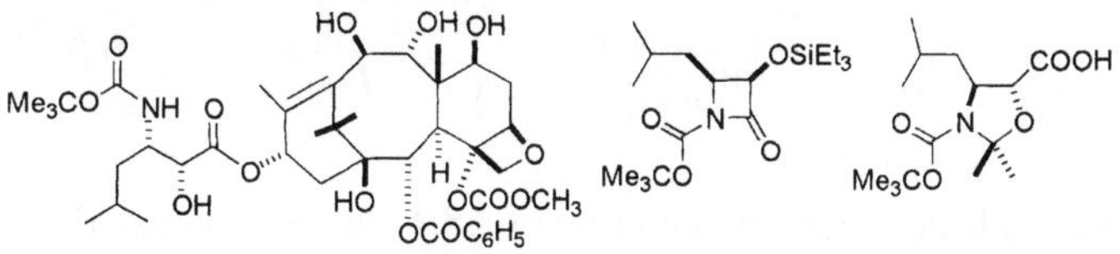

A7.1 A7.2 A7.3

A Korean group has developed a route to the β-lactam **A7.4** from diethyl L-tartrate (*535*). The conversion requires ten steps, but proceeds in reasonable overall yield. A different approach was taken by a Japanese group, who prepared racemic *syn* and *anti*-3-chloro-2-hydroxy-3-phenylpropanoates **A7.5** and **A7.6**, and then resolved them using a Lipase-catalyzed reaction. The resulting resolved esters were each converted stereospecifically to the taxol side chain (*536*). A final approach to the taxol side chain started from ethyl cinnamate by a route using the Sharpless hydroxylation reaction as the key step. Although this step has been used previously, the authors use an improved version of the cyclic iminocarbonate rearrangement to generate the key oxazolidinone **A7.7** with improved regiochemistry (*537*).

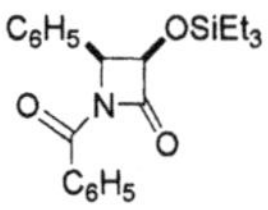

A7.4

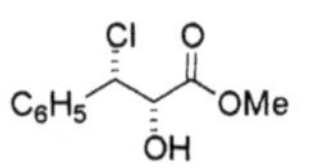

A7.5

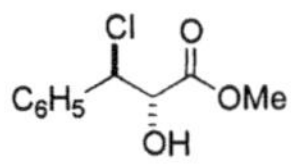

A7.6

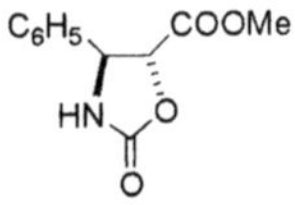

A7.7

Although not specifically related to the taxol side chain, two other publications deserve mention. Williams and his colleagues have developed a general stereocontrolled asymmetric synthesis of α-hydroxy-β-amino acids (*538*) while Verdine has published a concise stereoselective synthesis of α-hydroxy-α-methyl-β-amino acids (*539*).

A8. Synthesis of Taxol and Taxol Side Chain Analogs from Baccatin III

The preliminary publication by Appendino and his collaborators on the synthesis of conformationally restricted side chain analogs of taxol (*334*) such as **A8.1** has now been published as a full paper with additional theoretical studies (*540*). Although compound **A8.1** was almost as active as taxol in both tubulin-assembly and cytotoxicity assays, its epimer **A8.2** was very much less active.

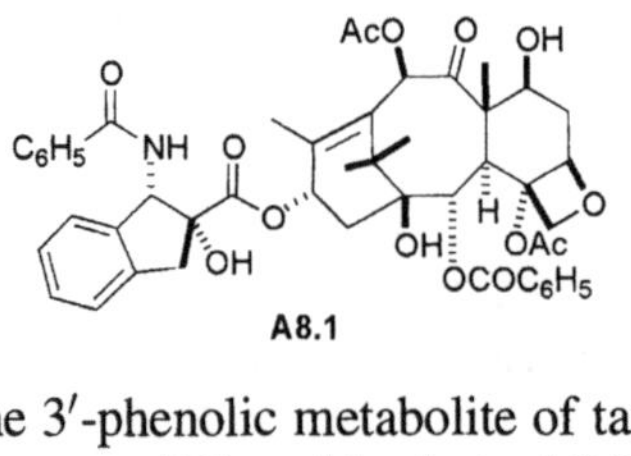

A8.1

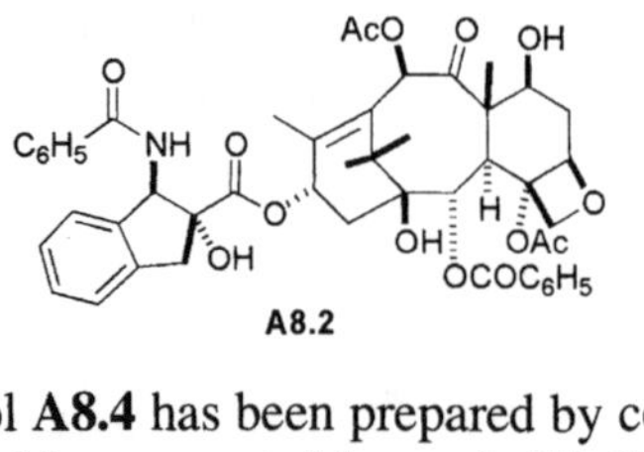

A8.2

The 3′-phenolic metabolite of taxol **A8.4** has been prepared by coupling the oxazolidine side chain **A8.3** with a protected baccatin III (*541*).

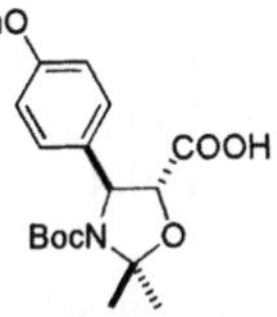

A8.3

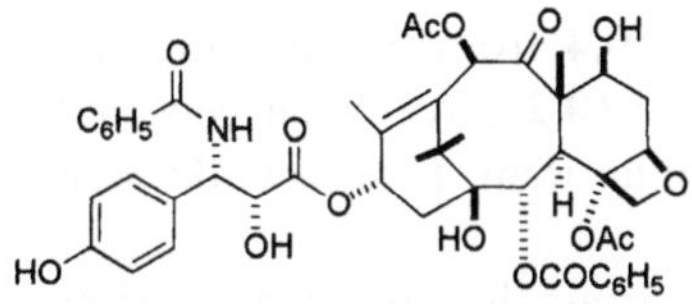

A8.4

One of the well-known features of the taxol side chain is that esters at the 2′-position are readily hydrolyzed, as was reported as early as 1984 (*542*). A recent study has explored the mechanistic reasons for this, using 2′-chloroacetates as a test case, and has concluded that the neighboring 3′-amide nitrogen atom catalyzes the reaction by serving as a nucleophilic hydrogen bond/proton acceptor (*543*).

A10. Taxol Analogs and Prodrugs

Work continues seemingly unabated on the synthesis of taxols with various substituents, either as prodrugs or as analogs with modified properties.

Conjugation of docosahexaenoic acid to the 2′-position of taxol gave a conjugate that showed improved tumor targeting and is currently in clinical trials (*544*, *545*). The syntheses of the cyclic AMP derivatives

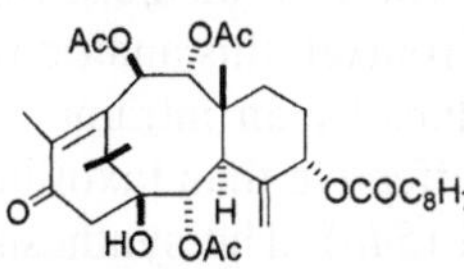

A10.1

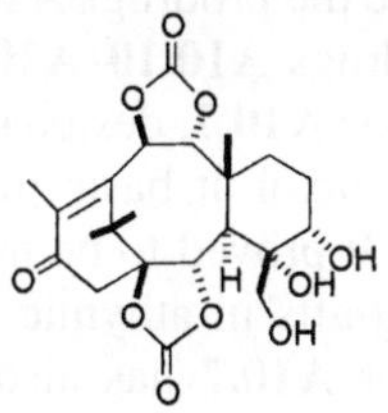

A10.2

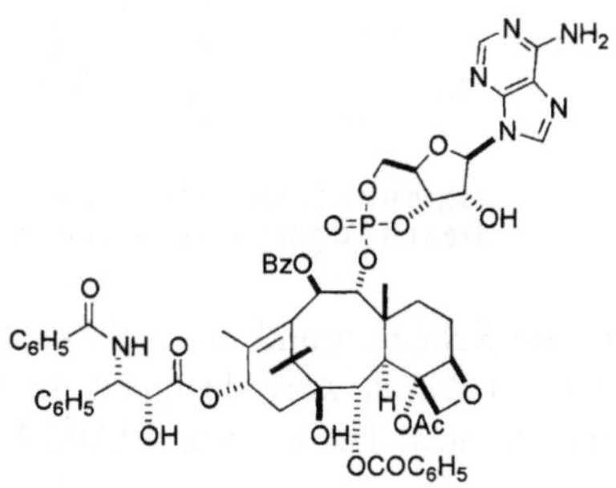

A10.3

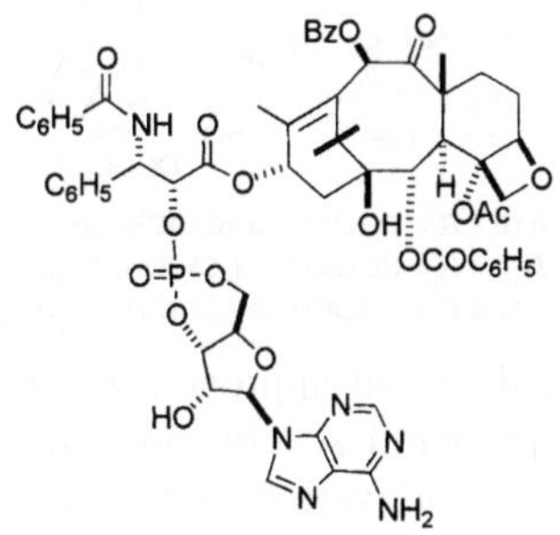

A10.4

A10.3 and **A10.4** from 5-cinnamoyltaxicin-I (**A10.1**) through the key intermediate **A10.2** have been reported (*546*). Conjugate **A10.3** was more cytotoxic than taxol, while conjugate **A10.4** had a reduced cytotoxicity.

The Lee group has prepared various analogs in which taxol or a related compound is conjugated with epipodophyllotoxin, with compounds **A10.5** and **A10.6** being representative examples. All of the analogs prepared were significantly less cytotoxic than taxol in most cell lines, with the KB cell line being the sole exception. Three of the analogs, including **A10.5** and **A10.6**, inhibited topoisomerase II as well as or better than etoposide (*547*).

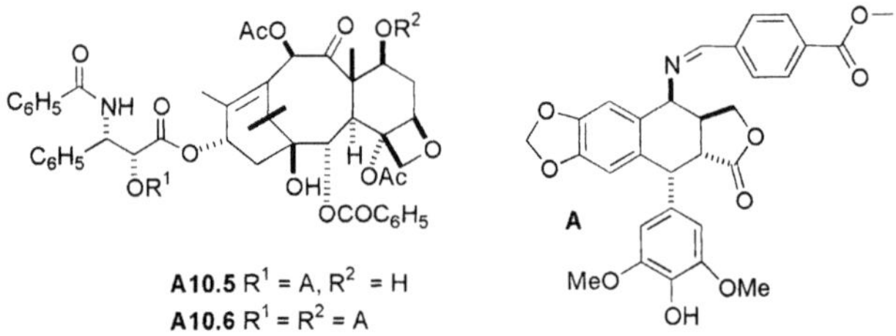

A10.5 R^1 = A, R^2 = H
A10.6 R^1 = R^2 = A

Interest in prodrugs of taxol has continued, with attachment at C-2′ or C-7 being favored. Various polyols were attached to the C-7 hydroxyl group to give the prodrugs **A10.7–A10.9**, and to the C-2′ hydroxyl group to give prodrugs **A10.10–A10.11**. The best of these from a therapeutic viewpoint was **A10.7**, designated Protaxel; this analog was stable in acid but released taxol at basic pH values by an intramolecular cyclization mechanism. It proved to be more effective than taxol in treating various tumor xenografts in athymic mice (*548*). The synthesis and preclinical evaluation of **A10.7** has also been reported by a collaborating group (*549*).

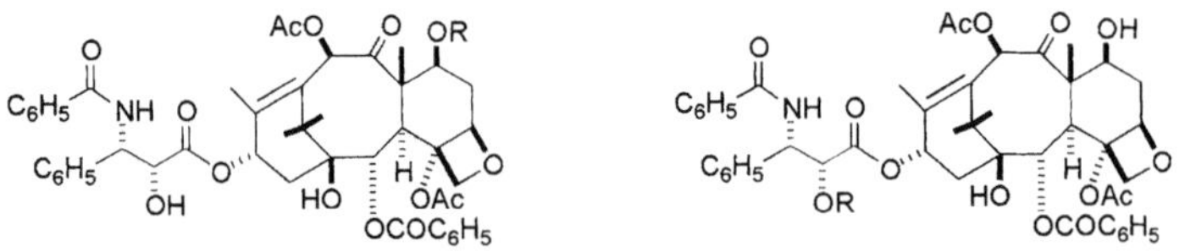

A10.7 R = $COOCH_2CH(OH)CH_2OH$
A10.8 R = $COOCH_2CH_2CH(OH)CH_2OH$
A10.9 R = $COOCH(CH_2CH(OH)CH_2OH)_2$

A10.10 R = $COOCH_2CH(OH)CH_2OH$
A10.11 R = $COOCH_2CH_2CH(OH)CH_2OH$

A different approach was taken by Scheeren and his collaborators, who prepared several prodrugs of taxol which were designed for bioreductive activation in hypoxic tumor tissues. The prodrugs **A10.12** and **A10.13** were found to be stable in buffer, to release taxol on reduction, and to have a reduced cytotoxicity under aerobic conditions, and were thus selected for further development. The remaining prodrugs prepared, which were all benzyl carbonate derivatives, did not show reduced cytotoxicity under aerobic conditions and were thus not of further interest (*550*).

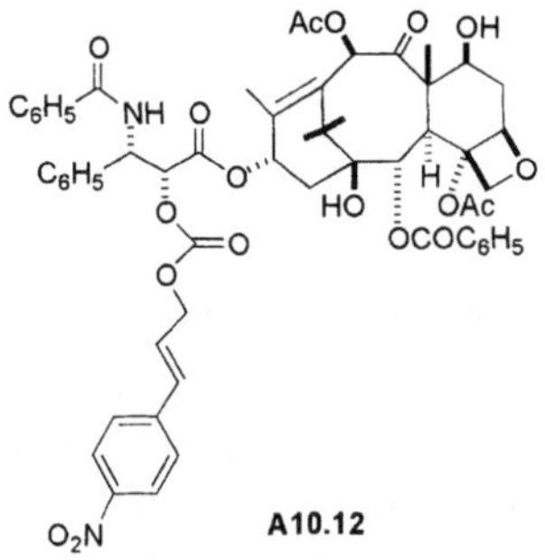

A10.12

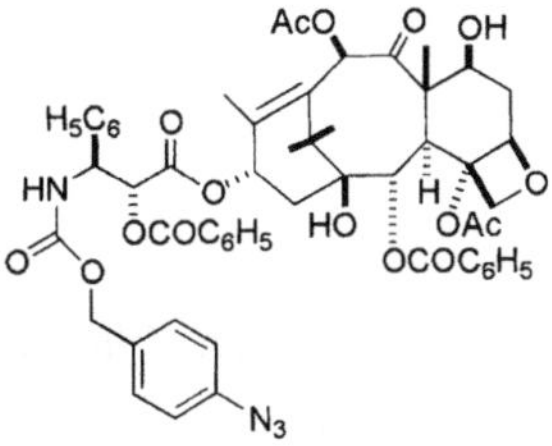

A10.13

The Zamir group has reported the design of the taxuspine analog **A10.14** based on a docking study with a refined $\alpha\beta$-tubulin structure. This analog was predicted to bind more strongly to $\alpha\beta$-tubulin than taxuspine, although not as strongly as taxol (*551*), but experimental data to support this prediction were not provided.

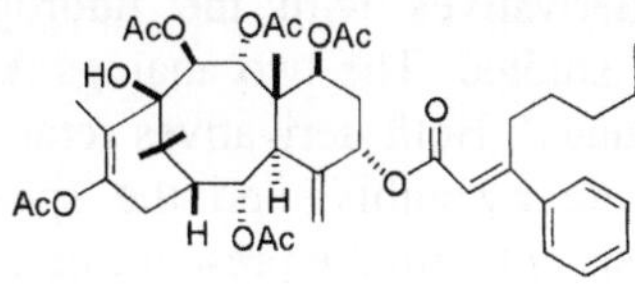

A10.14

A15. The Interaction of Taxol with Tubulin

The interaction of taxol and its derivatives with tubulin continues to be investigated by several groups, with the ultimate aim of understanding the interaction and designing improved analogs. Following an earlier report on the activity of a *m*-azidobenzoylbaccatin III derivative (*504*), Bane and her colleagues found that baccatin III itself will also assemble tubulin into normal microtubules and stabilize the polymers against cold-induced disassembly (*552*). The binding conformation of taxol in β-tubulin has been investigated using data from x-ray structures and NMR studies to generate potential conformers, and then docking these conformers into the experimental electron density map of the tubulin-taxol complex (*553*). This analysis led to the conclusion that taxol exists as a T-shaped or butterfly conformation (**A15.1**) when bound to tubulin.

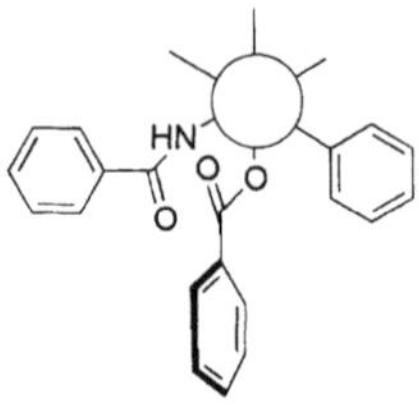

A15.1

A15.2. Fluorescence Spectroscopic Studies

The Andreu group has studied the binding of ligands to tubulin by measuring the displacement of the fluorescenyl taxol **11.3.13** from its binding site on microtubules. They found that baccatin III binds to microtubules with a binding constant about 200 times smaller than that of taxol, consistent with the results reported above (*554*). They conclude that the ring system of taxol provides about 75% of the free energy change of binding to microtubules. The same investigators also reported that centrosome and spindle pole microtubules are the main targets of taxoid-induced cell death (*555*).

Fluorescent taxol derivatives with the fluorophore at the C-10 position have also been studied. The two analogs **A15.2.1** and **A15.2.2** were prepared and evaluated. Both derivatives retained good activity as promoters of microtubule assembly, and the spectroscopic data for bound ligands were consistent with the presence of a charged amino acid in close proximity to the fluorophore (*556*).

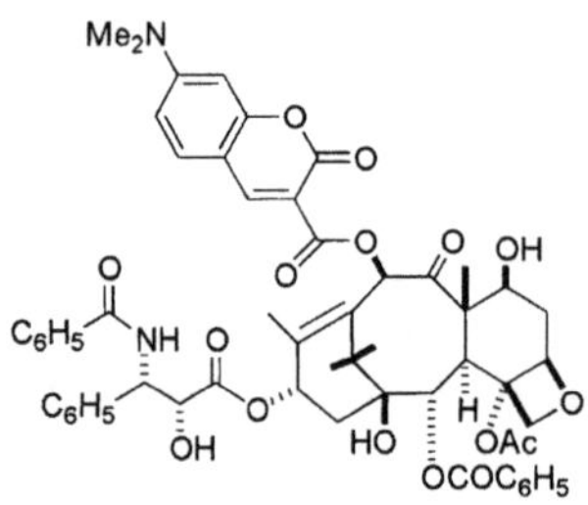

A15.2.1

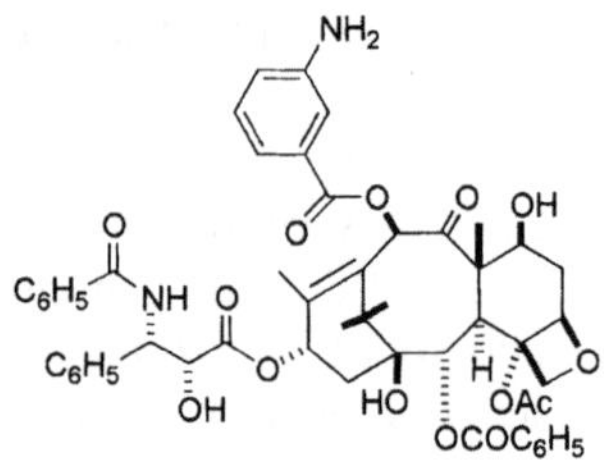

A15.2.2

A15.4. The Taxol Pharmacophore

Efforts continue to construct analogs of taxol that mimic its presumed binding conformation on the microtubule. The C-3′-phenyl/C-4

References, pp. 193–225

bridged analogs **A15.4.1** and **A15.4.2** were prepared and evaluated for their bioactivity; both were found to be active, with **A15.4.1** being about 10% as active as taxol. An analysis of the binding of these compounds using the DOCK protocol indicated that neither compound fit the binding pocket as snugly as "T-taxol", and in addition that an interaction between Phe272 of the protein and the tether occurred (*557*). The Ojima group has also reported an extension of their earlier work on the synthesis of bridged taxols by reporting on some bridged derivatives of simple scaffolds, such as **A15.4.3**. This compound was, however, inactive (*558*).

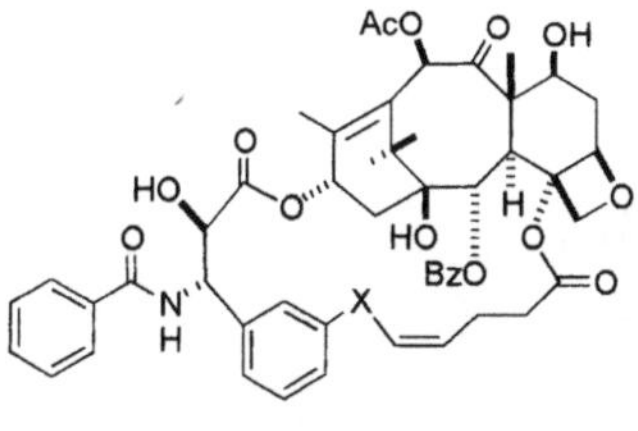

A15.4.1 X = OCH_2
A15.4.2 X = bond

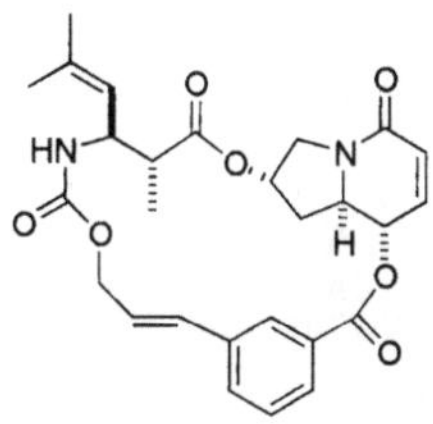

A15.4.3

A16. Miscellaneous Studies

The biosynthesis of taxol was not discussed in the body of this review, primarily because the subject had recently been reviewed. Some important new work has appeared during the past year, however, and so this area is briefly reviewed here.

The early steps in the overall biosynthetic pathway to taxol have become increasingly well understood, and the key enzymes have been cloned and expressed in *Escherichia coli*. The basic pathway begins with geranylgeranyl diphosphate (**A16.1**), which is cyclized by taxadiene synthase to taxa-4(5),11(12)-diene (**A16.2**). This then undergoes hydroxylation by cytochrome P450 taxadiene 5α-hydroxylase to give taxa-4(20),11(12)-dien-5α-ol (**A16.3**). Acetylation of **A16.3** by taxa-4(20), 11(12)-dien-5α-ol-*O*-acetyltransferase yields the acetate **A16.4**, and hydroxylation of **A16.4** by cytochrome P450 taxane 10β-hydroxylase gives the diol monoacetate **A16.5**. The conversion of **A16.5** to the 2-debenzoyltaxane **A16.6** is as yet poorly understood, but the benzoylation of **16.6** to 10-deacetylbaccatin III (**A16.7**) is brought about by a taxane 2α-*O*-benzoyltransferase. Similarly acetylation of **A16.7** to baccatin III (**A16.8**) utilizes 10-deacetylbaccatin III-10-*O*-acetyltransferase (*559*).

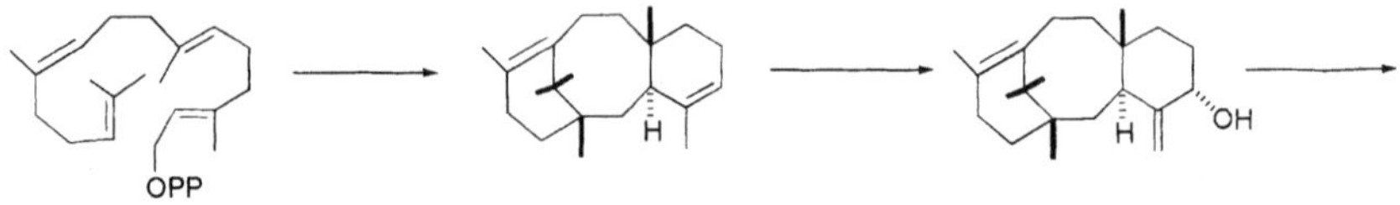

A16.1 **A16.2** **A16.3**

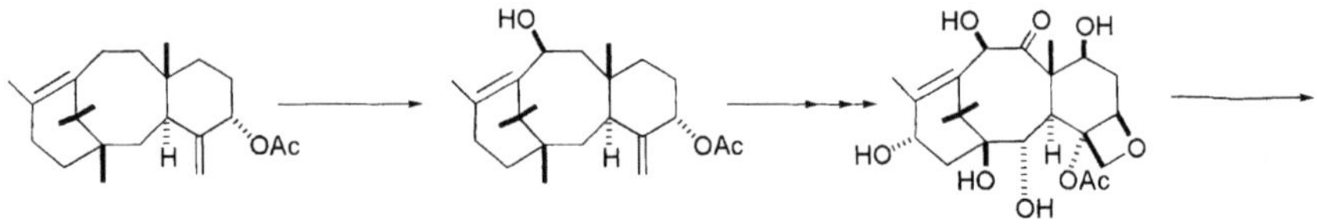

A16.4 **A16.5** **A16.6**

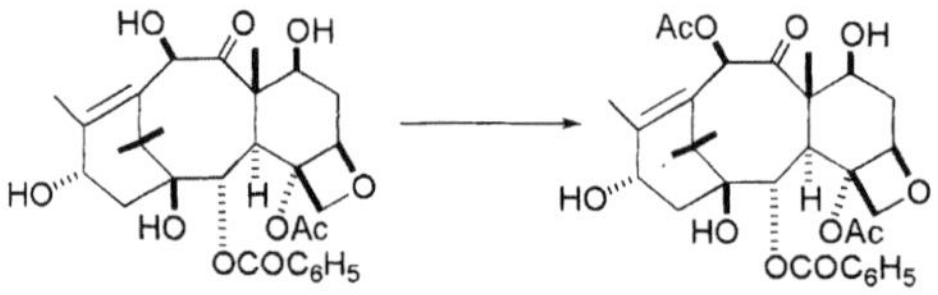

A16.7 **A16.8**

The genes for several of the key enzymes of this pathway have been cloned and expressed in *Escherichia coli*. This is the case for taxadiene synthase (*560*), taxadien-5α-ol-*O*-acetyltransferase (*561*), taxane 2α-*O*-benzoyltransferase (*562*) and 10-deacetylbaccatin III-10β-*O*-acetyltransferase (*563*). The gene for taxane 10β-hydroxylase has been cloned and expressed in yeast (*564*).

A practical application of these studies is that an engineered strain of *E. coli* has been developed that yields taxadiene (**A16.2**) in a reasonable yield of 1–3 mg per liter of cell culture (*565*). This successful bioproduction of a taxol precursor suggests that future work in this area will lead to improved bioproduction of taxol or baccatin III.

The cyclization of geranylgeranyl diphosphate (**A16.1**) to taxadiene **A16.2** has been thoroughly investigated, and it has been shown that the electrophilic cyclization involves initial closure of the A-ring, followed by an intramolecular transfer of the C-11 proton to the *re*-face of C-7 to promote closure of the B/C ring junction, followed by proton elimination from the β-face of C-5 (*566*).

Two new taxanes have entered clinical trials in addition to those mentioned in the main text. PNU166945 is a water-soluble polymer-conjugated prodrug of taxol with the general composition **A16.9**. It was well tolerated in a Phase I study and did not produce any dose-limiting toxicities up to doses of 196 mg/m^2 (*567*). The second analog is the cyclopropane analog **A16.10**; this compound also was well tolerated in a Phase I trial (*568*).

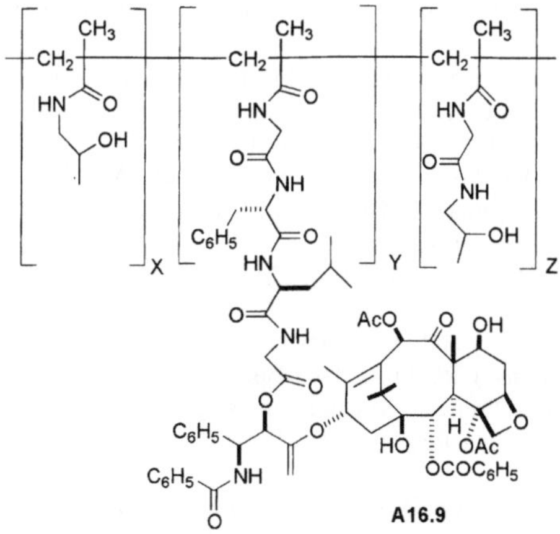

A16.9

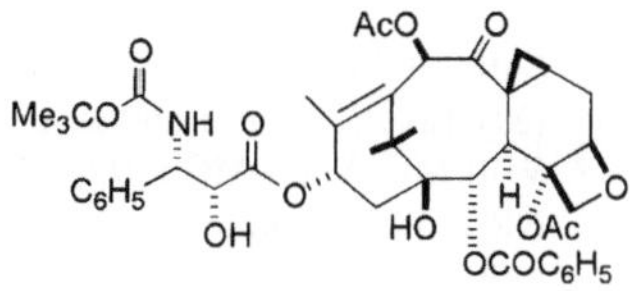

A16.10

Acknowledgements

We wish to acknowledge first and foremost three individuals without whose contributions taxol would not have become the important drug that it is. Drs. Monroe Wall and Mansukh Wani discovered taxol and its significant bioactivity; their recent receipt of the Charles F. Kettering Prize for cancer research is a fitting tribute to their contributions. Dr. Matthew Suffness of the National Cancer Institute championed the development of taxol within the Institute when it was far from clear that it would be a winner, and it is largely due to his persistence that we have the benefits of chemotherapy by taxol and docetaxel. Sadly his untimely death in 1995 prevented him from making any further contributions to taxol research, but his legacy lives on in the form of patients who have benefited from taxol chemotherapy.

We also wish to thank Drs. Gunda Georg (University of Kansas), Iwao Ojima (State University of New York at Stony Brook) and Giovanni Appendino (DiSCAFF, Università del Piemonte Orientale) for their care in reading the text and their helpful comments.

The senior author gratefully acknowledges research support from the National Cancer Institute and Bristol-Myers Squibb Pharmaceutical Research Institute.

References

1. Wani MC, Taylor HL, Wall ME, Coggon P, McPhail AT (1971) Plant Antitumor Agents. VI. The Isolation and Structure of Taxol, a Novel Antileukemic and Antitumor Agent from *Taxus brevifolia*. J Am Chem Soc **93**: 2325

2. Schiff PB, Fant J, Horwitz SB (1979) Promotion of microtubule assembly in vitro by taxol. Nature **277**: 665
3. McGuire WP, Rowinsky EK, Rosenshein NB, Grumbine FC, Ettinger DS, Armstrong DK, Donehower RC (1989) Taxol: A Unique Antineoplastic Agent with Significant Activity in Advanced Ovarian Epithelial Neoplasms. Ann Intern Med **111**: 273
4. Holmes FA, Walters RS, Theriault RL, Forman AD, Newton LK, Raber MN, Buzdar AU, Frye DK, Hortobagyi GN (1991) Phase II Trial of Taxol, an Active Drug in the Treatment of Metastatic Breast Cancer. J Nat Cancer Inst **83**: 1797
5. Thayer AM (2000) Busting Down A Blockbuster Drug. Chem & Eng News **78** (45): 20
6. Suffness M, Wall ME (1995) Discovery and Development of Taxol. In: Suffness M (ed) Taxol: Science and Applications. CRC Press, Inc., Boca Raton, Fl, p 3
7. Wall ME, Wani MC (1994) Taxol: Discovery to Clinic. In: Wagner H, Farnsworth NR (ed) Economic and Medicinal Plant Research. Academic Press, Ltd., London, p 299
8. Wall ME, Wani MC (1995) Paclitaxel: From Discovery to Clinic. In: Georg GI, Chen TT, Ojima I, Vyas DM (ed) Taxane Anticancer Agents: Basic Science and Current Status. American Chemical Society, Washington, DC, p 18
9. The chemical compound represented by structure **1.1.1** was known as taxol for over 20 years, until Bristol-Myers Squibb claimed the name Taxol as a trademark for their formulation of taxol; the chemical name paclitaxel was approved for the substance of structure **1.1.1**. Because of the historical nature of this review, and because many of the earlier papers cited use the name taxol rather than paclitaxel, the name taxol is retained for the substance **1.1.1**. No infringement of the Bristol-Myers Squibb trademark is intended or implied by this usage
10. Kingston DGI, Molinero AA, Rimoldi JM (1993) The Taxane Diterpenoids. In: Herz W, Kirby GW, Moore RE, Steglich W, Tamm C (ed) Progress in the Chemistry of Organic Natural Products. Springer-Verlag, New York, p 1
11. Appendino G (1995) The Phytochemistry of the Yew Tree. Nat Prod Rep 349
12. Appendino G (1995) Taxine. In: Pelletier SW (ed) Alkaloids: Chemical and Biological Perspectives. p 237
13. Appendino G (1995) Naturally Occurring Taxoids. In: Farina V (ed) The Chemistry and Pharmacology of Taxol and its Derivatives. Elsevier Science, New York, p 7
14. Parmar VS, Jha A (1998) Chemical Constituents of Taxus Species. In: Atta-ur-Rahman (ed) Studies in Natural Products Chemistry. Elsevier Science, Amsterdam, p 79
15. Das B, Rao SP, Das R (1995) Naturally Occurring Rearranged Taxoids. Planta Med **61**: 393
16. Das B, Rao SP (1996) Naturally Occurring Oxetane-type Taxoids. Ind J Chem **35B**: 883
17. Appendino G (1995) The Structural Elucidation of Taxoids. In: Farina V (ed) The Chemistry and Pharmacology of Taxol and its Derivatives. Elsevier Science, New York, p 55
18. Parmar VS, Jha A, Bisht KS, Taneja P, Singh SK, Kumar A, Poonam., Jain R, Olsen CE (1999) Constituents of the Yew Trees. Phytochemistry **50**: 1267
19. Baloglu E, Kingston DGI (1999) The Taxane Diterpenoids. J Nat Prod **62**: 1448
20. Kingston DGI (2000) Recent Advances in the Chemistry of Taxol. J Nat Prod **63**: 726
21. Floss HG, Mocek U (1995) Biosynthesis of Taxol. In: Suffness M (ed) Taxol: Science and Applications. CRC Press, Inc., Boca Raton, FL, p 191
22. Hezari M, Croteau R (1997) Taxol Biosynthesis: An Update. Planta Medica **63**: 291
23. Rohy J (1997) Biosynthesis of Taxol. Angew Chem Int Ed Engl **36**: 2190

24. Walker K, Croteau R (1999) Taxol Biosynthesis: a Review of Some Determinant Steps. Recent Adv Phytochem **33**: 31

25. Georg GI, Boge TC, Cheruvallath ZS, Clowers JS, Harriman GCB, Hepperle M, Park H (1995) The Medicinal Chemistry of Taxol. In: Suffness M (ed) Taxol: Science and Applications. CRC Press, Inc., Boca Raton, FL, p 317

26. Ojima I, Park YH, Fenoglio I, Duclos O, Sun C-M, Kuduk SD, Zucco M, Appendino G, Pera P, Veith JM, Bernacki RJ, Bissery M-C, Combeau C, Vrignaud P, Riou JG, Lavelle F (1995) Syntheses and Structure-Activity Relationships of New Taxoids. In: Georg GI, Chen TT, Ojima I, Vyas DM (ed) Taxane Anticancer Agents: Basic Science and Current Status. American Chemical Society, Washington, DC, p 262

27. Lin S, Ojima I (2000) Recent Strategies in the Development of Taxane Anticancer Drugs Exp Opin Ther Patents **10**: 1

28. Ojima I, Kuduk SD, Chakravarty S (1999) Recent Advances in the Medicinal Chemistry of Taxoid Anticancer Agents. Adv Med Chem JAI Press, Inc., p 69

29. Ter Haar E (1998) Taxanes and Other Microtubule Stabilizing Agents. Expert Opin Ther Pat **8**: 571

30. Rimoldi JM, Molinero AA, Chordia MD, Gharpure MM, Kingston DGI (1996) An Improved Method for the Separation of Paclitaxel and Cephalomannine. J Nat Prod **59**: 167

31. Kingston DGI, Samaranayake G, Ivey CA (1990) The Chemistry of Taxol, a Clinically Useful Anticancer Agent. J Nat Prod **53**: 1

32. Harriman GCB, Jalluri RK, Grunewald GL, Vander Velde DG, Georg GI (1995) The Chemistry of the Taxane Diterpene: Stereoselective Synthesis of 10-Deacetoxy-11,12-epoxypaclitaxel. Tetrahedron Lett **36**: 8909

33. Marder R, Dubois J, Guenard D, Gueritte-Voegelein F, Potier P (1995) Baccatin III Derivatives: Reduction of the C-11, C-12 Double Bond. Tetrahedron **51**: 1985

34. Appendino G, Jakupovic J, Cravotto G, Enriu R, Varese M, Bombardelli E (1995) Synthesis of Modified Baccatins *via* an Oxidation-Reduction Protocol. Tetrahedron Lett **36**: 3233

35. Marder R, Bricard L, Dubois J, Guenard D, Gueritte-Voegelein F (1996) Taxoids: 11,12-Dihydro-4-deacetyldocetaxel. Tetrahedron Lett **37**: 1777

36. Kelly RC, Wicnienski NA, Gebhard I, Qualls SJ, Han F, Dobrowolski PJ, Nidy EG, Johnson RA (1996) 12,13-Isobaccatin III. Taxane Enol Esters (12,13-Isotaxanes). J Am Chem Soc **118**: 919

37. Denis J-N, Greene AE (1996) Preparation of a Novel Paclitaxel/Docetaxel Derivative from Taxagifine. Nat Prod Lett **8**: 27

38. Magri NF, Kingston DGI, Jitrangsri C, Piccariello T (1986) Modified Taxols. 3. Preparation and Acylation of Baccatin III. J Org Chem **51**: 3239

39. Zheng QY, Murray CK (1995). Preparation of 10-deacetylbaccatin III and 7-protected-10-deacetylbaccatin III Derivatives from 10-deacetyltaxol A, 10-deacetyltaxol B, and 10-deacetyltaxol C. US Patent 5,449,790; Chem Abstr **132**: 340476e

40. Chatterjee A, Bhattacharyya S, McChesney JD, Elsohly HN, Williamson JS (1995) Preparation of 7-Triethylsilylbaccatin III using a Regioselective Reductive Deacetylation. Nat Prod Lett **6**: 139

41. Senilh V, Gueritte F, Guenard D, Colin M, Potier P (1984) Hemisynthese de nouveaux analogues du taxol. Etude de Leur Interaction Avec la Tubuline. C R Acad Sc Paris **299**: 1039

42. Appendino G, Fenoglio I, Cravotto G, Varese M (1994) The Chemistry and Occurrence of Taxane Derivatives. XIII. The Oxidation of 10-Deacetylbaccatin III. Gazz Chim Ital **124**: 253

43. Hoemann MZ, Vander Velde D, Aube J, Georg GI, Jayasinghe LR (1995) Synthesis of 13-*epi*-Taxol via a Transannular Delivery of a Borohydride Reagent. J Org Chem **60**: 2918
44. Chen S-H, Farina V, Vyas DM, Doyle TW, Long BH, Fairchild C (1996) Synthesis and Biological Evaluation of C-13 Amide-Linked Paclitaxel (Taxol) Analogs. J Org Chem **61**: 2065
45. Menichincheri M, Ceccarelli W, Fusar-Bassini D, Vanotti E, Ciomei M, Mongelli N, Pinciroli V, Vulpetti A (1996) A-Ring Modified Taxanes : 13-Aza Derivatives. Med Chem Res **6**: 264
46. Menichincheri M, Arlandini E, Ceccarelli W, Colombo M, Franzoi L, Fusar-Bassini D, Mongelli N, Pinciroli V, Vanotti E (1997) New 13-aza Baccatins. Tetrahedron Lett **38**: 3301
47. Appendino G, Gariboldi P, Gabetti B, Pace P, Bombardelli E, Viterbo D (1992) 14β-Hydroxy-10-deacetylbaccatin III, a new taxane from Himalayan yew (*Taxus wallichiana* Zucc.). J Chem Soc, Perkin Trans 1 **21**: 2925
48. Ojima I, Fenoglio I, Park YH, Pera P, Bernacki RJ (1994) Synthesis and Biological Activity of 14-Hydroxydocetaxel. Bioorg & Med Chem Lett **4**: 1571
49. Ojima I, Park YH, Sun C-M, Fenoglio I, Appendino G, Pera P, Bernacki RJ (1994) Structure-Activity Relationships of New Taxoids Derived from 14β-Hydroxy-10-deacetylbaccatin III. J Med Chem **37**: 1408
50. Ojima I, Fenoglio I, Park YH, Sun C-M, Appendino G, Pera P, Bernacki RJ (1994) Synthesis and Structure-Activity Relationships of Novel Nor-Seco Analogs of Taxol and Taxotere. J Org Chem **59**: 515
51. Ojima I, Lin S, Chakravarty S, Fenoglio I, Park YH, Sun C-M, Appendino G, Pera P, Veith JM, Bernacki RJ (1998) Syntheses and Structure-Activity Relationships of Novel Nor-seco Taxoids. J Org Chem **63**: 1637
52. Ojima I, Slater JC, Kuduk SD, Takeuchi CS, Gimi RH, Sun C-M, Park YH, Pera P, Veith JM, Bernacki RJ (1997) Syntheses and Structure-Activity Relationships of Taxoids Derived from 14β-Hydroxy-10-deacetylbaccatin III. J Med Chem **40**: 267
53. Nicoletti MI, Colombo T, Rossi C, Monardo C, Stura S, Zucchetti M, Riva A, Morazzoni P, Donati MB, Bombardelli E, D'Incalci M, Giavazzi R (2000) IDN5109, a Taxane with Oral Bioavailability and Potent Antitumor Activity. Cancer Res **60**: 842
54. Bombardelli E (1997) Preparation of Taxane Pyrazoline Derivatives for Use as Antitumor Agents, PCT Int. WO 97 43,291 (Cl. C07D491/08),20 Nov 1997, IT Appl 96/MI942,10 May 1996
55. Uoto K, Mitsui I, Terasawa H, Soga T (1997) First Synthesis and Cytotoxic Activity of Novel Docetaxel Analogs Modified at the C-18-Position. Bioorg & Med Chem Lett **7**: 2991
56. Fang WS, Fang QC, Liang XT (1998) The Ephemeral 10β-OH Trifluoromethanesulfonataes of 7-Epi Paclitaxel Analogs – a New Example for Anchimeric Assistance. Chin Chem Lett **9**: 29
57. Samaranayake G, Magri NF, Jitrangsri C, Kingston DGI (1991) Modified Taxols. 5. Reaction of Taxol with Electrophilic Reagents, and Preparation of a Rearranged Taxol Derivative with Tubulin Assembly Activity. J Org Chem **56**: 5114
58. Samaranayake G, Neidigh KA, Kingston DGI (1993) Modified Taxols. 8. Deacylation and Reacylation of Baccatin III. J Nat Prod **56**: 884
59. Johnson RA, Nidy EG, Dobrowolski PJ, Gebhard I, Qualls SJ, Wicnienski NA, Kelly RC (1994) Taxol Chemistry. 7-O-Triflates as Precursors to Olefins and Cyclopropanes. Tetrahedron Lett **35**: 7893

60. Datta A, Hepperle M, Georg GI (1995) Selective Deesterification Studies on Taxanes: Simple and Efficient Hydrazinolysis of C-10 and C-13 Ester Functionalities. J Org Chem **60**: 761
61. Zheng QY, Darbie LG, Cheng X, Murray CK (1995) Deacetylation of Paclitaxel and Other Taxanes. Tetrahedron Lett **36**: 2001
62. Kajiro H, Mitamura S, Mori A, Hiyama T (1999) Scandium Trifluoromethanesulfonate-Catalyzed Cleavage of Esters Bearing a Coordinative Group at a Vicinal Position. Bull Chem Soc Jpn **72**: 1553
63. Denis J-N, Greene AE, Guénard D, Guéritte-Voegelein F, Mangatal L, Potier P (1988) A Highly Efficient, Practical Approach to Natural Taxol. J Am Chem Soc **110**: 5917
64. Senilh V, Blechert S, Colin M, Guénard D, Picot F, Potier P, Varenne P (1984) Mise en Evidence de Nouveaux Analogues du Taxol Extraits de Taxus baccata. J Nat Prod **47**: 131
65. Sisti NJ, Zygmunt J, Brinkman HR, Chander MC, Liang X, McChesney JD (1999) U.S. US Patent 5,914,411
66. Holton RA, Zhang Z, Clarke PA, HN, Procter DJ (1998) Selective Protection of the C(7) and C(10) Hydroxyl Groups in 10-Deacetyl Baccatin III. Tetrahedron Lett **39**: 2883
67. Damen EWP, Braamer L, Scheeren HW (1998) Lanthanide Trifluoromethanesulfonate Catalysed Selective Acylation of 10-Deacetylbaccatin III. Tetrahedron Lett **39**: 6081
68. Damen EWP, Braamer L, Scheeren HW (1998) Direct Conversion of 10-Deacetylbaccatin III into Baccatin III as a Useful Synthetic Tool for a High Yield Paclitaxel Semisynthesis. Herba Pol **44**: 238
69. Cravallee C, Didier E, Pecquet P (1998) Methyleniminium Salts as Acylating Agent – One Step Synthesis of Baccatin III from 10-Deacetylbaccatin III with High Selectivity. Tetrahedron Lett **39**: 4263
70. Lee D, Kim K-C, Kim M-J (1998) Selective Enzymatic Acylation of 10-Deacetylbaccatin III. Tetrahedron Lett **39**: 9039
71. Bourzat JD, Commerçon A, Guenard D, Gueritte-Voegelein F, Potier P (1993) Preparation of O-Carbamoyl Taxol Analogs as Neoplasm Inhibitors, Eur Pat Appl Ep 524,093 (Cl. C007D305/14)FR Appl 91/8, 937,16 Jul 1991
72. Rao KV, Bhakuni RS, Johnson J, Oruganti RS (1995) Synthesis and Evaluation of Some 10-Mono-and 2′,10-Diesters of 10-Deacetylpaclitaxel. J Med Chem **38**: 3411
73. Kant J, O'Keeffe WS, Chen S-H, Farina V, Fairchild C, Johnston K, Kadow JF, Long BH, Vyas D (1994) A Chemoselective Approach to Functionalize the C-10 Position of 10-Deacetylbaccatin III. Synthesis and Biological Properties of Novel C-10 Taxol Analogues. Tetrahedron Lett **35**: 5543
74. Ojima I, Slater JC, Michaud E, Kuduk SC, Bounaud P-Y, Vrignaud P, Bissery M-C, Veith JM, Pera P, Bernacki RJ (1996) Syntheses and Structure-Activity Relationships of the Second-Generation Antitumor Taxoids: Exceptional Activity against Drug-Resistant Cancer Cells. J Med Chem **39**: 3889
75. Fang WS, Fang QC, Liang XT (1997) Preliminary Studies on Acylation of 10β-OH and 7α-OH in 7-Epi-10-deacetylpaclitaxel. Chinese Chemical Lett **8**: 847
76. Ojima I, Lin S, Slater JC, Wang T, Pera P, Bernacki RJ, Ferlini C, Scambia G (2000) Syntheses and Biological Activity of C-3′-Difluoromethyl-Taxoids. Bioorg Med Chem **8**: 1619
77. Bouchard H, Commerçon A, Combeau C, Riou JF, Lavelle F (1996) Synthesis of 7,10-Hydroxy-Modified Docetaxel Analogs, 20th IUPAC Symposium on the chemistry of Natural Products, Chicago, Sept. 15–20, Abstract No. SY-9

78. Hirofumi T, Tsunehiko S, Kouichi U (1996) Preparation of Taxol Derivatives as Antitumors, PCT, Int Appl WO 96 23,779 (CL. C07D305/14)JP Appl 95/15,652, 2 Feb 1995

79. Nakayama K, Terasawa H, Mitsui I, Ohsuki S, Uoto K, Iimura S, Soga T (1998) Synthesis and Cytotoxic Activity of Novel 10-Alkylated Docetaxel Analogs. Bioorg & Med Chem Lett **8**: 427

80. Iimura S, Ohsuki S, Uoto K, Nakayama K, Terasawa H, Soga T (2000) A Convenient Method for the Synthesis of C-10 Alkylated Docetaxel Analogs Using *Tris*-(Trimethylsilyl)silane as a Radical Mediator. Heterocycles **53**: 1579

81. Datta A, Aube J, Georg GI, Mitscher LA (1994) The First Synthesis of a C-9 Carbonyl Modified Baccatin III Derivative and its Conversion to Novel Taxol and Taxotere Analogues. Bioorg. Med Chem Lett **4**: 1831

82. Chen S-H, Fairchild C, Mamber SW, Farina V (1993) Taxol Structure-Activity Relationships: Synthesis and Biological Evaluation of 10-Deoxytaxol. J Org Chem **58**: 2927

83. Chaudhary AG, Kingston DGI (1993) Synthesis of 10-Deacetoxytaxol and 10-Deoxytaxotere. Tetrahedron Lett **34**: 4921

84. Chen S-H, Huang S, Kant J, Fairchild C, Wei J, Farina V (1993) Synthesis of 7-Deoxy- and 7,10-Dideoxytaxol via Radical Intermediates. J Org Chem **58**: 5028

85. Chen S-H, Huang S, Gao Q, Golik J, Farina V (1994) The Chemistry of Taxanes: Skeletal Rearrangements of Baccatin Derivatives via Radical Intermediates. J Org Chem **59**: 1475

86. Holton RA, Somoza C, Chai K-B (1994) A Simple Synthesis of 10-Deacetoxytaxol Derivatives. Tetrahedron Lett **35**: 1665

87. Georg GI, Cheruvallath ZS (1994) Samarium Diiodide-Mediated Deoxygenation of Taxol: A One-Step Synthesis of 10-Deacetoxytaxol. J Org Chem **59**: 4015

88. Georg GI, Cheruvallath ZS, Vander Velde DG, Himes RH (1995) Stereoselective Synthesis of 9β-Hydroxytaxanes via Reduction with Samarium Diiodide. Tetrahedron Lett **36**: 1783

89. Georg GI, Harriman GCB, Datta A, Ali S, Cheruvallath Z, Dutta D, Vander Velde DG, Himes RH (1998) The Chemistry of the Taxane Diterpene: Stereoselective Reductions of Taxanes. J Org Chem **63**: 8926

90. Py S, Khuong-Huu F (1995) Reactivity of Baccatin III Derivatives Towards Samarium II Reagents. Tetrahedron Lett **36**: 1661

91. Datta D, Hadd H, Vander Velde DG, Georg GI (1999) A One-Step Synthesis of a Deuterated Paclitaxel Analogue: 10-Deacetoxy-(10α-^{2}H)Paclitaxel. Bioorg Med Chem Lett **9**: 3277

92. Pulicani J-P, Bourzat J-D, Bouchard H, Commerçon A (1994) Electrochemical Reduction of Taxoids: Selective Preparation of 9-Dihydro-,10-Deoxy- and 10-Deacetoxy-taxoids. Tetrahedron Lett **35**: 4999

93. Bouchard H, Bourzat J-D, Commerçon A (1995) Preparation of Biologically Active Dioxoepoxytaxoids. PCT Int Appl WO 95 01,969, 19 Jan

94. Datta A, Vander Velde DG, Georg GI, Himes RH (1995) Syntheses of Novel C-9 and C-10 Modified Bioactive Taxanes. Tetrahedron Lett **36**: 1985

95. Appendino G, Jakupovic J, Varese M, Belloro E, Danieli B, Bombardelli E (1996) Synthesis of 7,9-Nitrogen-Substituted Paclitaxel Derivatives. Tetrahedron Lett **37**: 7837

96. Appendino G, Jakupovic J, Cravotto G, Enriu R, Varese M, Bombardelli E (1995) Synthesis of Modified Baccatins *via* an Oxidation-Reduction Protocol. Tetrahedron Lett **36**: 3233

97. Terasawa H, Soga T, Tsunehiko NK (1995) Preparation of Deacetoxytaxol Derivatives as Potential Antitumors, PCT Int Appl WO 95 33,740 (Cl. C07D305/14), JP Appl 94/127,334, 9 Jun 1994
98. Klein LL (1993) Synthesis of 9-Dihydrotaxol: A Novel Bioactive Taxane. Tetrahedron Lett **34**: 2047
99. Maring CJ, Grampovnik DJ, Yeung CM, Klein LL, Li L, Thomas SA, Plattner JJ (1994) C-3′-N-Acyl Analogs of 9(R)-Dihydrotaxol: Synthesis and Structure Activity Relationships. Bioorg & Med Chem Lett **4**: 1429
100. Li L, Thomas SA, Klein LL, Yeung CM, Maring CJ, Grampovnik DJ, Lartey PA, Plattner JJ (1994) Synthesis and Biological Evaluation of C-3′-Modified Analogs of 9(R)-Dihydrotaxol. J Med Chem **37**: 2655
101. Klein LL, Li L, Maring CJ, Yeung CM, Thomas SA, Grampovnik DJ, Plattner JJ (1995) Antitumor Activity of 9(R)-Dihydrotaxane Analogs. J Med Chem **38**: 1482
102. Klein LL, Yeung CM, Li L (1995) 9-Deoxo-taxanes, US 5,440,056 (Cl.549-510, C07D305/14), US Appl 870,509 17 Apr 1992
103. Margraff R, Bezard D, Bourzat JD, Commerçon A (1994) Synthesis of 19-Hydroxy Docetaxel from a Novel Baccatin. Bioorg Med Chem Lett **4**: 233
104. Chen S, Farina V, Wei J, Long B, Fairchild C, Mamber SW, Kadow JF, Vyas D, Doyle TW (1994) Structure-Activity Relationships of Taxol: Synthesis and Biological Evaluation of C2 Taxol Analogs. Bioorg Med Chem Lett **4**: 479
105. Datta A, Jayasinghe LR, Georg GI (1994) Internal Nucleophile Assisted Selective Deesterification Studies on Baccatin III. Synthesis of 2-Debenzoyl- and 4-Deacetylbaccatin III Analogues. J Org Chem **59**: 4689
106. Nicolaou KC, Nantermet PG, Ueno H, Guy RK (1994) Novel Chemistry of Taxol. Retrosynthetic and Synthetic Studies. J Chem Soc Chem Commun 295
107. Holton RA, Kim S (1995) Process for the Preparation of Baccatin III Analogs Bearing New C2 and C4 Functional Groups. U.S. Patent 5,399,726
108. Kim S-C, Chai K-B (1997) Migration Between C-2 and C-4 Hydroxyl Groups in Paclitaxel Core. Korean J Med Chem **7**: 2
109. Farina V, Huang S (1992) The chemistry of taxanes: unexpected rearrangement of baccatin III during chemoselective debenzoylation with Bu_3SnOMe/LiCl. Tetrahedron Lett **33**: 3979
110. Chaudhary AG, Gharpure MM, Rimoldi JM, Chordia MD, Gunatilaka AAL, Kingston DGI, Grover S, Lin CM, Hamel E (1994) Unexpectedly Facile Hydrolysis of the 2-Benzoate Group of Taxol and Synthesis of Analogs with Increased Activities. J Am Chem Soc **116**: 4097
111. Georg GI, Ali SM, Boge TC, Datta A, Falborg L, Park H, Mejillano M, Himes RH (1995) Synthesis of Biologically Active 2-Benzoyl Paclitaxel Analogues. Bioorg & Med Chem Lett **5**: 259
112. Kingston DGI, Chaudhary AG, Chordia MD, Gharpure M, Gunatilaka AAL, Higgs PI, Rimoldi JM, Samala L, Jagtap PG, Giannakakou P, Jiang YQ, Lin CM, Hamel E, Long BH, Fairchild CR, Johnston KA (1998) Synthesis and Biological Evaluation of 2-Acyl Analogues of Paclitaxel (Taxol). J Med Chem **41**: 3715
113. Georg GI, Ali SM, Boge TC, Datta A, Falborg L, Himes RH (1994) Selective C-2 and C-4 Deacylation and Acylation of Taxol: The First Synthesis of a C-4 Substituted Taxol Analogue. Tetrahedron Lett **35**: 8931
114. Pulicani J-P, Bezard D, Bourzat J-D, Bouchard H, Zucco M, Deprez D, Commerçon A (1994) Direct Access to 2-Debenzoyl Taxoids by Electrochemistry, Synthesis of 2-Modified Docetaxel Analogs. Tetrahedron Lett **35**: 9717

115. Nicolaou KC, Couladouros EA, Nantermet PG, Renaud J, Guy RK, Wrasidlo W (1994) Synthesis of C-2 Taxol Analogues. Angew Chem Int Ed Engl **33**: 1581

116. Kim S-C, Moon M-S, Choi K-M, Jun S-J, Kim H-K, Jung D-I, Lee K-S, Chai K-B (1998) Synthesis and Biological Evaluation of C-2 Modified Taxol Analogs. Bull Korean Chem Soc **19**: 1027

117. Nicolaou KC, Renaud J, Nantermet PG, Couladouros EA, Guy RK, Wrasidlo W (1995) Chemical Synthesis and Biological Evaluation of C-2 Taxoids. J Am Chem Soc **117**: 2409

118. Georg GI, Harriman GCB, Ali SM, Datta A, Hepperle M, Himes RH (1995) Synthesis of 2-O-Heteroaroyl Taxanes: Evaluation of Microtubule Assembly Promotion and Cytotoxicity. Bioorg & Med Chem Lett **5**: 115

119. Boge TC, Himes RH, Vander Velde DG, Georg GI (1994) The Effect of the Aromatic Rings of Taxol on Biological Activity and Solution Conformation: Synthesis and Evaluation of Saturated Taxol and Taxotere Analogues. J Med Chem **37**: 3337

120. Ojima I, Duclos O, Zucco M, Lavelle F (1994) Synthesis and Structure-Activity Relationships of New Antitumor Taxoids. Effects of Cyclohexyl Substitution at the C-3′ and/or C-2 of Taxotere (Docetaxel). J Med Chem **37**: 2602

121. Chordia MD, Kingston DGI (1996) Synthesis and Biological Evaluation of 2-*epi*-Paclitaxel. J Org Chem **61**: 799

122. Chen S-H, Wei J-M, Farina V (1993) Taxol Structure-Activity Relationships: Synthesis and Biological Evaluation of 2-Deoxytaxol. Tetrahedron Lett **34**: 3205

123. Chaudhary AG, Chordia MD, Kingston DGI (1995) A Novel benzoyl group migration: synthesis and biological evaluation of 1-benzoyl-2-des(benzoyloxy)paclitaxel. J Org Chem **60**: 3260

124. Chen S-H, Huang S, Gao Q, Golik J, Farina V (1994) The Chemistry of Taxanes: Skeletal Rearrangements of Baccatin Derivatives via Radical Intermediates. J Org Chem **59**: 1475

125. Yin DL, Sekiguchi Y, Kameo K (1998) Attempted Removal of Oxygen Function at 1-Position of 10-Deacetylbaccatin III. Chin Chem Lett **9**: 373

126. Kingston DGI, Chordia MD, Jagtap PG, Liang J, Shen Y-C, Long BH, Fairchild CR, Johnston KA (1999) Synthesis and Biological Evaluation of 1-Deoxypaclitaxel Analogues. J Org Chem **64**: 1814

127. Benigni D, Dillon J, Weaver D, Haines K, Shultis K, Faigle G, Lin H-Y, Rourick R, Crull G, Huang S, Baumes J, Long B, Fairchild C (2000) Isolation, Structure Elucidation and Biological Activity of a Novel Taxane, 1-Deoxypaclitaxel. 41st Annual Meeting of the American Society of Pharmacognosy, Seattle, WA, July 22–26, Abstract P-16

128. Holton RA, Tang S, Liang F, Somoza C (1997) Preparation of 1-Deoxybaccatin III, 1-Deoxytaxol and 1-Deoxytaxol Analogs. PCT Int Appl WO 97 42,181

129. Horiguchi T, Cheng Q, Oritani T (2000) Highly Regio- and Stereospecific Hydroxylation of C-1 Position of 2-Deacetoxytaxinine J Derivative with DMDO. Tetrahedron Lett **41**: 3907

130. Soto J, Mascarenas JL, Castedo L (1998) A Paclitaxel Analogue with a 2(3→20) *Abeo*taxane Skeleton: Synthesis and Biological Evaluation. Bioorg & Med Chem Lett **8**: 273

131. Py S, Pan J-W, Khuong-Huu F (1994) Cleavage Reactions of 10-Deacetylbaccatin III. Retrosynthetic Approach to the Total Synthesis of Taxol Derivatives. Tetrahedron **50**: 6881

132. Chordia MD, Gharpure MM, Kingston DGIK (1995) Facile AB Ring Cleavage Reactions of Taxoids. Tetrahedron **51**: 12963

133. Klein LL, Li L, Maring CJ, Yeung CM, Thomas SA, Grampovnik DJ, Plattner JJ (1995) Antitumor Activity of 9(R)-Dihydrotaxane Analogs. J Med Chem **38**: 1482

134. Grover S, Rimoldi JM, Molinero AA, Chaudhary AG, Kingston DGI, Hamel E (1995) Differential Effects of Paclitaxel (Taxol) Analogs Modified at Positions C-2, C-7, and C-3′ on Tubulin Polymerization and Polymer Stabilization: Identification of a Hyperactive Paclitaxel Derivative. Biochemistry **34**: 3927

135. Ojima I, Kuduk SD, Pera P, Veith JM, Bernacki RJ (1997) Syntheses and Structure-Activity Relationships of Nonaromatic Taxoids: Effects of Alkyl and Alkenyl Ester Groups on Cytotoxicity. J Med Chem **40**: 279

136. Kim S-C, Moon M-S, Cheong C-S (1999) C-2 Modified Taxol Analogs with Improved Aqueous Solubility. Bull Korean Chem Soc **20**: 1389

137. Ojima I, Wang T, Miller ML, Lin S, Borella CP, Geng X, Pera P, Bernacki RJ (1999) Synthesis and Structure-Activity Relationships of New Second-generation Taxoids. Bioorg & Med Chem Lett **9**: 3423

138. Uoto K, Takenoshita H, Yoshino T, Hirota Y, Ando S, Mitsui I, Terasawa H, Soga T (1998) Synthesis and Evaluation of Water-Soluble Non-Prodrug Analogs of Docetaxel Bearing *sec*-Aminoethyl Group at the C-10 Position. Chem Pharm Bull **46**: 770

139. Liu Y, Ali SM, Boge TC, Zygmunt J, Georg GI, Himes RH (2000) A Systematic Study of the Effects of *O*-10-Substitution on Paclitaxel Bioactivity: A Parallel Synthesis Approach. 219th ACS National Meeting, March 26–30. San Francisco, CA, Abstr. MEDI 72

140. Mellado W, Magri NF, Kingston DGI, Garcia-Arenas R, Orr GA, Horwitz SB (1984) Preparation and Biological Activity of Taxol Acetates. Biochem Biophys Res Commun **124**: 329

141. Lataste H, Senilh V, Wright M, Guenard D, Potier P (1984) Relationships between the Structures of Taxol and Baccatine III Derivatives and Their in vitro Action on the Disassembly of Mammalian Brain and *Physarum* Amoebal Microtubules. Proc Natl Acad Sci U.S.A. **81**: 4090

142. Kingston DGI, Samaranayake G, Ivey CA (1990) The Chemistry of Taxol, a Clinically Useful Anticancer Agent. J Nat Prod **53**: 1

143. Guéritte-Voegelein F, Guénard D, Lavelle F, Le Goff M-T, Mangatal L, Potier P (1991) Relationships between the Structure of Taxol Analogues and Their Antimitotic Activity. J Med Chem **34**: 992

144. Mathew AE, Mejillano MR, Nath JP, Himes RH, Stella VJ (1992) Synthesis and Evaluation of Some Water-Soluble Prodrugs and Derivatives of Taxol with Antitumor Activity. J Med Chem **35**: 145

145. Mongelli N, Biasoli G, Menichincheri M, Fusar Bassini D, Ciomei M (1995) Taxane Derivatives with Antitumor Activity. PCT Int Appl WO 95 25,728 (Chem Abstr **124**: 56359)

146. Bradley MO, Shashoua VE, Webb NL, Swindell CS (1997) Preparation of Conjugates of all-cis-docosahexaenoic Acid and Paclitaxel as Antitumor Agents. PCT Int Appl WO 97 44,336 (Chem Abstr **128**: 34910)

147. Ojima I, Bounaud P-Y, Takeuchi C, Pera P, Bernacki RJ (1998) New Taxanes as Highly Efficient Reversal Agents for Multi-Drug Resistance in Cancer Cells. Bioorg & Med Chem Lett **8**: 189

148. Chen S-H, Kant J, Mamber SW, Roth GP, Wei J-M, Marshall D, Vyas DM, Farina V (1994) Taxol Structure-Activity Relationships: Synthesis and Biological Evaluation of Taxol Analogs Modified at C-7. Bioorg Med Chem Lett **4**: 2223

149. Bhat L, Liu Y, Victory SF, Himes RH, Georg GI (1998) Synthesis and Evaluation of Paclitaxel C7 Derivatives: Solution Phase Synthesis of Combinatorial Libraries. Bioorg & Med Chem Lett **8**: 3181

150. Georg GI, Harriman GCB (1992) Taxol Photoaffinity Label: 7-(p-Azidobenzoyl)-taxol Synthesis and Biological Evaluation. Bioorg Med Chem Lett **2**: 735

151. Rimoldi JM, Kingston DGI, Chaudhary AG, Samaranayake G, Grover S, Hamel E (1993) Modified Taxols, 9. Synthesis and Biological Evaluation of 7-Substituted Photoaffinity Analogues of Taxol. J Nat Prod **56**: 1313

152. Georg GI, Liu Y, Boge TC, Himes RH (1997) 7-*O*-Acylpaclitaxel Analogues: Potential Probes to Map the Paclitaxel Binding Site. Bioorg & Med Chem Lett **7**: 1829

153. Wu Q, Bounaud P-Y, Kuduk SD, Yang C-PH, Ojima I, Horwitz SB, Orr GA (1998) Identification of the Domains of Photoincorporation of the 3′- and 7-Benzophenone Analogues of Taxol in the Carboxyl-Terminal Half of Murine *mdr*1b P-glycoprotein. Biochemistry **37**: 11272

154. Vyas DM, Wong H, Crosswell AR, Casazza AM, Knipe JO, Mamber SW, Doyle TW (1993) Synthesis and Antitumor Evaluation of Water Soluble Taxol Phosphates. Bioorg Med Chem Lett **3**: 1357

155. Vyas DM, Ueda Y, Wong H, Matiskella JD, Hauck S, Mikkilineni AB, Farina V, Rose WC, Casazza AM (1995) Phosphatase-Activated Prodrugs of Paclitaxel. In: Georg GI, Chen TT, Ojima I, Vyas DM (ed) Taxane Anticancer Agents: Basic Science and Current Status. American Chemical Society, Washington, DC, p 124

156. Ueda Y, Mikkilineni AB, Knipe JO, Rose WC, Casazza AM, Vyas DM (1993) Novel Water Soluble Phosphate Prodrugs of Taxol Possessing in Vivo Antitumor Activity. Bioorg Med Chem Lett **3**: 1761

157. Wittman MD (1997) 7-O-Methoxymethyl Paclitaxel. PCT Int Appl WO 97 42,948 (Chem Abstr 128:39533)

158. Wittman MD, Kadow JF (1996) Amino Acid Derivatives of Paclitaxel. US 5,489,589

159. Kadow JF, Chen S-H, Dextraze P, Fairchild CR, Golik J, Hansel SB, Johnston KA, Kramer RA, Lee FY, Long BH, Ouellet C, Perrone RK, Rose WC, Schulze GE, Xue M, Wei J-M, Wittman MD, Wong H, Wright JJK, Zoeckler ME, Vyas DM (2000) Discovery of More Efficacious Analogs of Paclitaxel (Taxol) for Human Clinical Evaluation. 219th ACS National Meeting. San Francisco, CA, MEDI 298

160. Fang W-S, Fang Q-C, Liang X-T (1997) Reinvestigation to the C-7 Epimerization of Paclitaxel and Related Taxoids under Basic Conditions. Synth Commun **27**: 2305

161. Chaudhary AG, Rimoldi JM, Kingston DGI (1993) Modified Taxols. 10. Preparation of 7-Deoxytaxol, a Highly Bioactive Taxol Derivative, and Interconversions of Taxol and 7-epi-Taxol. J Org Chem **58**: 3798

162. Chen S-H, Huang S, Kant J, Fairchild C, Wei J, Farina V (1993) Synthesis of 7-Deoxy- and 7,10-Dideoxytaxol via Radical Intermediates. J Org Chem **58**: 5028

163. Chen SH, Wei JM, Vyas DM, Doyle TW, Farina V (1993) A Facile Synthesis of 7,10-dideoxy Taxol and 7-epi-10-deoxy Taxol. Tetrahedron Lett **34**: 6845

164. Pulicani J-P, Bouchard H, Bourzat J-D, Commerçon A (1994) Preparation of 7-Modified Docetaxel Analogs Using Electrochemistry. Tetrahedron Lett **35**: 9709

165. Magri NF, Kingston DGI (1986) Modified Taxols. 2. Oxidation Products of Taxol. J Org Chem **51**: 797

166. Magri NF, Kingston DGI (1988) Modified Taxols. 4. Synthesis and Biological Activity of Taxols Modified in the Side Chain. J Nat Prod **51**: 298

167. Liang X, Kingston DGI, Lin CM, Hamel E (1995) Synthesis and Biological Evaluation of Paclitaxel Analogs Modified in Ring C. Tetrahedron Lett **36**: 2901

168. Menichincheri M, Botta M, Ceccarelli W, Ciomei M, Corelli F, D'Anello M, Fusar-Bassini D, Mongelli N, Pesenti E, Pinciroli V, Tafi A, Vanotti E (1995) Synthesis and Molecular Modeling of Taxanes Modified at Ring C. Med Chem Res **5**: 534

169. Chen S-H, Huang S, Wei J, Farina V (1993) Serendipitous Synthesis of a Cyclopropane-Containing Taxol Analog via Anchimeric Participation of an Unactivated Angular Methyl Group. J Org Chem **58**: 4520

170. Chen SH, Huang S, Farina V (1994) On the Reaction of Taxol with DAST. Tetrahedron Lett **35**: 41

171. Roth GP, Marshall DR, Chen S-H (1995) Reaction of Paclitaxel and 10-Desacetyl Baccatin III with Diethylamino Sulfurtrifluoride. Tetrahedron Lett **36**: 1609

172. Johnson RA, Dobrowolski PJ, Nidy EG, Gebhard I, Qualls SJ, Wicnienski NA, Kelly RC, Pitts TW, Lopes NM, DeKoning TF, Mattern SJ, Weed SD, McGovren JP, Chidester CG (1997) 12,13-Isotaxanes: Synthesis of New Potent Analogs and X-ray Crystallographic Confirmation of Structure. J Med Chem **40**: 2810

173. Pulicani J-P, Bouchard H, Bourzat J-D, Commerçon A (1994) Preparation of 7-Modified Docetaxel Analogs Using Electrochemistry. Tetrahedron Lett **35**: 9709

174. Klein LL, Maring CJ, Li L, Yeung CM, Thomas SA, Grampovnik DJ, Plattner JJ, Henry RF (1994) Synthesis of Ring B-Rearranged Taxane Analogs. J Org Chem **59**: 2370

175. Altstadt TJ, Gao Q, Wittman MD, Kadow JF, Vyas DM (1998) Crystallographic Determination of the Stereochemistry of C-6,7 Epoxy Paclitaxel. Tetrahedron Lett **39**: 4965

176. Wittman MD, Alstadt TJ, Kadow JF, Vyas DM, Johnson K, Fairchild C, Long B (1999) Stereospecific Synthesis of 7-Deoxy-6-hydroxy Paclitaxel. Tetrahedron Lett **40**: 4943

177. Yuan H, Kingston DGI (1998) Synthesis of 6α-Hydroxypaclitaxel, the Major Human Metabolite of Paclitaxel. Tetrahedron Lett **39**: 4967

178. Wittman MD, Kadow JF, Vyas DM (2000) Stereospecific Synthesis of the Major Human Metabolite of Paclitaxel. Tetrahedron Lett **41**: 4729

179. Liang X, Kingston DGI, Long BH, Fairchild CA, Johnston KA (1997) Paclitaxel Analogs Modified in Ring C: Synthesis and Biological Evaluation. Tetrahedron **53**: 3441

180. Chen S-H, Huang S, Roth GP (1995) An Interesting C-Ring Contraction in Paclitaxel (Taxol). Tetrahedron Lett **36**: 8933

181. Wender PA, Lee D, Lal TK, Horwitz SB, Rao S (1997) Synthesis of Novel Taxol Analogs and Evaluation of Their Biological Activities. Bioorg & Med Chem Lett **7**: 1941

182. Appendino G, Jakupovic J, Cravotto G, Enriu R, Varese M, Bombardelli E (1995) Synthesis of Modified Baccatins via an Oxidation-Reduction Protocol. Tetrahedron Lett **36**: 3233

183. Appendino G, Danieli B, Jakupovic J, Belloro E, Scambia G, Bombardelli E (1997) Synthesis and Evaluation of C-Seco Paclitaxel Analogues. Tetrahedron Lett **38**: 4273

184. Wang M, Cornett B, Nettles J, Liotta DC, Snyder JP (2000) The Oxetane Ring in Taxol. J Org Chem **65**: 1059

185. Neidigh KA, Gharpure MM, Rimoldi JM, Kingston DGI, Jiang YQ, Hamel E (1994) Synthesis and Biological Evaluation of 4-Deacetylpaclitaxel. Tetrahedron Lett **35**: 6839

186. Datta A, Jayasinghe LR, Georg GI (1994) Internal Nucleophile Assisted Selective Deesterification Studies on Baccatin III. Synthesis of 2-Debenzoyl- and 4-Deacetylbaccatin III Analogues. J Org Chem **59**: 4689

187. Datta A, Jayasinghe LR, Georg GI (1994) 4-Deacetyltaxol and 10-Acetyl-4-deacetyltaxotere: Synthesis and Biological Evaluation. J Med Chem **37**: 4258

188. Chen S-H, Kadow JF, Farina V, Fairchild CR, Johnston KA (1994) First Syntheses of Novel Paclitaxel(Taxol) Analogs Modified at the C4-Position. J Org Chem **59**: 6156

189. Kim S-C, Chai K-B (1997) Migration Between C-2 and C-4 Hydroxyl Groups in Paclitaxel Core. Korean J Med Chem **7**: 2

190. Georg GI, Ali SM, Boge TC, Datta A, Falborg L, Himes RH (1994) Selective C-2 and C-4 Deacylation and Acylation of Taxol: The First Synthesis of a C-4 Substituted Taxol Analogue. Tetrahedron Lett **35**: 8931

191. Chen S-H, Wei J-M, Long BH, Fairchild CA, Carboni J, Mamber SW, Rose WC, Johnston K, Casazza AM, Kadow JF, Farina V, Vyas D, Doyle TW (1995) Novel C-4 Paclitaxel (Taxol) Analogs: Potent Antitumor Agents. Bioorg & Med Chem Lett **5**: 2741

192. Chen S-H (1996) First Syntheses of C-4 Methyl Ether Paclitaxel Analogs and the Unexpected Reactivity of 4-Deacetyl-4-Methyl Ether Baccatin III. Tetrahedron Lett **37**: 3935

193. Chen S-H, Farina V, Vyas DM, Doyle TW (1998) Synthesis of a Paclitaxel Isomer: C-2-Acetoxy-C-4-Benzoate Paclitaxel. Bioorg & Med Chem Lett **8**: 2227

194. Chordia MD, Yuan H, Jagtap PG, Kadow JF, Long BH, Fairchild CR, Johnston KA, Kingston DGI (2001) Synthesis and Bioactivity of 2,4-Diacyl Analogues of Paclitaxel. Bioorg Med Chem **9**: 171

195. Uoto K, Takenoshita H, Ishiyama T, Terasawa H, Soga T (1997) A New Method to Modify the C-4 Position of 10-Deacetylbaccatin III. Chem Pharm Bull **45**: 2093

196. Chordia MD, Chaudhary AG, Kingston DGI, Jiang YQ, Hamel E (1994) Synthesis and Biological Evaluation of 4-Deacetoxypaclitaxel. Tetrahedron Lett **35**: 6843

197. Chen S, Huang S, Wei J, Farina V (1993) The Chemistry of Taxanes: Reaction of Taxol and Baccatin Derivatives with Lewis Acids in Aprotic and Protic Media. Tetrahedron **49**: 2805

198. Holton RA, Kim H-B, Somoza C, Liang F, Biediger RJ, Boatman PD, Shindo M, Smith CC, Kim S, Nadizadeh H, Suzuki Y, Tao C, Vu P, Tang S, Zhang P, Murthi KK, Gentile LN, Liu JH (1994) First Total Synthesis of Taxol. 2. Completion of the C and D Rings. J Am Chem Soc **116**: 1599

199. Farina V, Huang S (1992) The Chemistry of Taxanes: Unexpected Rearrangement of Baccatin III During Chemoselective Debenzoylation with $Bu_3SnOMe/LiCl$. Tetrahedron Lett **33**: 3979

200. Wahl A, Guéritte-Voegelein F, Guénard D, Le Goff M, Potier P (1992) Rearrangement Reactions of Taxanes: Structural Modifications of 10-Deacetylbaccatin III. Tetrahedron **48**: 6965

201. Marder-Karsenti R, Dubois J, Chiaroni A, Riche C, Guénard D, Guéritte F, Potier P (1998) New Skeletal Rearrangements of C and D Rings of a 13-Oxobaccatin III Derivative. Tetrahedron **54**: 15833

202. Fenoglio I, Nano GM, Vander Velde DG, Appendino G (1996) Synthesis of Azetidine-type Taxanes. Tetrahedron Lett **37**: 3203

203. Marder-Karsenti R, Dubois J, Bricard L, Guénard D, Guéritte-Voegelein F (1997) Synthesis and Biological Evaluation of D-Ring-Modified Taxanes: 5(20)-Azadocetaxel Analogs. J Org Chem. **62**: 6631

204. Payré C, Mourabit AA, Mercklé L, Ahond A, Poupat C, Potier P (2000) Semisynthesis of D-ring-modified Taxoids: Thietane Derivatives from Taxine B. Tetrahedron Lett **41**: 4891

205. Gunatilaka AAL, Ramdayal FD, Sarragiotto MH, Kingston DGI, Sackett DL, Hamel E (1999) Synthesis and Biological Evaluation of Novel Paclitaxel (Taxol) D-Ring Modified Analogues. J Org Chem **64**: 2694
206. Dubois J, Thoret S, Place E, Guéritte F, Guénard D, Potier P (2000) Synthesis and Biological Evaluation of Docetaxel Oxetane Modified Analogs. 219th ACS National Meeting, San Francisco, March 26–30, Abstract MEDI 066
207. Dubois J, Thoret S, Guéritte F, Guénard D (2000) Synthesis of 5(20)deoxydocetaxel, a New Active Docetaxel Analogue. Tetrahedron Lett **41**: 3331
208. Boge TC, Hepperle M, Vander Velde DG, Gunn CW, Grunewald GL, Georg GI (1999) The Oxetane Conformational Lock of Paclitaxel: Structural Analysis of D-secopaclitaxel. Bioorg & Med Chem Lett **9**: 3041
209. Guéritte-Voegelein F, Guénard D, Dubois J, Wahl A, Marder R, Muller R, Lund M, Bricard L, Potier P (1995) Chemistry and Structure-Activity Relationships of Taxoids with Modified Skeletons. In: Georg GI, Chen TT, Ojima I, Vyas DM (eds) Taxane Anticancer Agents: Basic Science and Current Status. American Chemical Society, Washington, DC, p 189
210. Appendino G, Varese M, Gariboldi P, Gabetta B (1994) The Chemistry (15) and Occurrence of Taxane Derivatives. XV. A New Rearrangement of Oxetane-Type Taxoids. Tetrahedron Lett **35**: 2217
211. Zamir LO, Zheng YF, Caron G, Sauriol F, Mamer O (1996) Rearrangement of the Major Taxane from Taxus canadensis. Tetrahedron Lett **37**: 6435
212. Zamir LO, Balachandran S, Zheng YF, Nedea ME, Caron G, Nikolakakis A, Vishwakarma RA, Sauriol F, Mamer O (1997) Acid Catalyzed Rearrangement and Acyl Migration Studies on 9-Dihydro-13-acetylbaccatin-III, A Major Taxane from Taxus canadensis. Tetrahedron **53**: 15991
213. Chordia MD, Kingston DGI, Hamel E, Lin CM, Long BH, Fairchild CA, Johnston KA, Rose WC (1997) Synthesis and Biological Activity of A-nor-Paclitaxel Analogues. Bioorg Med Chem **5**: 941
214. Georg GI, Cheruvalath ZS, Velde DV, Ye Q-M, Mitscher LA, Himes RH, Mejillano MR (1993) Semisynthesis and Biological Evaluation of Brevifoliol 13-[N-Benzoyl-(2′R,3′S)-3′-phenylisoserinate]. Bioorg Med Chem **3**: 1349
215. Chordia MD, Kingston DGI (1997) Synthesis and Biological Evaluation of Amide-Linked A-Norpaclitaxels. Tetrahedron **53**: 5699
216. Torregiani E, Barboni L, Rafaiani G, Lambertucci C, Appendino G (1996) The Chemistry and Occurrence of Taxane Derivatives. XXVI. A Reactivity Study on 11(15→1) Abeotaxanes. Gazz Chim Ital **126**: 809
217. Chen S-H, Farina V, Huang S, Gao Q, Jolik J, Doyle TW (1994) Studies on the Photochemistry of Taxol. Tetrahedron **50**: 8633
218. Chen S, Combs CM, Hill SE, Farina V, Doyle TW (1992) The Photochemistry of Taxol: Synthesis of a Novel Pentacyclic Taxol Isomer. Tetrahedron Lett **33**: 7679
219. Appendino G, Cravotto G, Gariboldi P, Gabetta B, Bombardelli E (1994) The Chemistry (10) and Occurrence of Taxane Derivatives X. The Photochemistry of Taxine B. Gazz Chim Ital **124**: 1
220. Chen S-H, Huang S, Gao Q, Golik J, Farina V (1994) The Chemistry of Taxanes: Skeletal Rearrangements of Baccatin Derivatives via Radical Intermediates. J Org Chem **59**: 1475
221. Klein LL, Maring CJ, Li L, Yeung CM, Thomas SA, Grampovnik DJ, Plattner JJ, Henry RF (1994) Synthesis of Ring B-Rearranged Taxane Analogs. J Org Chem **59**: 2370
222. Yu C-Z, Liu Z-Y (1997) Skeletal Rearrangement of A and B Ring of 13-Oxobaccatin III Derivative. Tetrahedron Lett **38**: 4133

223. Appendino G, Fenoglio I, Vander Velde DG (1997) Novel Base-Catalyzed Rearrangement of the Taxane Skeleton. J Nat Prod **60**: 464
224. van Rozendaal ELM, Veldhuis H, van Veldhuizen B, van Beek TA, de Groot A, Beusker PH, Scheeren HW (2000) Rearrangement of O-Cinnamoyltaxicin I to a Novel C-13 Spiro-Taxane. J Nat Prod **63**: 179
225. Appendino G, Jakupovic J, Cravotto G, Varese M (1994) The Chemistry(18) and Occurrence of Taxane Derivatives. XVIII. The Synthesis of A-Nor-B-homobaccatin III Derivatives. Tetrahedron Lett **35**: 6547
226. Pinciroli V, Ceccarelli W, Fusar-Bassini D, Menichincheri M, Mongelli N, Vanotti E (1996) New Rearranged Products from the Methylation of 13-Oxobaccatin III. Tetrahedron Lett **37**: 9365
227. Lee D, Kim M-J (1999) Enzymatic Selective Dehydration and Skeleton Rearrangement of Paclitaxel Precursors. Org Lett **1**: 925
228. Zamir LO, Cherestes AD, Nikolakakis A, Sauriol F, Mamer O (1999) Rearrangement of the Major Taxane from *Taxus canadensis*. Tetrahedron Lett **40**: 7917
229. Appendino G, Jakupovic J, Varese M, Bombardelli E (1996) Acid and Base Catalysed Rearrangements of 9,10-Dioxotaxanes. Tetrahedron Lett **37**: 727
230. Nidy EG, Wicnienski NA, Chidester CG, Dobrowolski PJ, Mizsak SA, Toy A, Runge TA, Qualls SJ, Kelly RC, Johnson RA (1997) Intramolecular 2 + 2 Cycloaddition in a Baccatin. From a Tetracyclic to a Hexacyclic Molecule. J Org Chem **62**: 4900
231. Hosoyama H, Shigemori H, Kobayashi J (2000) Further Unexpected Boron Trifluoride-catalyzed Reactions of Taxoids with α- and β-4,20-epoxides. J Chem Soc Perkin Trans 1 449
232. Kant J (1995) The Chemistry of the Taxol Side Chain: Synthesis Modifications and Conformational Studies. In: Farina V (ed) The Chemistry and Pharmacology of Taxol and its Derivatives. Elsevier Science B V, Amsterdam, p 255
233. Ojima I (1993) β-Lactam Synthon Method: Enantiomerically Pure β-Lactams as Synthetic Intermediates. In: Georg GI (ed) The Organic Chemistry of β-Lactams. VCH Publishers, Inc., New York, p 197
234. Ojima I (1995) Recent Advances in the β-Lactam Synthon Method. Acc Chem Res **28**: 383
235. Andruszkiewicz R (1998) Recent Developments in the Synthesis of α-Hydroxy-β-amino Acids. Polish J Chem **72**: 1
236. Ojima I, Lin S, Wang T (1999) Recent Advances in the Medicinal Chemistry of Taxoids with Novel β-amino Acid Side Chains. Curr Med Chem **6**: 927
237. Ojima I, Habus I, Zhao M, Zucco M, Park YH, Sun CM, Brigaud T (1992) New and efficient approaches to the semisynthesis of taxol and its C-13 side chain analogs by means of β-lactam synthon method. Tetrahedron **48**: 6985
238. Georg GI, Cheruvallath ZS, Harriman GCB, Hepperle M, Park H (1993) An Efficient Semisynthesis of Taxol from (3R,4S)-N-Benzoyl-3-[(t-butyldimethylsilyl)oxy]-4-phenyl-2-azetidinone and 7-(Triethylsilyl)baccatin III. Bioorg Med Chem Lett **3**: 2467
239. Farina V, Hauck SI, Walker DG (1992) A Simple Chiral Synthesis of the Taxol Side Chain. Synlett **1**: 761
240. Bourzat JD, Commerçon A (1993) A Practical Access to Chiral Phenylisoserinates, Preparation of Taxotere Analogs. Tet Lett **34**: 6049
241. Palomo C, Aizpurua JM, Miranda JI, Mielgo A, Odriozola JM (1993) Asymmetric Synthesis of α-keto-β-lactams via [2+2] Cycloaddition Reaction – A Concise Approach to Optically Active α-hydroxy β-lactams and β-alkyl(aryl)isoserines. Tetrahedron Lett **34**: 6325

242. Holton RA, Liu JH (1993) A Novel Asymmetric Synthesis of Cis-3-Hydroxy-4-Aryl Azetidin-2-Ones. Bioorg. Med Chem Lett **3**: 2475

243. Kayser MM, Mihovilovic MD, Kearns J, Feicht A, Stewart JD (1999) Baker's Yeast-Mediated Reductions of α-Keto Esters and an α-Keto-β-Lactam. Two Routes to the Paclitaxel Side Chain. J Org Chem **64**: 6603

244. Shimizu M, Ishida T, Fujisawa T (1994) Highly Stereocontrolled Construction of 3-Alkoxyazetidin-2-ones Using Ester Enolate-Imine Condensation. Chem Lett 1403

245. Brieva R, Crich JZ, Sih CJ (1993) Chemoenzymatic Synthesis of the C-13 side Chain of Taxol: Optically-active 3-Hydroxy-4-phenyl β-lactam Derivatives. J Org Chem **58**: 1068

246. Ojima I, Sun CM, Zucco M, Park YH, Duclos O, Kuduk S (1993) A Highly Efficient Route to Taxotere by the β-Lactam Synthon Method. Tetrahedron Lett **34**: 4149

247. Ojima I, Zucco M, Duclos O, Kuduk SD, Sun CM, Park YH (1993) N-Acyl-3-Hydroxy-β-Lactams as Key Intermediates for Taxotere and its Analogs. Bioorg Med Chem Lett **3**: 2479

248. Georg GI, Harriman GCB, Park H, Himes RH (1994) Taxol Photoaffinity Labels 2. Synthesis and Biological Evaluation of N-(4-Azidobenzoyl)-N-debenzoyltaxol, N-(4-Azido-2,3,5,6-tetrafluorobenzoyl)-N-debenzoyltaxol, and 7-(4-Azido-2,3,5,6-tetrafluorobenzoyl)taxol. Bioorg Med Chem Lett **4**: 487

249. Song CE, Lee SW, Roh EJ, Lee S-G, Lee W-K (1998) A New Synthetic Route to (3R,4S)-3-hydroxy-4-phenylazetidin-2-one as a Taxol Side Chain Precursor. Tetrahedron: Asymmetry **9**: 983

250. Banik BK, Manhas MS, Bose AK (1994) Studies on Lactams. 96. Stereospecific Glycosylation via Ferrier Rearrangement for Optical Resolution. J Org Chem **59**: 4714

251. Lin S, Geng C, Qu C, Tynebor R, Gallagher D, Pollina E, Rutter J, Ojima I (2000) Synthesis of Highly Potent Second-Generation Taxoids Through Effective Kinetic Resolution Coupling of Racemic β-Lactams with Baccatins. Chirality **12**: 431

252. Denis J-N, Greene AE, Serra AA, Luche M-J (1986) An Efficient, Enantioselective Synthesis of the Taxol Side Chain. J Org Chem **51**: 46

253. Bonini C, Righi G (1994) Enantio- and Stereo-selective Route to the Taxol Side Chain *via* Asymmetric Epoxidation of *trans*-Cinnamyl Alcohol and Subsequent Epoxide Ring Opening. J Chem Soc Chem Comm 2767

254. Righi G, Rumboldt G (1996) Stereoselective Preparation of Syn α-Hydroxy-β-amino Ester Units via Regioselective Opening of α,β-Epoxy Esters: Enantioselective Synthesis of Taxol C-13 Side Chain and Cyclohexylnorstatine. J Org Chem **61**: 3557

255. Srivastava RP, Zjawiony JK, Peterson JR, McChesney JD (1994) N-Benzoyl-(2R,3S)-Phenylisoserine Methyl Ester; A Facile and Convenient Synthesis and Resolution by Entrainment. Tetrahedron: Asymmetry **5**: 1683

256. Srivastava RP, McChesney JD (1995) A Practical and Inexpensive Synthesis of the Taxol C-13 Side Chain; N-Benzoyl-(2R,3S)-3-Phenylisoserine. Nat Prod Lett **6**: 147

257. Cabon O, Buisson D, Larcheveque M, Azerad R (1995) Stereospecific Preparation of Glycidic Esters from 2-Chloro-3-Hydroxyesters. Application to the Synthesis of (2*R*,3*S*)-3-Phenylisoserine. Tetrahedron: Asymmetry **6**: 2211

258. Deng L, Jacobsen EN (1992) A Practical, Highly Enantioselective Synthesis of the Taxol Side Chain via Asymmetric Catalysis. J Org Chem **57**: 4320

259. Koskinen AMP, Karvinen EK, Siirila JP (1994) Enantioselective Synthesis of the Taxol and Taxotere Side Chains. J Chem Soc Chem Comm **1**: 21

260. Lohray BB, Bhushan V (1995) Enantioselective Synthesis of Taxol Side Chain via Regio- and Stereoselective Reaction of 4(R)-carbethoxy-5(S)-phenyl-1,3,2-dioxathiolane-2-oxide. Indian J Chem Sect B: Org Chem Incl Med Chem **34B**: 471

261. Nandanan E, Phukan P, Sudalai A (1999) Regio and Stereoselective Opening of Chiral Cyclic Sulfates with MgBr2-Et2O: a Practical Strategy for the Synthesis of (−)-N-Benzoyl-3-phenylisoserine Ethyl Ester (Taxol Side Chain). Indian J Chem Sect B: Org Chem Incl Med Chem **38B**: 283

262. Li G, Sharpless KB (1996) Catalytic Asymmetric Aminohydroxylation Provides a Short Taxol Side-chain Synthesis. Acta Chem Scand **50**: 649

263. Song CE, Oh CR, Roh EJ, Lee S-G, Choi JH (1999) One-Step Synthesis of Paclitaxel Side-Chain Precursor: Benzamide-based Asymmetric Aminohydroxylation of Isopropyl *trans*-cinnamate. Tetrahedron: Asymmetry **10**: 671

264. Bushey ML, Haukaas MH, O'Doherty GA (1999) Asymmetric Aminohydroxylation of Vinylfuran. J Org Chem **64**: 2984

265. Raatz D, Innertsberger C, Reiser O (1999) Asymmetric Aminohydroxylation of Heteroaromatic Acrylates. Synlett 1907

266. Chandrasekhar S, Mohapatra S (1998) Neighbouring Group Assisted Sulfonamide Cleavage of Sharpless Aminols under Acetonation Conditions. Tetrahedron Lett **39**: 695

267. Hattori K, Miyata M, Yamamoto H (1993) Highly selective and operationally simple synthesis of enantiomerically pure β-amino esters via double stereodifferentiation. J Am Chem Soc **115**: 1151

268. Hattori K, Yamamoto H (1994) Practical Preparation of α-Hydroxy-β-Amino Ester Units; Stereoselective Synthesis of Taxol Side Chain and Norstatine. Tetrahedron **50**: 2785

269. Kobayashi S, Ishitani H, Ueno M (1998) Catalytic Asymmetric Synthesis of Both Syn- and Anti-β-Amino Alcohols. J Am Chem Soc **120**: 431

270. Kanazawa AM, Denis J-N, Greene AE (1994) Highly Stereocontrolled and Efficient Preparation of the Protected, Esterification-Ready Docetaxel (Taxotere) Side Chain. J Org Chem **59**: 1238

271. Kanazawa AM, Denis J-N, Greene AE (1994) Direct, Stereoselective Synthesis of the Protected Paclitaxel (Taxol) Side Chain and High-yield Transformation to Paclitaxel. J Chem Soc Chem Comm 2591

272. Ha H-J, Park G-S, Ahn Y-G, Lee GS (1998) Practical Synthesis of Taxol Side Chain. Bioorg Med Chem Lett **8**: 1619

273. Schade W, Reibig H-U (1999) Stereocontrolled Synthesis of the Taxol C-13 Side Chain: Methyl (2R,3S)-3-Benzoylamino-2-hydroxy-3-phenylpropanoate. J Prakt Chem **341**: 685

274. Merino P, Castillo E, Franco S, Merchan FL, Tejero T (1998) Nucleophilic Additions of Grignard Reagents to N-benzyl-2,3-O-isopropylidene-D-glyceraldehyde Nitrone (BIGN). Synthesis of (2S,3R) and (2S,3S)-3-phenylisoserine. Tetrahedon **54**: 12301

275. Kanazawa AM, Correa A, Denis J, Luche M, Greene A (1993) A short synthesis of the taxotere side chain through dilithiation of boc-benzylamine. J Org Chem **58**: 255

276. Mukai C, Kim I, Hanaoka M (1992) An Optically Active Chromium(0)-Complexed Benzaldehyde in Organic Synthesis: a Highly Stereocontrolled Asymmetric Synthesis of (2R,3S)-(−)-N-Benzoyl-3-phenylisoserine Methyl Ester, the Taxol C-13 Side Chain analogue. Tetrahedron: Asymmetry **3**: 1007

277. Mukai C, Kim IJ, Furu E, Hanaoka M (1993) Highly Stereocontrolled Asymmetric Syntheses of Taxol and Taxotere C-13 Side Chain Analogues. Tetrahedron **49**: 8323

278. Mukaiyama T, Shiina I, Iwadare H, Saitoh M, Nishimura T, Ohkawa N, Sakoh H, Nishimura K, Tani Y-I, Hasegawa M, Yamada K, Saitoh K (1999) Asymmetric Total Synthesis of Taxol. Chem Eur J **5**: 121

279. Dondoni A, Perrone D, Semola T (1995) Synthesis of Taxol and Taxotere Side chains by 2-(Trimethylsilyl)thiazole Based Homologation of L-Phenylglycine. Synthesis 181

280. Ishizuka T, Morooka K, Ishibuchi S, Kunieda T (1996) Synthetic Application of Chiral 4-Methoxy-2-oxazolidinone Synthons to 2-Amino Alcohols of Biological Interest. Heterocycles **42**: 837

281. Kingston DGI, Chaudhary AG, Gunatilaka AAL, Middleton ML (1994) Synthesis of Taxol from Baccatin III via an Oxazoline Intermediate. Tetrahedron Lett **35**: 4483

282. Tiecco M, Testaferri L, Temperini A, Bagnoli L, Marini F, Santi C (1998) Electrophilic Azido Selenenylation of Alkenes. A Simple Synthetic Route to Racemic Taxol Side Chain Syn Comm **28**: 2167

283. Tiecco M, Testaferri L, Temperini A, Marini F, Bagnoli L, Santi C (1999) Selenium Promoted Stereospecific One-Pot Conversion of Cinnamyl Derivatives into Oxazolines. A Simple Synthetic Route to Racemic Taxol Side Chain Syn Comm **29**: 1773

284. Lee K-Y, Kim Y-H, Park M-S, Ham W-H (1998) A Highly Stereocontrolled Asymmetric Synthesis of the Taxol C-13 Side Chain; (4*S*,5*R*)-2,4-Diphenyloxazoline-5-carboxylic Acid. Tetrahedron Lett **39**: 8129

285. Lee S-H, Yoon J, Nakamura K, Lee Y-S (2000) Efficient Syntheses and Ring-Opening Reactions of *trans*- and *cis*-Oxazoline-5-carboxylates. Org Lett **2**: 1243

286. Swindell CS, Krauss NE, Horwitz SB, Ringel I (1991) Biologically Active Taxol Analogues with Deleted A-Ring Side Chain Substituents and Variable C-2′ Configurations. J Med Chem **34**: 1176

287. Swindell CS, Tao M (1993) Chiral Auxiliary-Mediated Asymmetric Induction in a Thermal Inverse Electron Demand Hetero-Diels-Alder Reaction – Enantioselective Synthesis of the Taxol A-Ring Side Chain. J Org Chem **58**: 5889

288. Patel RN, Banerjee A, Howell JM, McNamee CG, Brzozowski D, Nanduri V, Thottathil JK, Szarka LJ (1995) Stereoselective Microbial Reduction of 2-Keto-3-(N-benzoylamino)-3-phenyl Propionic Acid Ethyl Ester: Synthesis of Taxol Side-Chain Synthon. Ann NY Acad Sci **750**: 166

289. Kearns J, Kayser MM (1994) Application of Yeast-Catalyzed Reductions to Synthesis of (2R,3S)-Phenylisoserine. Tetrahedron Lett **35**: 2845

290. Lee D, Kim M-J (1998) Lipase-Catalyzed Transesterification as a Practical Route to Homochiral syn-1,2-Diols. The Synthesis of the Taxol Side Chain. Tetrahedron Lett **39**: 2163

291. Wuts PGM, Gu RL, Northuis JM (2000) Synthesis of (2R, 3S)-Isobutyl Phenylisoserinate, the Taxol Side Chain, from Ethyl Benzoylacetate. Tetrahedron: Asymmetry **11**: 2117

292. Wroblewski AE, Piotrowska DG (1998) Phosphonate Analogs of *N*-benzoyl- and *N*-Boc-3-phenylisoserine, the Taxol C-13 Side Chain. Tetrahedron **54**: 8123

293. Wroblewski AE, Piotrowska DG (1999) Enantiomeric Phosphonate Analogs of the Paclitaxel C-13 Side Chain. Tetrahedron: Asymmetry **10**: 2037

294. Wroblewski AE, Piotrowska DG (2000) Enantiomeric Phosphonate Analogs of the Docetaxel C-13 Side Chain. Tetrahedron: Asymmetry **11**: 2615

295. Wuts PGM (1998) Semisynthesis of Taxol. Curr Opin Drug Disc Devel **1**: 329

296. Didier E, Fouque E, Commerçon A (1994) Expeditious Semisynthesis of Docetaxel Using 2-Trichloromethyl-1,3-Oxazolidine as Side-Chain Protection. Tetrahedron Lett **35**: 3063

297. Denis J-N, Kanazawa AM, Greene AE (1994) Taxotere by Esterification with Stereochemically "Wrong" (2S,3S)-Phenylisoserine Derivatives. Tetrahedron Lett **35**: 105

298. Baloglu E, Kingston DGI (1999) A New Semisynthesis of Paclitaxel from Baccatin III. J Nat Prod **62**: 1068

299. Gennari C, Vulpetti A, Donghi M, Mongelli N, Vanotti E (1996) Semisynthesis of Taxol: A Highly Enantio- and Diastereoselective Synthesis of the Side Chain and a New Method for Ester Formation at C13 Using Thioesters. Angew Chem Int Ed Engl **35**: 1723

300. Gennari C, Carcano M, Donghi M, Mongelli N, Vanotti E, Vulpetti A (1997) Taxol Semisynthesis: A Highly Enantio- and Diastereoselective Synthesis of the Side Chain and a New Method for Ester Formation at C-13 Using Thioesters. J Org Chem **62**: 4746

301. Yamaguchi T, Harada N, Ozaki K, Hayashi M, Arakawa H, Hashiyama T (1999) Synthesis of Taxoids 4. Novel and Versatile Methods for Preparation of New Taxoids by Employing cis- or trans-Phenyl Glycidic Acid. Tetrahedron **55**: 1005

302. Mandai T, Kuroda A, Okumoto H, Nakanishi K, Mikuni K, Hara K-J, Hara K-Z (2000) Synthesis of New Baccatin III Derivatives through Transesterification of α-keto Esters with a Protected 10-deacetylbaccatin III. Tetrahedron Lett **41**: 239

303. Mandai T, Kuroda A, Okumoto H, Nakanishi K, Mikjuni K, Hara K-J, Hara K-Z (2000) A Semisynthesis of Paclitaxel via a 10-deacetylbaccatin III Derivative Bearing a β-keto Ester Appendage. Tetrahedron Lett **41**: 243

304. Wuts PGM, Kelly RC (1997) Process to Prepare Taxol. WO-09742167. Cited in ref 295

305. Guénard D, Guéritte-Voegelein F, Potier P (1993) Taxol and Taxotere: Discovery, Chemistry, and Structure-Activity Relationships. Acc Chem Res **26**: 160

306. Georg GI, Boge TC, Cheruvallath ZS, Harriman GCB, Hepperle M, Park H, Himes RH (1994) Topliss Approach to the Synthesis of Biologically Active Substituted *N*-Benzoyl Taxol Analogues. Bioorg & Med Chem Lett **4**: 1825

307. Roh EJ, Song CE, Kim D, Pae H-O, Chung H-T, Lee K-S, Chai K-B, Lee CO, Choi SU (1999) Synthesis and Biology of 3′-*N*-Acyl-*N*-debenzoylpaclitaxel Analogues. Bioorg & Med Chem **7**: 2115

308. Xue M, Long BH, Fairchild C, Johnston K, Rose WC, Kadow JF, Vyas DM, Chen S-H (2000) Structure-Activity Relationships Study at the 3′-*N* Position of Paclitaxel. Part 2: Synthesis and Biological Evaluation of 3′-*N*-Thiourea- and 3′-*N*-Thiocarbamate-Bearing Paclitaxel Analogues. Bioorg Med Chem Lett **10**: 1327

309. Holton RA, Nadizadeh H, Beidiger RJ, Kim S (1993) Preparation of Substituted Taxanes as Antitumor Agents. Eur Pat Appl 534,709 (Chem Abstr 119, 49694)

310. Holton RA, Nadizadeh H, Beidiger RJ (1993) Preparation of Furyl- and Thienyl-substituted Taxanes as Antitumor Agents. Eur Pat Appl 534,708 (Chem Abstr **119**: 49693)

311. Georg GI, Harriman GCB, Hepperle M, Himes RH (1994) Heteroaromatic Taxol Analogues: the Chemistry and Biological Activities of 3′-Furyl and 3′-Pyridyl Substituted Taxanes. Bioorg & Med Chem Lett **4**: 1381

312. Holton RA, Rengan K (1996) Preparation and Pharmaceutical Compositions of Taxanes Having a Pyridyl Substituted Side-Chain. US 5,489,601 (Chem Abstr **124**: 317534)

313. Ojima I, Duclos O, Kuduk SD, Sun C-M, Slater JC, Lavelle F, Veith JM, Bernacki RJ (1994) Synthesis and Biological Activity of 3′-Alkyl- and 3′-Alkenyl-3′-dephenyldocetaxels. Bioorg Med Chem Lett **4**: 2631

314. Ojima I, Lin S (1998) Efficient Asymmetric Syntheses of β-Lactams Bearing a Cyclopropane or an Epoxide Moiety and their Application to the Syntheses of Novel Isoserines and Taxoids. J Org Chem **63**: 224

315. Ojima I, Kuduk SD, Slater JC, Gimi RH, Sun CM (1996) Syntheses of New Fluorine-Containing Taxoids by Means of β-Lactam Synthon Method. Tetrahedron **52**: 209

316. Ojima I, Slater JC, Pera P, Veith JM, Abouabdellah A, Begue J-P, Bernacki RJ (1997) Synthesis and Biological Activity of Novel 3′-Trifluoromethyl Taxoids. Bioorg & Med Chem Lett **7**: 133

317. Ojima I, Slater JC (1997) Synthesis of Novel 3′-Trifluoromethyl Taxoids Through Effective Kinetic Resolution of Racemic 4-CF_3-β-Lactams with Baccatins. Chirality **9**: 487

318. Ojima I, Kuduk SD, Slater JC, Gimi RH, Sun CM, Chakravarty S, Ourevitch M, Abouabdellah A, Bonnet-Delpon D, Begue J-P, Veith JM, Pera P, Bernacki RJ (1996) Syntheses, Biological Activity, and Conformational Analysis of Fluorine-Containing Taxoids. In: Ojima I, McCarthy J, Welch JT (ed) Biomedical Frontiers of Fluorine Chemistry. American Chemical Society, Washington, DC, p 228

319. Bourzat J-D, Lavelle F, Commerçon A (1995) Synthesis and Biological Activity of *Para*-Substituted 3′-Phenyl Docetaxel Analogs. Bioorg & Med Chem Lett **5**: 809

320. Tao C, Desai NP, Soon-Shiong P, Sandford PA (1996) Dual Functional Cytotoxic/Radiosensitizer Compounds and Methods for the Preparation Thereof. PCT Int Appl WO 96 40,091 (Chem Abstr **126**: 131664)

321. Swindell CS, Heerding JM, Krauss NE, Horwitz SB, Ringel I (1994) Characterization of the Taxol Structure-Activity Profile for the Locus of the A-Ring Side Chain. Bioorg Med Chem Lett **4**: 1531

322. Holton RA, Nadizadeh H, Biediger RJ, Rengan K, Suzuki Y, Tao C, Chai K-B, Idmoumaz H (1994) Preparation of Taxanes Having Furyl or Thienyl Substituted Side-Chain as Antileukemia and Antitumor Agents. PCT Int Appl WO 94,21,651

323. Ali SM, Hoemann MZ, Aube J, Mitscher LA, Georg GI (1995) Novel Cytotoxic 3′-(*tert*-Butyl) 3′-Dephenyl Analogs of Paclitaxel and Docetaxel. J Med Chem **38**: 3821

324. Ali SM, Hoemann MZ, Aube J, Georg GI, Mitscher LA (1997) Butitaxel Analogues: Synthesis and Structure-Activity Relationships. J Med Chem **40**: 236

325. Georg GI, Cheruvallath ZS, Harriman GCB, Hepperle M, Park H, Himes RH (1994) Synthesis and Biology of Substituted 3′-Phenyl Taxol Analogues. Bioorg & Med Chem Lett **4**: 2331

326. Georg GI, Harriman GCB, Hepperle M, Clowers JS, Vander Velde DG, Himes RH (1996) Synthesis, Conformational Analysis, and Biological Evaluation of Heteroaromatic Taxanes. J Org Chem **61**: 2664

327. Cabri W, Curini M, Marcotullio MC, Rosati O (1996) A High Yield Semisynthetic Approach to 2′-epi-Taxol. Tetrahedron Lett **37**: 4785

328. Kant J, Huang S, Wong H, Fairchild C, Vyas D, VF (1993) Studies toward Structure-Activity Relationships of Taxol: Synthesis and Cytotoxicity of Taxol Analogues with C-2′ Modified Phenylisoserine Side Chains. Bioorg Med Chem Lett **3**: 2471

329. Uoto K, Ohsuki S, Takenoshita H, Ishiyama T, Iimura S, Hirota Y, Mitsui I, Terasawa H, Soga T (1997) Synthesis and Structure-Activity Relationships of Novel 2′,2′-Difluoro Analogues of Docetaxel. Chem Pharm Bull **45**: 1793

330. Jayasinghe LR, Datta A, Ali SM, Zygmunt J, Vander Velde DG, Georg GI (1994) Structure-Activity Studies of Antitumor Taxanes: Synthesis of Novel C-13 Side Chain Homologated Taxol and Taxotere Analogs. J Med Chem **37**: 2981

331. Denis J-N, Fkyerat A, Gimbert Y, Coutterez C, Mantellier P, Jost S, Greene AE (1995) Docetaxel (Taxotere) Derivatives: Novel $NbCl_3$-based Stereoselective Approach to 2′-Methyldocetaxel. J Chem Soc Perkin Trans 1, 1811

332. Kant J, Schwartz WS, Fairchild C, Gao Q, Huang S, Long BH, Kadow JF, Langley DR, Farina V, Vyas D (1996) Diastereoselective Addition of Grignard Reagents to Azetidine-2,3-dione: Synthesis of Novel Taxol Analogues. Tetrahedron Lett **37**: 6495

333. Ojima I, Wang T, Delaloge F (1998) Extremely Stereoselective Alkylation of 3-Siloxy-β-lactams and Its Applications to the Asymmetric Syntheses of Novel 2-Alkylisoserines, Their Dipeptides, and Taxoids. Tetrahedron Lett **39**: 3663

334. Barboni L, Lambertucci C, Ballini R, Appendino G, Bombardelli E (1998) Synthesis of a Conformationally Restricted Analogue of Paclitaxel. Tetrahedron Lett **39**: 7177

335. Chen S-H, Xue M, Huang S, Long BH, Fairchild CA, Rose WC, Kadow JF, Vyas D (1997) Structure-Activity Relationships Study at the 3′-N Position of Paclitaxel-Part 1: Synthesis and Biological Evaluation of the 3′-(*t*)-Butylaminocarbonyloxy Bearing Paclitaxel Analogs. Bioorg & Med Chem Lett **7**: 3057

336. Gunatilaka AAL, Chordia MD, Kingston DGI (1997) Efficient Conversion of Cephalomannine to Paclitaxel and 3′-N-Acyl-3′-N-debenzoylpaclitaxel Analogs. J Org Chem **62**: 3775

337. Kingston DGI, Gunatilaka AAL, Ivey CA (1992) Modified Taxols, 7. A Method for the Separation of Taxol and Cephalomannine. J Nat Prod **55**: 259

338. Beckvermit JT, Anziano DJ, Murray CK (1996) An Improved Method for Separating Paclitaxel and Cephalomannine Using Ozone and Girard Reagents. J Org Chem **61**: 9038

339. Rimoldi JM, Molinero AA, Chordia MD, Gharpure MM, Kingston DGI (1996) An Improved Method for the Separation of Paclitaxel and Cephalomannine. J Nat Prod **59**: 167

340. Jagtap PG, Kingston DGI (1999) A Facile N-Debenzoylation of Paclitaxel: Conversion of Paclitaxel to Docetaxel. Tetrahedron Lett **40**: 189

341. Rowinsky EK, Wright M, Monsarrat B, Lesser GJ, Donehower RC (1993) Taxol: Pharmacology, Metabolism and Clinical Implications. Cancer Surv **17**: 283

342. Vuilhorgne M, Gaillard C, Sanderink GJ, Royer I, Monsarrat B, Dubois J, Wright M (1995) Metabolism of Taxoid Drugs. In: Georg GI, Chen TT, Ojima I, Vyas DM (eds) Taxane Anticancer Agents: Basic Science and Current Status. American Chemical Society, Washington, DC, p 98

343. Crommentuyn KML, Schellens JHM, Van Den Berg JD, Beijnen JH (1998) In-vitro Metabolism of Anticancer Drugs, Methods and Applications: Paclitaxel, Docetaxel, Tamoxifen and Ifosfamide. Cancer Treat Rev **24**: 345

344. Harris JW, Katki A, Anderson LW, Chmurny GN, Paukstelis JV, Collins JM (1994) Isolation, Structural Determination, and Biological Activity of 6α-Hydroxytaxol, the Principal Human Metabolite of Taxol. J Med Chem **37**: 706

345. Sparreboom A, Huizing MT, Boesen JJB, Nooijen WJ, van Tellingen O, Beijnen JH (1995) Isolation, Purification, and Biological Activity of Mono- and Dihydroxylated Paclitaxel Metabolites from Human Feces. Cancer Chemother Pharmacol **36**: 299

346. Sottani C, Minoia C, Colombo A, Zucchetti M, D'Incalci M, Fanelli R (1997) Structural Characterization of Mono- and Dihydroxylated Metabolites of Paclitaxel in Rat Bile Using Liquid Chromatography/Ion Spray Tandem Mass Spectrometry. Rapid Comm in Mass Spec **11**: 1025

347. Park H, Hepperle M, Boge TC, Himes RH, Georg GI (1996) Preparation of Phenolic Paclitaxel Metabolites. J Med Chem **39**: 2705

348. Monegier B, Gaillard C, Sable S, Vuilhorgne M (1994) Structures of the Major Human Metabolites of Docetaxel (RP 56976-Taxotere). Tetrahedron Lett **35**: 3715

349. Commerçon A, Bourzat J-D, Bezard D, Vuilhorgne M (1994) Partial Synthesis of Major Human Metabolites of Docetaxel. Tetrahedron **50**: 10289

350. Deutsch HM, Glinski JA, Hernandez M, Haugwitz RD, Narayanan VL, Suffness M, Zalkow LH (1989) Synthesis of Congeners and Prodrugs. 3. Water-Soluble Prodrugs of Taxol with Potent Antitumor Activity. J Med Chem **32**: 788
351. Mathew AE, Mejillano MR, Nath JP, Himes RH, Stella VJ (1992) Synthesis and Evaluation of Some Water-Soluble Prodrugs and Derivatives of Taxol with Antitumor Activity. J Med Chem **35**: 145
352. Zhao Z, Kingston DGI, Crosswell AR (1991) Modified taxols, 6. Preparation of Water-Soluble Prodrugs of Taxol. J Nat Prod **54**: 1607
353. Shimizu T, Kobayashi R, Ohmori H, Nakata T (1995) Synthesis of Dicarboxylic Monoesters with Cyclic Anhydrides under High Pressure. Synlett 650
354. Harada N, Ozaki K, Yamaguchi T, Arakawa H, Ando A, Oda K, Nakanishi N, Ohashi M, Hashiyama T, Tsujihara K (1997) Synthesis of Taxoids II. Synthesis and Antitumor Activity of Water-Soluble Taxoids. Heterocycles **46**: 241
355. Khmelnitsky YL, Budde C, Arnold JM, Usyatinsky A, Clark DS, Dordick JS (1997) Synthesis of Water-Soluble Paclitaxel Derivatives by Enzymatic Acylation. J Am Chem Soc **119**: 11554
356. Kingston DGI, Liang J (1995) Water Soluble Analogs and Prodrugs of Taxol. U.S. Patent 5,411,984
357. Damen EWP, Wiegerinck PHG, Braamer L, Sperling D, de Vos D, Scheeren HW (2000) Paclitaxel Esters of Malic Acid as Prodrugs with Improved Water Solubility. Bioorg & Med Chem **8**: 427
358. Mayhew E, Franklin JC, Bhatia S, Harmon PA, Janoff AS (1996) Preparation of Hydrophobic Taxane Derivatives for Treatment of Cancer and Pharmaceutical Compositions Containing Them. PCT Int Appl WO 96 21,658 (Chem Abstr 125:196057)
359. Ueda Y, Wong H, Matiskella JD, Mikkilineni AB, Farina V, Fairchild C, Rose WC, Mamber SW, Long BH, Kerns EH, Casazza AM, Vyas DM (1994) Synthesis and Antitumor Evaluation of 2′-Oxycarbonylpaclitaxels (Paclitaxel-2′-Carbonates). Bioorg & Med Chem Lett **4**: 1861
360. Vyas DM, Wong H, Crosswell AR, Casazza AM, Knipe JO, Mamber SW, Doyle TW (1993) Synthesis and Antitumor Evaluation of Water Soluble Taxol Phosphates. Bioorg Med Chem Lett **3**: 1357
361. Ueda Y, Mikkilineni AB, Knipe JO, Rose WC, Casazza AM, Vyas DM (1993) Novel Water Soluble Phosphate Prodrugs of Taxol Possessing in Vivo Antitumor Activity. Bioorg Med Chem Lett **3**: 1761
362. Ueda Y, Matiskella JD, Mikkilineni AB, Farina V, Knipe JO, Rose WC, Casazza AM, Vyas DM (1995) Novel, Water-soluble Phosphate Derivatives of 2′-Ethoxy Carbonylpaclitaxel as Potential Prodrugs of Paclitaxel: Synthesis and Antitumor Evaluation. Bioorg & Med Chem Lett **5**: 247
363. Golik J, Wong HSL, Chen S-H, Doyle TW, Wright JJK, Knipe J, Rose WC, Casazza AM, Vyas DM (1996) Synthesis and Antitumor Evaluation of Paclitaxel Phosphonooxymethyl Ethers: a Novel Class of Water Soluble Paclitaxel Pro-Drugs. Bioorg & Med Chem Lett **6**: 1837
364. Dischino DD, Chen SH, Golik J, Walker DW, Wong HSL (1997) Synthesis of the Monosodium Salt of Carbon-14 Labeled Paclitaxel (Taxol)2′-Ethyl Carbonate 7-Phosphonooxymethyl Ether, a Potential Prodrug of Paclitaxel. J Labelled Compd and Radiopharm **39**: 173
365. Hwu JR, Hakimelahi GH, Sambaiah T, Patel HV, Tsay S-C, Lai Y-K, Lieu C-H (1997) Design and Synthesis of New Taxol-Containing Aminophosphates as Protaxols. Bioorg & Med Chem Lett **7**: 545

366. Greenwald RB, Pendri A, Bolikal D, Gilbert CW (1994) Highly Water Soluble Taxol Derivatives: 2′-Polyethyleneglycol Esters as Potential Prodrugs. Bioorg & Med Chem Lett **4**: 2465
367. Greenwald RB, Gilbert CW, Pendri A, Conover CD, Xia J, Martinez A (1996) Drug Delivery Systems: Water Soluble Taxol 2′-Poly(ethylene glycol) Ester Prodrugs – Design and *in Vivo* Effectiveness. J Med Chem **39**: 424
368. Li C, Yu D, Inoue T, Yang DJ, Milas L, Hunter NR, Kim EE, Wallace S (1996) Synthesis and Evaluation of Water-Soluble Polyethylene Glycol-Paclitaxel Conjugate as a Paclitaxel Prodrug. Anti-Cancer Drugs **7**: 642
369. Pendri A, Conover CD, Greenwald RB (1998) Antitumor Activity of Paclitaxel-2′-glycinate Conjugated to Poly(ethylene glycol): a Water-Soluble Prodrug. Anti-Cancer Drug Design **13**: 387
370. Mongelli N, Biasoli G, Lombardi Borgia A, Ciomei M, Pesenti E, Angelucci F. (1996: 6 Sep) Preparation of Conjugates of Biologically Active Compounds with Polypyrrolecarboxamidonaphthalene Derivatives with Increased Bioavailability. PCT Int Appl WO 96 26,950 (Chem Abstr **125**: 300690)
371. Rodrigues ML, Carter P, Wirth C, Mullins S, Lee A, Blackburn B (1995) Synthesis and β-lactamase-mediated Activation of a Cephalosporion-taxol Prodrug. Chem Biol **2**: 223
372. de Bont DBA, Leenders RGG, Haisma HJ, van der Meulen-Muileman I, Scheeren HW (1997) Synthesis and Biological Activity of β-Glucuronyl Carbamate-Based Prodrugs of Paclitaxel as Potential Candidates for ADEPT. Bioorg Med Chem **5**: 405
373. Safavy A, Raisch KP, Khazaeli MB, Buchsbaum DJ, Bonner JA (1999) Paclitaxel Derivatives for Targeted Therapy of Cancer: Toward the Development of Smart Taxanes. J Med Chem **42**: 4919
374. Huang C-M, Wu Y-T, Chen S-T (2000) Targeting Delivery of Paclitaxel into Tumor Cells via Somatostatin Receptor Endocytosis. Chem Biol **7**: 453
375. de Groot FMH, van Berkom LWA, Scheeren HW (2000) Synthesis and Biological Evaluation of 2′-Carbamate-Linked and 2′-Carbonate-Linked Prodrugs of Paclitaxel: Selective Activation by the Tumor-Associated Protease Plasmin. J Med Chem **43**: 3093
376. Taylor GF, Thornton SS, Tallent CR, Kepler JA (1993) Synthesis of [3″-^{3}H]Taxol and [13-^{3}H]Taxol. J Labelled Cpd Radiopharm **33**: 501
377. Shu AYL, Heys JR (2000) Direct, Efficient and Selective Tritiations of Paclitaxel and Photoaffinity Taxoids. Tetrahedron Lett **41**: 9015
378. Walker DG, Swigor JE, Kant J, Schroeder DR (1994) Synthesis of Carbon-14 Labeled Taxol (Paclitaxel). J Labelled Cpd Radiopharm **34**: 973
379. Walker DG, Standridge RT, Swigor JE (1995) Synthesis of Paclitaxel-C3′-^{14}C. J Labelled Cpd Radiopharm **36**: 479
380. deSuzzoni S, Boullais C, Vandais A, Noel JP (1995) Hemisynthesis of [3′-^{14}C]-Taxotere. J Labelled Cpd Radiopharm **36**: 739
381. Rodriguez I, Voges R (2000) Synthesis of [$^{14}C_2$]Taxol: Application of (+)/(−)-[$^{13,14}C_n$]BABS, part 3. J Labelled Cpd Radiopharm **43**: 169
382. Rao CS, Chu J-J, Liu R-S, Lai Y-K (1998) Synthesis and Evaluation of [^{14}C]-Labelled and Fluorescent-Tagged Paclitaxel Derivatives as New Biological Probes. Bioorg Med Chem **6**: 2193
383. Giribone D, Fontana E (2000) Synthesis of the Taxane Analogue PNU-105298 Labelled with Deuterium and Carbon-14. J Labelled Comp Radiopharm **43**: 933
384. Roh EJ, Park YH, Song CE, Oh S-J, Choe YS, Kim B-T, Chi DY, Kim D (2000) Radiolabeling of Paclitaxel with Electrophilic ^{123}I. Bioorg & Med Chem **8**: 65

385. Swindell CS, Heerding JM, Krauss NE, Horwitz SB, Rao S, Ringel I (1994) Characterization of Two Taxol Photoaffinity Analogues Bearing Azide and Benzophenone-Related Photoreactive Substituents in the A-Ring Side Chain. J Med Chem **37**: 1446

386. Rao S, Krauss NE, Heerding JM, Swindell CS, Ringel I, Orr GA, Horwitz SB (1994) 3′-(*p*-Azidobenzamido)taxol Photolabels the N-Terminal 31 Amino Acids of β-Tubulin. J Biol Chem **269**: 3132

387. Dasgupta D, Park H, Harriman GCB, Georg GI, Himes RH (1994) Synthesis of a Photoaffinity Taxol Analogue and Its Use in Labeling Tubulin. J Med Chem **37**: 2976

388. Georg GI, Boge TC, Park H, Himes RH (1995) Paclitaxel and Docetaxel Photoaffinity Labels. Bioorg Med Chem Lett **5**: 615

389. Ojima I, Duclos O, Dorman G, Simonot B, Prestwich GD, Rao S, Lerro KA, Horwitz SB (1995) A New Paclitaxel Photoaffinity Analog with a 3-(4-Benzoylphenyl)propanoyl Probe for Characterization of Drug-Binding Sites on Tubulin and P-Glycoprotein. J Med Chem (Communications to the Editor) **38**: 3891

390. Ojima I, Bounaud P-Y, Ahern DG (1999) New Photoaffinity Analogs of Paclitaxel. Bioorg & Med Chem Lett **9**: 1189

391. Lin S, Fang K, Hashimoto M, Nakanishi K, Ojima I (2000) Design and Synthesis of a Novel Photoaffinity Taxoid as a Potential Probe for the Study of Paclitaxel-microtubules Interactions. Tetrahedron Lett **41**: 4287

392. Rimoldi JM, Kingston DGI, Chaudhary AG, Samaranayake G, Grover S, Hamel E (1993) Modified Taxols. 9. Synthesis and Biological Evaluation of 7-Substituted Photoaffinity Analogues of Taxol. J Nat Prod **56**: 1313

393. Rao S, Orr GA, Chaudhary AG, Kingston DGI, Horwitz SB (1995) Characterization of the Taxol Binding Site on the Microtubule. J Biol Chem **270**: 20235

394. Sengupta S, Boge TC, Georg GI, Himes R (1995) Interaction of a Fluorescent Paclitaxel Analogue with Tubulin. Biochemistry **34**: 11889

395. Li Y, Edsall J, R, Jagtap PG, Kingston DGI, Bane S (2000) Equilibrium Studies of a Fluorescent Paclitaxel Derivative Binding to Microtubules. Biochemistry **39**: 616

396. Sengupta S, Boge TC, Liu Y, Hepperle M, Georg GI, Himes RH (1997) Probing the Environment of Tubulin-Bound Paclitaxel Using Fluorescent Paclitaxel Analogues. Biochemistry **36**: 5179

397. Han Y, Chaudhary AG, Chordia MD, Sackett DL, Perez-Ramirez B, Kingston DGI, Bane S (1996) Interaction of a Fluorescent Derivative of Paclitaxel (Taxol) with Microtubules and Tubulin-Colchicine. Biochemistry **35**: 14173

398. Kingston DGI (1991) The Chemistry of Taxol. Pharmac Ther **52**: 1

399. Dubois J, LeGoff M-T, Guéritte-Voegelein F, Guénard D, Tollon Y, Wright M (1995) Fluorescent and Biotinylated Analogues of Docetaxel: Synthesis and Biological Evaluation. Bioorg Med Chem **3**: 1357

400. Evangelio JA, Abal M, Barasoain I, Souto AA, Lillo MP, Acuna AU, Amat-Guerri F, Andreu JM (1998) Fluorescent Taxoids as Probes of the Microtubule Cytoskeleton. Cell Motility and the Cytoskeleton **39**: 73

401. Souto AA, Acuna AU, Andreu JM, Barasoain I, Abal M, Amat-Guerri F (1995) New Fluorescent Water-Soluble Taxol Derivatives. Angew Chem Int Ed Engl **34**: 2710

402. Guy RK, Scott ZA, Sloboda RD, Nicolaou KC (1996) Fluorescent Taxoids. Chem Biol **3**: 1021

403. Nicholov R, Kingston DGI, Chordia MC, DiCosmo F (1997) Molecular Mobility of Nitroxyl-labelled Taxol During Tubulin Assembly. FEBS Lett **405**: 73

404. Rao KV (1997) Semi-Synthesis of Paclitaxel from Naturally Occurring Glycosidic Precursors. J Heterocyclic Chem **34**: 675

405. Nikolakakis A, Caron G, Cherestes A, Sauriol F, Mamer O, Zamir LO (2000) *Taxus canadensis* Abundant Taxane: Conversion to Paclitaxel and Rerrangements. Bioorg & Med Chem **8**: 1269
406. Wiegerinck PHG, Fluks L, Hammink JB, Mulders SJE, de Groot FMH, van Rozendaal HLM, Scheeren HW (1996) Semisynthesis of Some 7-Deoxypaclitaxel Analogs from Taxine B. J Org Chem **61**: 7092
407. Mourabit AA, Poujol H, Poupat C, Ahond A, Potier P (1996) Chemistry of Taxoids: One-step Functionalization of the Positions 1, 2, 9 and 10 from the Selectively Protected 5-O-Cinnamoyltaxicine I. Tetrahedron Lett **37**: 9189
408. Poujol H, Ahond A, Mourabit AA, Chiaroni A, Poupat C, Riche C, Potier P (1997) Taxoides: Nouveaux Analogues du 7-Deshydroxydocetaxel Prepares a Partir des Alcaloides de l'If. Tetrahedron **53**: 5169
409. Poujol H, Mourabit AA, Ahond A, Poupat C, Potier P (1997) Taxoides: 7-Deshydroxy-10-acetyldocetaxel et Nouveaux Analogues Prepares a partir des Alcaloides de l'If. Tetrahedron **53**: 12575
410. Matovic R, Saicic RN (1998) An Efficient Semisynthesis of 7-Deoxypaclitaxel from Taxine. Chem Comm 1745
411. Saicic RN, Matovic R (2000) An Efficient Semisynthesis of 7-deoxypaclitaxel from Taxine. J Chem Soc, Perkin Trans I 59
412. Cheng Q, Oritani T, Horiguchi T, Yamada T, Mong Y (2000) Synthesis and Biological Evaluation of Novel 9-Functional Heterocyclic Coupled 7-Deoxy-9-Dihydropaclitaxel Analogue. Bioorg & Med Chem Lett **10**: 517
413. Liu RW, Guo JY, Liang XT (1997) Construction of the Oxetane Ring from 5α-Acyloxy-Δ-4(20)-Taxoids. Chin Chem Lett **8**: 99
414. Liu R-W, Lin D-L, Guo J-Y, Liang X-T (2000) A Facile Approach for the Construction of the Oxetane Ring from 5α-Acyloxy-Δ-4(20)-Taxoids. Chin J Chem **18**: 236
415. Yin D, Liu R-W, Wang D-H, Guo J-Y, Liang X-T, Yoshinori S, Kazuya K (1999) Synthesis of Analogues of Paclitaxel with 14β-Side Chain from Sinenxan. A J Chin Pharm Sci **8**: 191
416. Blechert S, Kleine-Klausing A (1991) Synthesis of a Biologically Active Taxol Analogue. Angew Chem Int Ed Engl **30**: 412
417. Blechert S, Jansen R, Velder J (1994) Synthesis of New Taxoids. Tetrahedron **50**: 9649
418. Fuji K, Watanabe Y, Ohtsubo T, Nuruzzaman M, Hamajima Y, Kohno M (1999) Synthesis of Extremely Simplified Compounds Possessing the Key Pharmacophore Units of Taxol, Phenylisoserine and Oxetane Moieties. Chem Pharm Bull **47**: 1334
419. Gentile G, Fattori D, Botta M, Corelli F, Fusar-Bassini D, Lamba D (2000) Synthesis of Dimeric and Tetrameric Macrolactams with Cytotoxic Activity. Can J Chem **78**: 925
420. Morihira K, Nishimori T, Kusama H, Horiguchi Y, Kuwajima I, Tsuruo T (1998) Synthesis of C-Ring Aromatic Taxoids and Evaluation of their Multi-Drug Resistance Reversing Activity. Bioorg & Med Chem Lett **8**: 2973
421. Klar U, Graf H, Schenk O, Rohr B, Schulz H (1998) New Synthetic Inhibitors of Microtubule Depolymerization. Bioorg & Med Chem Lett **8**: 1397
422. Soto J, Mascarenas JL, Castedo L (1998) A Paclitaxel Analogue with a 2(3$\rightarrow$20) *Abeo*taxane Skeleton: Synthesis and Biological Evaluation. Bioorg Med Chem Lett **8**: 273
423. Appendino G, Belloro E, Jakupovic S, Danieli B, Jakupovic J, Bombardelli E (1999) Synthesis of Paclitaxel (Docetaxel)/2-Deacetoxytaxinine J Dimers. Tetrahedron **55**: 6567

424. Cheng Q, Oritani T, Horiguchi T, Yamada T, Hassner A (2000) Novel Dimeric Taxoids via Highly Regio- and Stereospecific Diels-Alder Cycloadditions of Taxinine B and Taxicine I Derivatives. Tetrahedron **56**: 1181

425. Kar AK, Braun PD, Wandless TJ (2000) Synthesis and Evalution of Daunorubicin-Paclitaxel Dimers. Bioorg Med Chem Lett **10**: 261

426. Holton RA, Somoza C, Kim H-B, Liang F, Biediger RJ, Boatman PD, Shindo M, Smith CC, Kim S, Nadizadeh H, Suzuki Y, Tao C, Vu P, Tang S, Zhang P, Murthi KK, Gentile LN, Liu JH (1994) First Total Synthesis of Taxol. 1. Functionalization of the B Ring. J Am Chem Soc **116**: 1597

427. Holton RA, Kim H-B, Somoza C, Liang F, Biediger RJ, Boatman PD, Shindo M, Smith CC, Kim S, Nadizadeh H, Suzuki Y, Tao C, Vu P, Tang S, Zhang P, Murthi KK, Gentile LN, Liu JH (1994) First Total Synthesis of Taxol. 2. Completion of the C and D Rings. J Am Chem Soc **116**: 1599

428. Holton RA, Juo RR, Kim HB, Williams AD, Harusawa S, Lowenthal RE, Yogai S (1988) A Synthesis of Taxusin. J Am Chem Soc **110**: 6558

429. Nicolaou KC, Yang Z, Liu JJ, Ueno H, Nantermet PG, Guy RK, Claiborne CF, Renaud J, Couladouros EA, Paulvannan K, Sorensen EJ (1994) Total Synthesis of Taxol. Nature **367**: 630

430. Nicolaou KC, Nantermet PG, Ueno H, Guy RK, Couladouros EA, Sorensen EJ (1995) Total Synthesis of Taxol. 1. Retrosynthesis, Degradation, and Reconstitution. J Am Chem Soc **117**: 624

431. Nicolaou KC, Liu J-J, Yang Z, Ueno H, Sorensen EJ, Claiborne CF, Guy RK, Hwang C-K, Nakada M, Nantermet PG (1995) Total Synthesis of Taxol. 2. Construction of A and C Ring Intermediates and Initial Attempts to Construct the ABC Ring System. J Am Chem Soc **117**: 634

432. Nicolaou KC, Yang Z, Liu J-J, Nantermet PG, Claiborne CF, Renaud J, Guy RK, Shibayama K (1995) Total Synthesis of Taxol. 3. Formation of Taxol's ABC Ring Skeleton. J Am Chem Soc **117**: 645

433. Nicolaou KC, Ueno H, Liu J-J, Nantermet PG, Yang Z, Renaud J, Paulvannan K, Chadha R (1995) Total Synthesis of Taxol. 4. The Final Stages and Completion of the Synthesis. J Am Chem Soc **117**: 653

434. Danishefsky SJ, Masters JJ, Young WB, Link JT, Snyder LB, Magee TV, Jung DK, Isaacs RCA, Bornmann WG, Alaimo CA, Coburn CA, Di Grandi MJ (1996) Total Synthesis of Baccatin III and Taxol. J Am Chem Soc **118**: 2843

435. Wender PA, Badham NF, Conway SP, Floreancig PE, Glass TE, Granicher C, Houze JB, Janichen J, Lee D, Marquess DG, McGrane PL, Meng W, Mucciaro TP, Muhlebach M, Natchus MG, Paulsen H, Rawlins DB, Satkofsky J, Shuker AJ, Sutton JC, Taylor RE, Tommoka K (1997) The Pinene Path to Taxanes. 5. Stereocontrolled Synthesis of a Versatile Taxane Precursor. J Am Chem Soc **119**: 2755

436. Wender PA, Badham NF, Conway SP, Floreancig PE, Glass TE, Houze JB, Krauss NE, Lee D, Marquess DG, McGrane PL, Meng W, Natchus MG, Shuker AJ, Sutton JC, Taylor RE (1997) The Pinene Path to Taxanes. 6. A Concise Stereocontrolled Synthesis of Taxol. J Am Chem Soc **119**: 2757

437. Morihira K, Hara R, Kawahara S, Nishimori T, Nakamura N, Kusama H, Kuwajima I (1998) Enantioselective Total Synthesis of Taxol. J Am Chem Soc **120**: 12980

438. Kusama H, Hara R, Kawahara S, Nishimori T, Kashima H, Nakamura N, Morihira K, Kuwajima I (2000) Enantioselective Total Synthesis of (−)-Taxol. J Am Chem Soc **122**: 3811

439. Mukaiyama T, Shiina I, H. I, Sakoh H, Tani Y-I, Hasegawa M, Saitoh K (1997) Asymmetric Total Synthesis of Taxol. Proc Japan Acad **73B**: 95

440. Shiina I, Iwadare H, Sakoh H, Hasegawa M, Tani Y-I, Mukaiyama T (1998) A New Method for the Synthesis of Baccatin III. Chem Lett 1
441. Shiina I, Saitoh K, Frechard-Ortuno I, Mukaiyama T (1998) Total Asymmetric Synthesis of Taxol by Dehydration Condensation between 7-TES Baccatin III and Protected *N*-Benzoylphenylisoserines Prepared by Enantioselective Aldol Reaction. Chem Lett 3
442. Manfredi JJ, Horwitz SB (1984) Taxol: An Antimitotic Agent with a New Mechanism of Action. Pharmac Ther **25**: 83
443. Horwitz SB, Davidson MW, Holton RA (1992) Mechanism of action of Taxol. Trends Pharmacol Sci **13**: 134
444. Milross CG, Mason KA, Hunter NR, Chung W-K, Peters LJ, Milas L (1996) Relationship of Mitotic Arrest and Apoptosis to Antitumor Effect of Paclitaxel. J Nat Cancer Inst **88**: 1308
445. Schimming R, Mason KA, Hunter N, Weil M, Kishi K, Milas L (1999) Lack of Correlation Between Mitotic Arrest or Apoptosis and Antitumor Effect of Docetaxel. Cancer Chemother Pharmacol **43**: 165
446. Fan W (1999) Possible Mechanisms of Paclitaxel-Induced Apoptosis. Biochem Pharmacol **57**: 1215
447. Wang T-H, Wang H-S, Soong Y-K (2000) Paclitaxel-Induced Cell Death. Cancer **88**: 2619
448. Rodi DJ, Janes RW, Sanganee HJ, Holton RA, Wallace BA, Makowski L (1999) Screening of a Library of Phage-displayed Peptides Identifies Human Bcl-2 as a Taxol-binding Protein. J Mol Biol **285**: 197
449. Scatena CD, Stewart ZA, Mays D, Tang LJ, Keefer CJ, Leach SD, Pietenpol JA (1998) Mitotic Phosphorylation of Bcl-2 During Normal Cell Cycle Progression and Taxol-Induced Growth Arrest. J Biol Chem **273**: 30777
450. Blagosklonny MV, Giannakakou P, El-Deiry WS, Kingston DGI, Higgs PI, Neckers L, Fojo T (1997) Raf-1/bcl-2 Phosphorylation: A Step from Microtubule Damage to Cell Death. Cancer Res **57**: 130
451. Nogales E, Wolf SG, Downing KH (1997) Visualizing the Secondary Structure of Tubulin: Three-Dimensional Map at 4 A. J Struct Bio **118**: 119
452. Nogales E, Whittaker M, Milligan RA, Downing KH (1999) High-Resolution Model of the Microtubule. Cell **96**: 79
453. Michaelis ML, Ranciat N, Chen Y, Bechtel M, Ragan R, Hepperle M, Liu Y, Georg G (1998) Protection Against β-Amyloid Toxicity in Primary Neurons by Paclitaxel (Taxol). J Neurochem **70**: 1623
454. Morihira K, Nishimori T, Kusama H, Horiguchi Y, Kuwajima I, Tsuruo T (1998) Synthesis and Evaluation of Artificial Taxoids with Antitumor and Multi-Drug Resistance Reversing Activities. Bioorg & Med Chem Lett **8**: 2977
455. Morihira K, Nishimori T, Kusama H, Horiguchi Y, Kuwajima I, Tsuruo T (1998) Synthesis of C-Ring Aromatic Taxoids and Evaluation of their Multi-Drug Resistance Reversing Activity. Bioorg & Med Chem Lett **8**: 2973
456. Ojima I, Bounaud P-Y, Bernacki RJ (1998) New Weapons in the Fight Against Cancer. CHEMTECH 31
457. Ojima I, Bounaud P-Y, Bernacki RJ (1999) Designing Taxanes to Treat Multidrug-resistant Tumors. Modern Drug Discovery 45
458. Kirkiae T, Ojima I, Kirikae F, Ma Z, Kuduk SD, Slater JC, Takeuchi CS, Bonaud P-Y, Nakano M (1996) Structural Requirements of Taxoids for Nitric Oxide and Tumor Necrosis Factor Production by Murine Macrophages. Biochem Biophys Res Commun **227**: 227
459. Kirikae T, Ojima I, Ma Z, Kirikae F, Hirai Y, Nakano M (1998) Structural Significance of the Benzoyl Group at the C-3′-N Position of Paclitaxel for Nitric

Oxide and Tumor Necrosis Factor Production by Murine Macrophages. Biochem & Biophys Res Comm **245**: 698

460. Burkhart CA, Berman JW, Swindell CS, Horwitz SB (1994) Relationship between the Structure of Taxol and Other Taxanes on Induction of Tumor Necrosis Factor-α-Gene Expression and Cytotoxicity. Cancer Research **54**: 5779

461. Kirikae T, Ojima I, Fuero-Oderda C, Lin S, Kirikae F, Hashimoto M, Nakano M (2000) Structural Significance of the Acyl Group at the C-10 Position and the A-ring of the Taxane Core of Paclitaxel for Inducing Nitric Oxide and Tumor Necrosis Factor Production by Murine Macrophages. FEBS Lett **478**: 221

462. Ding A, Sanchez E, Nathan CF (1993) Taxol Shares the Ability of Bacterial Lipopolysaccharide to Induce Tyrosine Phosphorylation of Microtubule-associated Protein Kinase. J Immunol **151**: 5596

463. Liu Y, Bhalla K, Hill C, Priest DG (1994) Evidence for Involvement of Tyrosine Phosphorylation in Taxol-induced Apoptosis in a Human Ovarian Tumor Cell Line. Biochem Pharmacol **48**: 1265

464. Carboni JM, Singh C, Tepper MA (1993) Taxol and Lipopolysaccharide Activation of a Murine Macrophage Cell Line and Induction of Similar Tyrosine Phosphoproteins. J Natl Cancer inst Monogr **15**: 95

465. Yang C-PH, Horwitz SB (2000) Taxol Mediates Serine Phosphorylation of the 66-kDa Shc Isoform. Cancer Res **60**: 5171

466. Rao S, Horwitz SB, Ringel I (1992) Direct photoaffinity labeling of tubulin with taxol. J Natl Cancer Inst **84**: 785

467. Combeau C, Commercon A, Mioskowski C, Rousseau B, Aubert F, Goeldner M (1994) Predominant Labeling of β- over α-Tubulin from Porcine Brain by a Photo-activatable Taxoid Derivative. Biochemistry **33**: 6676

468. Rao S, He L, Chakravarty S, Ojima I, Orr GA, Horwitz SB (1999) Characterization of the Taxol Binding Site on the Microtubule. J Biol Chem **274**: 37990

469. Han Y, Malak AG, Chaudhary AG, Chordia MD, Kingston DGI, Bane S (1998) Distances Between the Paclitaxel, Colchicine, and Exchangeable GTP Binding Sites on Tubulin. Biochemistry **37**: 6636

470. Diaz JF, Strobe R, Engelborghs Y, Souto AA, Andreu JM (2000) Molecular Recognition of Taxol by Microtubules. J Biol Chem **275**: 26265

471. Baker JK (1992) Nuclear Overhauser Effect Spectroscopy (Noesy) and Dihedral Angle Measurements in the Determination of the Conformation of Taxol in Solution. Spectroscopy Letters **25**: 31

472. Hilton BD, Chmurny GN, Muschik GM (1992) Taxol: Quantitative internuclear proton-proton distances in deuterated chloroform solution form nOe data: 2D NMR ROESY buildup rates at 500 MHz. J Nat Prod **55**: 1157

473. Falzone CJ, Benesi AJ, J. LT (1992) Characterization of Taxol in Methylene Chloride by NMR Spectroscopy. Tetrahedron Lett **33**: 1169

474. Dubois J, Guénard D, Guéritte-Voeglein F, Guedira N, Potier P, Gillet B, Betoeil J-C (1993) Conformation of Taxotere and Analogues Determined by NMR Spectroscopy and Molecular Modeling Studies. Tetrahedron **49**: 6533

475. Williams HJ, Scott AI, Dieden RA, Swindell CS, Chirlian LE, Francl MM, Heerding JM, Krauss NE (1993) NMR and Molecular Modeling Study of the Conformations of Taxol and of its Side Chain Methylester in Aqueous and Non-Aqueous Solution. Tetrahedron **49**: 6545

476. Williams HJ, Scott AI, Dieden RA, Swindell CS, Chirlian LE, Francl MM, Heerding JM, Krauss NE (1994) NMR and Molecular Modeling Study of Active and Inactive Taxol Analogues in Aqueous and Nonaqueous Solution. Can J Chem **72**: 252

477. Balasubramanian SV, Alderfer JL, Straubinger RM (1994) Solvent- and Concentration-Dependent Molecular Interactions of Taxol (Paclitaxel). J Pharm Sci **83**: 1470
478. Cachau RE, Gussio R, Beutler JA, Chmurny GN, Hilton BD, Muschik GM, Erickson JW (1994) Solution Structure of Taxol Determined Using a Novel Feedback-Scaling Procedure for Noe-Restrained Molecular Dynamics. Supercomput Appl High Perform Comput **8**: 24
479. Ojima I, Kuduk SD, Chakravarty S, Ourevitch M, Begue J-P (1997) A Novel Approach to the Study of Solution Structures and Dynamic Behavior of Paclitaxel and Docetaxel Using Fluorine-Containing Analogs as Probes. J Am Chem Soc **119**: 5519
480. Milanesio M, Ugliengo P, Viterbo D, Appendino G (1999) *ab Initio* Conformational Study of the Phenylisoserine Side Chain of Paclitaxel. J Med Chem **42**: 291
481. Ojima I, Chakravarty S, Inoue T, Lin S, He L, Horwitz SB, Kuduk SC, Danishefsky SJ (1999) A Common Pharmacophore for Cytotoxic Natural Products that Stabilize Microtubules. Proc Natl Acad Sci U.S.A. **96**: 4256
482. Vander Velde DG, Georg GI, Grunewald GL, Gunn CW, Mitscher LA (1993) "Hydrophobic Collapse" of Taxol and Taxotere Solution Conformations in Mixtures of Water and Organic Solvent. J Am Chem Soc **115**: 11650
483. Paloma LG, Guy RK, Wrasidlo W, Nicolaou KC (1994) Conformation of a Water-Soluble Derivative of Taxol in Water by 2D-NMR Spectroscopy. Chem Biol **1**: 107
484. Jimenez-Barbero J, Souto AA, Abal M, Barasoain I, Evangelio JA, Acuna AU, Andreu JM, Amat-Guerri F (1998) Effect of 2′-OH Acetylation on the Bioactivity and Conformation of 7-*O*-[*N*-(4′-Fluoresceincarbonyl)-L-alanyl]taxol. A NMR-fluorescence Microscopy Study. Bioorg Med Chem **6**: 1857
485. Snyder JP, Nevins N, Cicero DO, Jansen J (2000) The Conformations of Taxol in Chloroform. J Am Chem Soc **122**: 724
486. Moyna G, Williams HJ, Scott AI, Ringel I, Gorodetsky R, Swindell CS (1997) Conformational Studies of Paclitaxel Analogs Modified at the C-2′ Position in Hydrophobic and Hydrophilic Solvent Systems. J Med Chem **40**: 3305
487. Ojima I, Inoue T, Slater JC, Lin S, Kuduk SD, Chakravarty S, Walsh JJ, Cresteil T, Monsarrat B, Pera P, Bernacki RJ. Synthesis of Enantiopure Fluorine-Containing Taxoids and Their Use as Anticancer Agents as well as Probes for Biomedical Problems. In: Ramachandran PV (2000) Asymmetric Fluoroorganic Chemistry: Synthesis, Applications, and Future Developments. American Chemical Society, Washington, DC, p 158
488. Ojima I, Inoue T, Chakravarty S (1999) Enantiopure Fluorine-containing Taxoids: Potent Anticancer Agents and Versatile Probes for Biomedical Problems. J Fluorine Chem **97**: 3
489. Li Y, Poliks B, Cegelski L, Poliks M, Cryczynski Z, Piszczek G, Jagtap P, G, Studelska DR, Kingston DGI, Schaefer J, Bane S (2000) Conformation of Microtubule-Bound Paclitaxel Determined by Fluorescence Spectroscopy and REDOR NMR. Biochemistry **39**: 281
490. Bollag DM, McQueney PA, Zhu J, Hensens O, Koupal L, Liesch J, Goetz M, Lazarides E, Woods CM (1995) Epothilones, a New Class of Microtubule-stabilizing Agents with a Taxol-like Mechanism of Action. Cancer Res **55**: 2325
491. Kowalski RJ, Giannakakou P, Hamel E (1997) Activities of the Microtubule-Stabilizing Agents Epothilones A and B with Purified Tubulin and in Cells Resistant to Paclitaxel (Taxol). J Biol Chem **272**: 2534
492. ter Haar E, Kowalski RJ, Hamel E, Lin CM, Longley RE, Gunasekera SP, Rosenkranz HS, Day BW (1996) Discodermolide, A Cytotoxic Marine Agent that Stabilizes Microtubules More Potently Than Taxol. Biochemistry **35**: 243

493. Lindel T, Jensen PR, Fenical W, Long BH, Casazza AM, Carboni J, Fairchild CR (1997) Eleutherobin, a New Cytotoxin that Mimics Paclitaxel (Taxol) by Stabilizing Microtubules. J Am Chem Soc **119**: 8744
494. Dupont C, Guénard D, Tchertanov L, Thoret S, Guéritte F (1999) D-Ring Substituted Rhazinilam Analogues: Semisynthesis and Evalluation of Antitubulin Activity. Bioorg Med Chem **7**: 2961
495. Mooberry SL, Tien G, Hernandez AH, Plubrukarn A, Davidson BS (1999) Laulimalide and Isolaulimalide, New Paclitaxel-Like Microtubule-Stabilizing Agents. Cancer Res **59**: 653
496. Vanderwal CD, Vosburg DA, Weiler S, Sorensen EJ (1999) Postulated Biogenesis of WS9885B and Progress Toward an Enantioselective Synthesis. Org Lett **1**: 645
497. Rouz D, Hadi HA, Thoret S, Guénard D, Thoison O, Päis M, Sévenet T (2000) Structure-Activity Relationship of Polyisoprenyl Benzophenones from *Garcinia pyrifera* on the Tubulin/Microtubule System. J Nat Prod **63**: 1070
498. Miglietta A, Bocca C, Gadoni E, Gabriel L, Rampa A, Bisi A, Valenti P, Da Re P (1996) Interaction of Geiparvin and Related Compounds with Purified Microtubular Protein. Anti-Cancer Drug Design **11**: 35
499. Winkler JD, Axelsen PH (1996) A Model for the Taxol (Paclitaxel)/Epothilone Pharmacophore. Bioorg Med Chem Lett **6**: 2963
500. Ojima I, Lin S, Inoue T, Miller ML, Borella CP, Geng X, Walsh JJ (2000) Macrocycle Formation by Ring-Closing Metathesis. Application to the Syntheses of Novel Macrocyclic Taxoids. J Am Chem Soc **122**: 5343
501. Boge TC, Wu Z-J, Himes RH, Vander Velde DG, Georg GI (1999) Conformationally Restricted Paclitaxel Analogues: Macrocyclic Mimics of the "Hydrophobic Collapse" Conformation. Bioorg Med Chem Lett **9**: 3047
502. Yuan H, Kingston DGI (1999) Synthesis and Biological Activity of a Novel C4–C6 Bridged Paclitaxel Analog. Tetrahedron **55**: 9707
503. Giannakakou P, Gussio R, Nogales E, Downing KH, Zaharevitz D, Bollbuck B, Poy G, Sackett D, Nicolaou KC, Fojo T (2000) A Common Pharmacophore for Epothilone and Taxanes: Molecular Basis for Drug Resistance Conferred by Tubulin Mutations in Human Cancer Cells. Proc Natl Acad Sci U.S.A. **97**: 2904
504. He L, Jagtap PG, Kingston DGI, Shen H-J, Orr GA, Horwitz SB (2000) A Common Pharmacophore for Taxol and the Epothilones Based on the Biological Activity of a Taxane Molecule Lacking a C-13 Side Chain. Biochemistry **39**: 3972
505. Day BW (2000) Mutants Yield a Pharmacophore Model for the Tubulin-paclitaxel Binding Site. TiPS **21**: 321
506. Wang M, Xia X, Kim Y, Hwang D, Jansen JM, Botta M, Liotta DC, Snyder JP (1999) A Unified and Quantitative Receptor Model for the Microtubule Binding of Paclitaxel and Epothilone. Org Lett **1**: 43
507. Reference 503, footnote 14
508. Czaplinski K-HA, Grunewald GL (1994) A Comparative Molecular Field Analysis Derived Model of the Binding of Taxol Analogues to Microtubules. Biorg Med Chem Lett **4**: 2211
509. Morita H, Gonda A, Wei L, Takeya K, Itokawa H (1997) 3D QSAR Analysis of Taxoids from *Taxus cuspidata* var. *nana* by Comparative Molecular Field Approach. Bioorg & Med Chem Lett **7**: 2387
510. Zhu Q, Guo Z, Huang N, Wang M, Chu F (1997) Comparative Molecular Field Analysis of a Series of Paclitaxel Analogues. J Med Chem **40**: 4319
511. Cheng CC, Gordon MA (2000) An Interesting Observation of the Position of Some Electronegative Atoms in the Taxol Molecule. Med Hypoth **54**: 172

512. Kingston DGI (2001) Taxol, a Molecule for All Seasons. Chem Commun 867
513. Guéritte F (2001) General and Recent Aspects of the Chemistry and Structure-Activity Relationships of Taxoids. Curr Pharm Design **7**: 1229
514. Goodman J, Walsh V (2001) The Story of Taxol. Nature and Politics in Pursuit of an Anti-Cancer Drug. Cambridge University Press, Cambridge
515. Michaud LB, Valero V, Hortobagyi G (2000) Risks and Benefits of Taxanes in Breast and Ovarian Cancer. Drug Safety **23**: 401
516. Wilson CR, Sauer J-M, Hooser SB (2001) Taxines: A Review of the Mechanism and Toxicity of Yew (*Taxus spp.*) Alkaloids. Toxicon **39**: 175
517. Fang WS, Fang QC, Liang XT (2001) Synthesis of 14-Bromo and 14-Hydroxy Baccatin III Deriviatives. Chin Chem Lett **12**: 675
518. Iimura S, Uoto K, Ohsuki S, Chiba J, Yoshino T, Iwahana M, Jimbo T, Terasawa H, Soga T (2001) Orally Active Docetaxel Analogue: Synthesis of 10-Deoxy-10-*C*-morpholinoethyl Docetaxel Analogues. Bioorg & Med Chem Lett **11**: 407
519. Park H (2001) Synthesis of Novel Taxol Analogues with Heterocyclic Amine Moieties. Yakhak Hoeji **45**: 321
520. Mandai T, Okumoto H, Oshitari T, Nakanishi K, Mikuni K, Hara K-J, Hara K-Z, Iwatani W, Amano T, Nakamura K, Tsuchiya Y (2001) Synthesis and Biological Evaluation of Water Soluble Taxoids Bearing Sugar Moieties. Heterocycles **54**: 561
521. Walker MA, Johnson TD, Huang S, Vyas DM, Kadow JF (2001) Synthesis of a Novel C-10 Spiro-Epoxide of Paclitaxel. Bioorg & Med Chem Lett **11**: 1683
522. Gibson FS, Wei J, Vemishetti P, Gao Q, Dillon JL (2000) Synthesis, Utility, and X-ray Crystal Structure of Novel Complexes of Baccatin III with Imidazole and 2-Propanol. Org Lett **2**: 3269
523. Fang W-S, Fang Q-C, Liang X-T (2001) Synthesis of the 2α-benzoylamido Analogue of Docetaxel. Tetrahedron Lett **42**: 1331
524. Mastalerz H, Zhang G, Kadow J, Fairchild C, Long B, Vyas DM (2001) Synthesis of 7β-Sulfur Analogues of Paclitaxel Utilizing a Novel Epimerization of the 7α-Thiol Group. Org Lett **3**: 1613
525. Wittman MD, Kadow JF, Vyas DM, Lee FL, Rose WC, Long BH, Fairchild C, Johnston K (2001) Synthesis and Antitumor Activity of Novel Paclitaxel-Chlorambucil Hybrids. Bioorg & Med Chem Lett **11**: 811
526. Wittman MD, Altstadt TJ, Fairchild C, Hansel S, Johnston K, Kadow JF, Long BH, Rose WC, Vyas DM, Wu M-J, Zoeckler ME (2001) Synthesis of Metabolically Blocked Paclitaxel Analogues. Bioorg & Med Chem Lett **11**: 809
527. Beusker PH, Veldhuis H, van den Bossche BAC, Scheeren HW (2001) Semisynthesis of Flexible 5,7-Dideoxypaclitaxel Derivatives from Taxine B. Eur J Org Chem 1761
528. Horiguchi T, Nagura M, Cheng Q, Oritani T, Kudo T (2000) Chemical Oxidation of Taxoids with *m*-CPBA and Dimethyl Dioxirane: Regioselective Epoxidation of Taxinine J Derivatives. Heterocycles **53**: 2629
529. Barboni L, Datta A, Dutta D, Georg GI, Vander Velde DG, Himes RH, Wang M, Snyder JP (2001) Novel D-Seco Paclitaxel Analogues: Synthesis, Biological Evaluation, and Model Testing. J Org Chem **66**: 3321
530. Merckle L, Dubois J, Place E, Thoret S, Guéritte F, Guénard D, Poupat C, Ahond A, Potier P (2001) Semisynthesis of D-Ring Modified Taxoids: Novel Thia Derivatives of Docetaxel. J Org Chem **66**: 5058
531. Cheng Q, Horiguchi T, Yamaguchi M, Oritani T (2000) High Regio- and Stereo-selectivity in Facile One-pot Converison of Taxoids to the Primary Alcohols: Precursors for the Synthesis of 4-Deacetoxypaclitaxel and Analogues. Heterocycles **53**: 2759

532. Sun D-A, Nikolakakis A, Sauriol F, Mamer O, Zamir LO (2001) Microbial and Reducing Agents Catalyze the Rearrangement of Taxanes. Bioorg & Med Chem **9**: 1985

533. DeMattei JA, Leanna MR, Li W, Nichols PJ, Rasmussen MW, Morton HE (2001) An Efficient Synthesis of the Taxane-Derived Anticancer Agent ABT-271. J Org Chem **66**: 3330

534. Leanna MR, DeMattei JA, Li W, Nichols PJ, Rasmussen M, Morton HE (2000) Synthesis of the C-13 Side Chain Precursors of the 9-Dihydrotaxane Analogue ABT-271. Org Lett **2**: 3627

535. Kim S-C, Kim H-K (2000) Synthesis of New Taxol Side Chain Precursor from L-Tartaric Ester. Bull Korean Chem Soc **21**: 1047

536. Hamamoto H, Mamedov VA, Kitamoto M, Hayashi N, Tsuboi S (2000) Chemoenzymatic Synthesis of the C-13 Side Chain of Paclitaxel (Taxol) and Docetaxel (Taxotere). Tetrahedron: Asymmetry **11**: 4485

537. Cho GY, An KM, Ko SY (2001) Improving the Regioselectivity in the Cyclic Iminocarbonate Rearrangement: Enantioselective Synthesis of the Paclitaxel Side Chain. Bull Korean Chem Soc **22**: 432

538. Aoyagi Y, Jain RP, Williams RM (2001) Stereocontrolled Asymmetric Synthesis of α-Hydroxy-β-amino Acids. A Stereodivergent Approach. J Am Chem Soc **123**: 3472

539. Roers R, Verdine GL (2001) Concise Enantio- and Diastereoselective Synthesis of α-Hydroxy-α-methyl-β-amino Acids. Tetrahedron Lett **42**: 3563

540. Barboni L, Lambertucci C, Appendino G, Vander Velde DG, Himes RH, Bombardelli E, Wang M, Snyder JP (2001) Synthesis and NMR-Driven Conformational Analysis of Taxol Analogues Conformationally Constrained on the C13 Side Chain. J Med Chem **44**: 1576

541. Reddy SHK, Lee S, Datta A, Georg GI (2001) Efficient Synthesis of the 3′-Phenolic Metabolite of Paclitaxel. J Org Chem **66**: 8211

542. Mellado W, Magri NF, Kingston DGI, Garcia-Arenas R, Orr GA, Horwitz SB (1984) Preparation and Biological Activity of Taxol Acetates. Biochem Biophys Res Commun **124**: 329

543. Klis WA, Sarver JG, Erhardt PW (2001) Mechanistic Considerations Pertaining to the Solvolysis of Paclitaxel Analogs Bearing Ester Groups at the C2′ Position. Tetrahedron Lett **42**: 7747

544. Bradley MO, Swindell CS, Anthony FH, Witman PA, Devanesan P, Webb NL, Baker SD, Wolff AC, Donehower RC (2001) Tumor Targeting by Conjugation of DHA to Paclitaxel. J Controlled Release **74**: 233

545. Bradley MO, Webb NL, Anthony FH, Devanesan P, Witman PA, Hemamalini S, Chander MC, Baker SD, He L, Horwitz SB, Swindell CS (2001) Tumor Targeting by Covalent Conjugation of a Natural Fatty Acid to Paclitaxel. Clin Cancer Res **7**: 3229

546. Cheng Q, Oritani T, Horiguchi T (2000) The Synthesis and Biological Activity of 9- and 2′-cAMP 7-Deoxypaclitaxel Analogues from 5-Cinnamoyltriacetyltaxicin-I. Tetrahedron **56**: 1667

547. Shi Q, Wang H-K, Bastow KF, Tachibana Y, Chen K, Lee F-Y, Lee K-H (2001) Antitumor Agents 210. Synthesis and Evaluation of Taxoid-Epipodophyllotoxin Conjugates as Novel Cytotoxic Agents. Bioorg & Med Chem **9**: 2999

548. Seligson AL, Terry RC, Bressi JC, Douglass JG III, Sovak M (2001) A New Prodrug of Paclitaxel: Synthesis of Protaxel. Anti-Cancer Drugs **12**: 305

549. Niethammer A, Gaedicke G, Lode HN, Wrasidlo W (2001) Synthesis and Preclinical Characterization of a Paclitaxel Prodrug with Improved Antitumor Activity and Water Solubility. Bioconjugate Chem **12**: 414

550. Damen EWP, Nevalainen TJ, van den Bergh TJM, de Groot FMH, Scheeren HW (2002) Synthesis of Novel Paclitaxel Prodrugs Designed for Bioreductive Activation in Hypoxic Tumour Tissue. Bioorg & Med Chem **10**: 71
551. Wu JH, Zamir LO (2000) A Rational Design of Bioactive Taxanes with Side Chains Situated Elsewhere Than on C-13. Anti-Cancer Drug Design **15**: 73
552. Chatterjee SK, Barron DM, Vos S, Bane S (2001) Baccatin III Induces Assembly of Purified Tubulin into Long Microtubules. Biochemistry **40**: 6964
553. Snyder JP, Nettles JH, Cornett B, Downing KH, Nogales E (2001) The Binding Conformation of Taxol in Beta Tubulin: A Model Based on the Electron Crystallographic Density. Proc Natl Acad Sci USA **98**: 5312
554. Andreu JM, Barasoain I (2001) The Interaction of Baccatin III with the Taxol Binding Site of Microtubules Determined by a Homogeneous Assay with Fluorescent Taxoid. Biochemistry **40**: 11975
555. Abal M, Souto AA, Amat-Guerri F, Acuna AU, Andreu JM, Barasoain I (2001) Centrosome and Spindle Pole Microtubules are Main Targets of a Fluorescent Taxoid Inducing Cell Death. Cell Motil Cytoskeleton **49**: 1
556. Baloglu E, Kingston DGI, Patel P, Chatterjee SK, Bane SL (2001) Synthesis and Microtubule Binding of Fluorescent Paclitaxel Derivatives. Bioorg & Med Chem Lett **11**: 2249
557. Metaferia BB, Hoch J, Glass TE, Bane SL, Chatterjee SK, Snyder JP, Lakdawala A, Cornett B, Kingston DGI (2001) Synthesis and Biological Evaluation of Novel Macrocyclic Paclitaxel Analogues. Org Lett **3**: 2461
558. Geng X, Vorella CP, Li D, Lin S, Pera P, Bernacki RJ, Ojima I (2001) Design and Synthesis of *de novo* Macrocyclic Hybrids as Potential Anticancer Agents. 221st ACS National Meeting. San Diego, CA, April 1–5
559. Walker K, Croteau R (2001) Molecules of Interest: Taxol Biosynthetic Genes. Phytochemistry **58**: 107
560. Wildung MR, Croteau R (1996) A cDNA Clone for Taxadiene Synthase, the Diterpene Cyclase That Catalyzes the Committed Step of Taxol Biosynthesis. J Biol Chem **271**: 9201
561. Walker K, Schoendorf A, Croteau R (2000) Molecular Cloning of a Taxa-4(20),11(12)-dien-5a-ol-*O*-Acetyl Transferase cDNA from Taxus and Functional Expression in Escherichia coli. Arch Biochem Biophys **374**: 371
562. Walker K, Croteau R (2000) Taxol Biosynthesis: Molecular Cloning of a Benzoyl-CoA:Taxane 2α-*O*-benzoyltransferase cDNA from Taxus and Functional Expression in *Escherichia coli*. PNAS **97**: 13591
563. Walker K, Croteau R (2000) Molecular Cloning of a 10-Deacetylbaccatin III-10-*O*-acetyl Transferase cDNA from Taxus and Functional Expression in *Escherichia coli*. Proc Natl Acad Sci USA **97**: 583
564. Schoendorf A, Rithner CD, Williams RM, Croteau RB (2001) Molecular Cloning of a Cytochrome P450 Taxane 10β-hydroxylase cDNA from *Taxus* and Functional Expression in Yeast. Proc Natl Acad Sci USA **98**: 1501
565. Huang Q, Roessner CA, Croteau R, Scott AI (2001) Engineering *Escherichia coli* for the Synthesis of Taxadiene, a Key Intermediate in the Biosynthesis of Taxol. Bioorg & Med Chem **9**: 2237
566. Williams DC, Carroll BJ, Jin Q, Rithner CD, Lenger SR, Floss HG, Coates RM, Williams RM, Croteau R (2000) Intramolecular Proton Transfer in the Cyclization of Geranylgeranyl Diphosphate to the Taxadiene Precursor of Taxol Catalyzed by Recombinant Taxadiene Synthase. Chem & Biol **7**: 969

567. Terwogt JMM, Huinink WWTB, Schellens JHM, Schot M, Mandjes IAM, Zurio MG, Rocchetti M, Rosing H, Koopman FJ, Beijnen JH (2001) Phase I Clinical and Pharmacokinetic Study of PNU166945, a Novel Water-soluble Polymer-Conjugated Prodrug of Paclitaxel. Anti-Cancer Drugs **12**: 315

568. Gelmon KA, Latreille J, Tolcher A, Genier L, Fisher B, Forand D, D'Aloisio S, Vernillet L, Daigneault L, Lebecq A, Besenval M, Eisenhauer E (2000) Phase I Dose-Finding Study of a New Taxane, RPR 109881A, Administered as a One-Hour Intravenous Infusion Days 1 and 8 to Patients with Advanced Solid tumors. J Clin Oncol **18**: 4098

Author Index

Page numbers printed in *italics* refer to References

Subject Index

MIX
Papier aus verantwortungsvollen Quellen
Paper from responsible sources
FSC® C105338

If you have any concerns about our products,
you can contact us on
ProductSafety@springernature.com

In case Publisher is established outside the EU,
the EU authorized representative is:
Springer Nature Customer Service Center GmbH
Europaplatz 3, 69115 Heidelberg, Germany

Printed by Libri Plureos GmbH
in Hamburg, Germany